NEW TECHNOLOGIES AND THE ARMS RACE

Also edited by David Carlton and Carlo Schaerf

STUDIES IN DISARMAMENT AND CONFLICTS

THE DYNAMICS OF THE ARMS RACE
INTERNATIONAL TERRORISM AND WORLD SECURITY
ARMS CONTROL AND TECHNOLOGICAL INNOVATION
*CONTEMPORARY TERROR: *Studies in Sub-State Violence*
*THE HAZARDS OF THE INTERNATIONAL ENERGY CRISIS
*THE ARMS RACE IN THE 1980s
*SOUTH-EASTERN EUROPE AFTER TITO: *A Powder-Keg for the 1980s?*
*REASSESSING ARMS CONTROL
*THE ARMS RACE IN THE ERA OF STAR WARS
*PERSPECTIVES ON THE ARMS RACE

*Also published by Palgrave Macmillan

New Technologies and the Arms Race

Edited by

Carlo Schaerf
Professor of Physics
University of Rome

Brian Holden Reid
Lecturer in War Studies
King's College, London

and

David Carlton
Lecturer in International Studies
University of Warwick

MACMILLAN

First published 1989

Published by
THE MACMILLAN PRESS LTD
Houndmills, Basingstoke, Hampshire RG21 2XS
and London
Companies and representatives
throughout the world

British Library Cataloguing in Publication Data
New technologies and the arms race. —
1. Warfare. Effects of technological
innovation
I. Schaerf, Carlo   II. Reid, Brian Holden
III. Carlton, David, *1938–*   IV. Series
355'.02
ISBN 978-1-349-10617-2   ISBN 978-1-349-10615-8 (eBook)
DOI 10.1007/978-1-349-10615-8

Contents

Foreword ix

Message to Participants from the President
of the Republic of Italy x

List of Abbreviations xi

Notes on the Contributors xv

List of Conference Participants xix

PART I CIVILIAN VERSUS MILITARY TECHNOLOGY

1 Do Civilian Spin-offs Justify Investments in Military
Technology?
Umberto Colombo and Guiseppe Lanzavecchia 3

2 Non-Military Justification of Military R and D:
Rationality or Rationalisation?
Lloyd J. Dumas 17

3 Non-Military Justifications for Investments in
Military R and D
Roberto Fieschi 29

4 Non-Military Justifications for Investments in
Military Technologies
Shalheveth Freier 50

PART II THE STRATEGIC DEFENSE INITIATIVE: MILITARY AND CIVILIAN SPACE TECHNOLOGY

5 Scientists and the Strategic Defense Initiative
Sidney D. Drell 57

6 Defensive and Offensive Weapons in Space and
Civilian Space Technologies
Richard L. Garwin 66

7 Clarifying ABM Treaty Ambiguities: Threshold Limits
John E. Pike 80

8   Cosmic Space and the Role of Europe
    Rolf Linkohr   96

9   A Technical Assessment of Potential Threats to NATO
    from Non-Nuclear Soviet Tactical Ballistic Missiles
    Benoit Morel and Theodore A. Postol   106

PART III   CIVILIAN NUCLEAR TECHNOLOGIES AND
           NUCLEAR WEAPONS PROLIFERATION

10 Civilian Nuclear Technologies and Nuclear Weapons
    Proliferation
    John P. Holdren   161

11 Nuclear Disarmament and Peaceful Nuclear
    Technology – Can We Have Both?
    Theodore B. Taylor   199

12 Pakistan's Nuclear Weapons Programme and its
    Implications
    Jorma K. Miettinen   210

PART IV   PROBLEMS OF COMMAND AND CONTROL

13 Routine Nuclear Operations
    Paul Bracken   219

14 Arms, Decision-Making and Information Technology
    Allan M. Din   226

15 Controlling Theatre Nuclear War
    Desmond Ball   237

16 Problems in European Command and Control
    George W. Rathjens   275

PART V   ARMS CONTROL: SUCCESSES AND
         FAILURES

17 The Purposes, Achievements and Priorities of Arms
    Control
    Paul S. Brown   285

18 Progress and Failure in Arms Control
    Edy Korthals Altes   310

19 Steps to Constrain Arms Races
    F. A. Long   316

20 Progress and Failure in Disarmament
    Maj Britt Theorin   324

21 Technology, the Arms Race and Disarmament
    Joseph Rotblat   332

22 Experimental Nuclear Explosions and the Arms Race
    Francesco Lenci   345

23 The US National Resources Defense Council/Soviet
    Academy of Sciences Nuclear Test Ban Verification
    Project
    Thomas B. Cochran   354

Index   363

Foreword

This volume contains the proceedings of the International Conference on 'Technology, the Arms Race and Arms Control' which took place at Castiglioncello, Rosignano Marittimo, in the Province of Livorno, Italy, from 25 to 30 September 1987. The Conference was organised by the Union of Scientists for Disarmament (Unione Scienziati Per Il Disarmo – USPID), thanks to the decisive contribution of funds and the valuable co-operation of the Town Council of Rosignano Marittimo.

We are glad to express our thanks here to every person who helped to make the initiative a success.

The Accademia Nazionale dei Lincei, the Istituto di Biofisica of CNR in Pisa, the Regione Toscana and the Amministrazione Provinciale di Livorno sponsored the Conference. The President of the Senate and the President of the Chamber of Deputies both supported the Conference.

The President of the Republic, Senator Francesco Cossiga, sent a goodwill message for the opening of the Congress. The keen attention and appreciative esteem he expressed – the full message is reproduced below – greatly encouraged all concerned.

The President of the Senate, Senator Giovanni Spadolini, the President of the Chamber of Deputies, on. Nilde Jotti, and the Minister of Foreign Affairs, on. Giulio Andreotti, sent greetings telegrams expressing their great interest in the work of the Conference.

The objective of the meeting was, on the one hand, to promote discussions and exchanges among experts and, on the other hand, to better inform political circles and, through the Press, public opinion in general, about matters that are all too often handled lightly or used as mere propaganda, depending on ideological standpoints and preconceived opinions. We sincerely hope that this volume will contribute to the international debate – both between the two great blocs and within them – about matters that are of crucial importance to all mankind.

THE SCIENTIFIC COUNCIL OF USPID
(Carlo Bernardini, Bruno Bertotti,
Francesco Calogero, Paolo Cotta-Ramusino,
Michelangelo De Maria, Roberto
Fieschi, Francesco Lenci, Carlo Schaerf)

Message to Participants from the President of the Republic of Italy

On the occasion of the Conference on 'Technology, the Arms Race and Arms Control' I am happy to send hearty greetings to the distinguished scientists taking part in the meeting and a warm welcome to our foreign guests.

Affirming once more its active commitment to themes of great general relevance and of particular importance for the peaceful development of humankind, USPID will afford with this initiative a useful moment of reflection and a great opportunity for an exchange of ideas and experiences within the international scientific community.

In the hope that these goals will be reached, I express my best wishes for a successful conference to all participants.

FRANCESCO COSSIGA

List of Abbreviations

ABM	Anti-ballistic Missile
ACE	Allied Command Europe
ACCHAN	Allied Command Channel
ACLANT	Allied Command Atlantic
ADA	Airborne Optical Adjunct Program
AGI	Auxiliary General Intelligence
AI	Artificial Intelligence
ALB	Airland Battle
AOC	Air Officer Commanding
ASAT	Anti-satellite
ASW	Anti-submarine Warfare
ATBM	Anti Tactical Ballistic Missile
AUTODIN	Automatic Digital Network
AUTOVON	Automatic Voice Network
AWACS	Airborne Warning and Control System
BAOR	British Army on the Rhine
BMC^3	Battle Management Command, Control and Communications
BMD	Ballistic Missile Defence
BMEWS	Ballistic Missile Early Warning System
CCIS	Command and Control Information System
CEP	Circle of Equal Probability
CINCAFCE	Commander-in-Chief Allied Forces Central Europe
CINCEASTLANT	Commander-in-Chief of the Eastern Atlantic
CINCHAN	Commander-in-Chief Channel
CINCLANT	Commander-in-Chief Atlantic
CINCSAC	Commander-in-Chief of the Strategic Air Command (US)
CINCUKAIR	Commander-in-Chief UK Air
CISC	Commanders-in-Chief Committee
COMCENTLANT	Commander of the Central Atlantic
COMMAIRCHAN	Commander of the Air Channel
COMMAIREASTLANT	Commander of the Air East Atlantic
COMNORCHAN	Commander of the North Channel

COMNORLANT	Commander of the North Atlantic
COMPLYMCHAN	Commander of Plymouth Channel
COMSEC	Communications Security
CONUS	Continental United States
CPXs	Command Post Exercises
CSCE	Conference on Security and Co-operation in Europe
CTB	Comprehensive Test Ban
C^3I	Command, Control, Communications and Intelligence
DARPA	Defense Advanced Research Projects Agency (US)
DMAIN	Deployed Main Divisional Command Post
DPC	Defence Planning Committee (NATO)
DSCS	Defense Satellites Communications System (US)
DSP	Defense Support Program (US)
EAM	Emergency Action Message
ECM	Electronic Counter-measures
ELINT	Electronic Intelligence
EMP	Electromagnetic Pulse
EMSI	European Manned Space Infrastructure
ESA	European Space Agency
FAE	Fuel Air Explosive
FAS	Federation of American Scientists
FBM	Fleet Ballistic Missile
FOFA	Follow-On-Forces-Attack
GDR	German Democratic Republic
GIUK	Greenland/Iceland/UK
GPS	Global Positioning System
GWe	Gigawatts of Electricity
HTGR	High-temperature Gas-cooled Reactor
HTOL	Horizontal Take-Off and Landing
IAEA	International Atomic Energy Agency
ICBM	Intercontinental Ballistic Missile
IDE	Institute of Physics of the Earth (Soviet)
IDEA	Innovative Dimension in Energy and Agriculture
IFF	Interrogation Friend or Foe
INF	Intermediate Nuclear Forces

IPS	Instrument Pointing System
ISMA	Internationl Satellite Monitoring Agency
IVSN	Initial Voice Switched Network
KKV	Kinetic Kill Vehicle
kt	Kiloton
LEO	Low Earth Orbit
LMFBR	Liquid Metal Fast Breeder Reactor
LTBT	Limited Test Ban Treaty
MAS	Mutual Assured Security
MCP	Mobile Command Post
MGT	Mobile Ground Terminal
MIRV	Multiple Independently-targetable Re-entry Vehicle
NARS	North Atlantic Relay System
NASA	National Aeronautics and Space Administration (US)
NATO	North Atlantic Treaty Organisation
NCA	National Command Authority (US)
NCF	Network Control Facility
NICS	NATO's Integrated Communications System
NPT	Non-Proliferation Treaty
NRDC	Natural Resources Defense Council (US)
NTB	National Test Bed
NTEM	Nuclear Test Experts' Meeting
NTM	National Technical Means
OSI	On-Site Inspection
OTA	Office of Technology Assessment (US)
PALs	Permissive Action Links
PAR	Perimeter Acquisition Radar
PNET	Peaceful Nuclear Explosions Treaty
PSAC	President's Science Advisory Committee (US)
PWR	Pressurised-water Reactor
R and D	Research and Development
RAF	Royal Air Force
REC	Radio-electronic Combat
RN	Royal Navy
RV	Re-entry Vehicle
SACEUR	Supreme Allied Commander Europe

SACLANT	Supreme Allied Commander Atlantic
SALT	Strategic Arms Limitation Talks
SATCOM	Satellite Communications
SBI	Space-based Interceptor
SCC	Standing Consultative Commission
SCF	Satellite Control Facility
SCI	Strategic Computing Initiative
SDI	Strategic Defense Initiative
SDIO	Strategic Defense Initiative Office
SHAPE	Supreme Headquarters Allied Power Europe
SIGINT	Signals Intelligence
SLBM	Submarine-launched Ballistic Missile
SLCM	Sea-launched Cruise Missile
SNDV	Strategic Nuclear Delivery Vehicle
SRAM	Short-range Attack Missile
START	Strategic Arms Reduction Talks
TBM	Tactical Ballistic Missile
TRADOC	Training and Doctrine Command (US Army)
TTBT	Threshold Test-Ban Treaty
USAEUR	US Army Europe
USAFE	US Air Forces Europe
USCINCEUR	US Commander-in-Chief Europe
USEUCOM	US European Command
USNAVEUR	US Navy Europe
WHSC	White House Science Council (US)
2ATAF	Second Allied Tactical Air Force

Notes on the Contributors

Desmond Ball (*Australian*) was Research Associate at the International Institute for Strategic Studies from 1979 to 1980. Since 1984 he has been Head of the Strategic and Defence Studies Center at the Australian National University, Canberra. He is author of *Politics and Force Levels: The Strategic Missile Program of the Kennedy Administration*.

Paul Bracken (*US*) has since 1983 been Professor of Public Policy at Yale University where he works in the fields of national security policy and arms control. He was previously at the Hudson Institute, New York. He is author of *The Command and Control of Nuclear Forces*.

Paul S. Brown (*US*) has worked since 1966 in the Lawrence Livermore Laboratory, where he is at present Assistant Director for Arms Control. During 1986 he served as US Department of Energy Delegate to the Nuclear Test Experts Meeting held in Geneva.

David Carlton (*British*) (*co-editor*) is Lecturer in International Studies at the University of Warwick. He is author of *Anthony Eden: A Biography*; and co-editor of *The Nuclear Arms Race Debated* and of *The Cold War Debated*.

Thomas B. Cochran (*US*) is a Senior Staff Scientist and Director of the Nuclear Weapons Databook Project at the National Resources Defense Council. He has served as a consultant to numerous US governmental agencies and non-governmental organisations on energy and nuclear non-proliferation matters. He holds a PhD in Physics from Vanderbilt University.

Umberto Colombo (*Italian*) is President of the Italian Atomic Energy Commission. He was formerly Director of the G. Donegani Research Insitute.

Allan M. Din (*Danish*) read Physics at the University of Gothenberg. He has held research posts at CERN, Geneva, and at the University of Lausanne. He is at present Senior Research Fellow at the Stockholm International Peace Research Institute.

Sidney D. Drell (*US*) is Professor and Deputy Director of the Stanford Linear Accelerator Center, Stanford University; and Co-Director of the Stanford Center for International Security and Arms Control.

Lloyd J. Dumas (*US*) is Professor of Political Economy at the University of Texas at Dallas. He was formerly Associate Professor of Industrial Engineering at Columbia University. His publications include *The Overburdened Economy*.

Roberto Fieschi (*Italian*) is Professor of Physics at the University of Parma. His field of interest is Solid State Physics. He is a member of the USPID Scientific Council.

Shalheveth Freier (*Israeli*) is a Physicist. He was Director of the Research and Development Authority of the Ministry of Defence of Israel (1953–6); Deputy Director-General of the Weizmann Institute of Sciençe (1967–70); Director-General of the Israeli Atomic Energy Commission (1971–6); and Chairman of the Presidential Council for Science Policy (1977–8).

Richard L. Garwin (*US*) is a Physicist at the IBM T. J. Watson Research Center, Yorktown Heights, New York. He is a member of the US National Academy of Sciences and of the National Academy of Engineering; and a consultant of several US government departments and agencies.

John P. Holdren (*US*) is Professor of Energy and Resources at the University of California, Berkeley. He is Chairman of the US Pugwash Group and a former Chairman of the Federation of American Scientists. He is also Faculty Consultant in Magnetic Fusion Energy at the Lawrence Livermore National Laboratory.

Edy Korthals Altes (*Dutch*) is a former diplomat. He served as his country's Ambassador in Warsaw and in Madrid. He was also a Permanent Delegate to the European Community in Brussels; and Director for the Foreign Service in The Hague.

Giuseppe Lanzavecchia (*Italian*) is Assistant to the President of the Italian Atomic Energy Commission. His principal interest is in strategies for energy and for technological innovation. He is involved in technological forecasting and assessment. He was formerly Head of the Materials Department and then Scientific Director of the G. Donegani Research Institute.

Francesco Lenci (*Italian*) is a Physicist. He is a researcher at the Consiglio Nationale delle Ricerche (CNR) at the Institute of Biophysics at Pisa. He is Secretary-General of USPID; and was a member of the 'Initiative Group' that promoted the International Forums in Moscow in July 1986 and February 1987.

Rolf Linkohr (*West German*) has been a Member of the European Parliament since 1979. He is Co-ordinator of the Socialist Group for Energy, Research and Technology.

F. A. Long (*US*) is Professor Emeritus of Chemistry and of Science and Society at Cornell University. He was Assistant Director for Science and Technology of the US Arms Control and Disarmament Agency and participated in the negotiations in Moscow that led to the Limited Test Ban Treaty of 1963.

Jorma K. Miettinen (*Finnish*) is Research Director of Military Science at the Institute of Military Science in Helsinki. He was formerly Head of the Department of Radiochemistry at the University of Helsinki.

Benoit Morel (*Swiss-French*) is a faculty member of the Department of Engineering and Public Policy at Carnegie-Mellon University, Pittsburgh. He was formerly a Science Fellow at the Stanford Center for International Security and Arms Control; a Senior Research Fellow at the California Institute of Technology; and an Adjunct Professor of Physics at the University of Southern California.

John E. Pike (*US*) is the Associate Director for Space Policy at the Federation of American Scientists. He co-ordinates the Federation's research, public education and lobbying on space policy, including the Strategic Defense Initiative.

Theodore A. Postol (*US*) is a Senior Research Associate at the Stanford Center for International Security and Arms Control. Previously he served in the US Department of Defense as a scientific adviser to the Chief of Naval Operations; in the Office of Technology Assessment of the US Congress; and on the scientific staff of the Argonne National Laboratory.

George W. Rathjens (*US*) is Professor of Political Science at the Massachusetts Institute of Technology. He was previously associated with the US Arms Control and Disarmament Agency; and has also been Chief Scientist and Deputy Director of the Advanced Research Projects Agency of the US Department of Defense.

Brian Holden Reid (*British*) (*co-editor*) is Lecturer in War Studies at King's College, London, and Resident Historian at the Staff College, Camberley. He was formerly editor of the *Journal of the Royal United Services Institute*. He is author of *J. F. C. Fuller: Military Thinker*. He is a Fellow of the Royal Historical Society.

Joseph Rotblat (*British*) is Emeritus Professor of Physics at the University of London. He started his career in Poland as a nuclear physicist and eventually participated in atomic bomb work at Los Alamos during the Second World War. As a result of this experience he switched to medical physics. And for the same reason he became a founder of the Pugwash Conferences on Science and World Affairs of which he was Secretary General for 17 years.

Carlo Schaerf (*Italian*) (*co-editor*) is Professor of Physics at the University of Rome I. He was previously a Research Associate at Stanford University and on the staff of the Italian Atomic Energy Commission. With Professor Eduardo Amaldi he founded in 1966 the International School on Disarmament and Research on Conflicts (ISODARCO) of which he has been a Director since 1970.

Theodore B. Taylor (*US*) is a consulting physicist working mainly on nuclear disarmament options and non-proliferation. He gained a PhD in theoretical physics from Cornell University. He was formerly engaged in nuclear weapon design at Los Alamos National Laboratory.

Maj Britt Theorin (*Swedish*) has been a Member of the Swedish Parliament (Social Democrat) since 1971. Since 1982 she has been Chairman of the Swedish Disarmament Commission with the rank of ambassador. She was a founder and first Chairperson of the World Women Parliamentarians for Peace.

List of Conference Participants

Rosignano Marittimo, 25–30 September 1987

ITALY

Agnese, Angelo Gino, Dipartimento di Fisica, Viale Dodecaneso, 3, 16132 Genova.

Aiello, Santi, Dipartimento Fisica/Spazio, Via L. Pancaldi, 3/45, 50127 Firenze.

Alberti, Maria Alberta, Ist. di Cibernetica, Via Viotti, 5, 20100 Milano.

Amaldi, Eduardo, Dip. di Fisica Facolta' di Science MFN, P. le A. Moro 5, 00185 Roma.

Anderlini, Luigi, Archivio Disarmo, Via di Torre Argentina, 18, 00186 Roma.

Ascoli, Cesare, CNR-Ist. di Biofisica, Via S. Lorenzo, 26, 56100 Pisa.

Bagnoli, Ali', Via Tagliamento, 1, 57012 Castiglioncello (LI).

Baldocchi, Maria Antonia, CNR-Ist. di Biofisica, Via S. Lorenzo, 26, 56100 Pisa.

Bangone, Gianfranco, Il Manifesto, Via Tomacelli, 146, 00186 Roma.

Barbieri, Giacomo, Fiom-Cgil, Corso Trieste, 36, 00198 Roma.

Bassi, Luigi, Via Vasco de Gama, 13, Bologna.

Batani, Bragagia, Silvia, Via Rossini, 49, 57013 Rosignano Solvay (LI)

Batani, Dimitri Dino, Via Rossini, 49, 57013 Rosignano Solvay (LI).

Bernardini, Carlo, Dipartimento di Fisica, Fac. Scienze MFN, P. le Aldo Moro, 5, 00185 Roma.

Bernini, Bruno, Cespi, Via Della Vite, 13, 00187 Roma.

Bertoli, Franco, Istituto Professionale, 57013 Rosignano Solvay (LI).

Bisi, Fulvio, CNR Istituto di Biofisica, Via San Lorenzo, 26, 56100 Pisa.

Boba, Silvia, Cgil, Corso di Italia, 25, 00198 Roma.

Bochicchio, Francesco, Dip. Fisica, P. le A. Moro, 2, 00186 Roma.

Boffa, Giuseppe, Cespi, Via Della Vite, 13, 00187 Roma.

Bovet, Daniel, P. za S. Apollinare, 33, 00186 Roma.

Caccio', Ornella, Archivio Disarmo, Via Torre Argentina, 18, 00187 Roma.

Calogero, Francesco, Dipartimento di Fisica, Fac. Scienze MFN, P. le Aldo Moro, 5, 00185 Roma.

Carossini, Paolo, Dip. Fisica, Largo Fermi, 2, 50125 Firenze.

Castelli, Antonio, La Giudecca 204/B, 30133 Venezia.

Cavaglia', Antonio, Agi, Via Nomentana, 92, 00161 Roma.

Ceragioli, Paola, C. So Lodi, 65, 20151 Milano.

Cerrina, Simone, E.L.E.A., Firenze.

Cesari, Anna, Via Delle Porte Nuove 20, 50144 Firenze.

Chahoud, Joseph, Dip. di Fisica, Via Irnerio, 46, 40126 Bologna.

Ciampa, Francesco, Enea – V.E.L. Syst. Prolog, C.P. 2400, Casaccia, 00100 Roma.

Codrignani, Giancarla, Via Milazzo, 5, 40100 Bologna.

Colombetti, Giuliano, CNR-Istituto di Biofisica, Via S. Lorenzo, 26, 56100 Pisa.

Conti, Maurizio, Dip. Fisica, Piazza Torricelli, 2, 56100 Pisa.

Corona, Eliseo, Consigliere Comunale, Comune di Rosignano, 57011 Rosignano M.Mo (LI).

Corridoni, Fabio, Il Tirreno, Viale Alfieri, 9, 57100 Livorno.

Cortesi, Luigi, P.za Tarquinia, 5d, 00183, Roma.

Cotta-Ramusino, Paolo, Dip. di Fisica, Via Celoria, 16, 20133 Milano.

Croce, Lucia, Capogruppo P.S.I., Comune di Rosignano, 57011 Rosignano M.Mo (LI).

Cuffaro, Antonino, P.C.I., Via Botteghe Oscure, 00186 Roma.

D'Amico, Arnaldo, Repubblica, Roma.

Dassu', Marta, Cespi, Via Della Vite, 13, 00187 Roma.

De Andreis, Marco, Iai, Viale Mazzini, 88, 00180 Roma.

De Giorgi, Ennio, Scuola Normale Superiore, Piazza dei Cavalieri, 56100 Pisa.

De Marchi, Maria Vittoria, Rinascita, Via dei Taurini, 19, 00185 Roma.

De Maria, Michelangelo, Dipartimento di Fisica, Facolta Scienze MFN, P. le Aldo Moro, 5, 00185 Roma.

De Murtas, Angelo, Scientia, Via Fratelli Bronzetti, 20129 Milano.

Del Giudice, Paolo, Dip. di Fisica, P. le A. Moro, 2, 00185 Roma.

Dello Sbarba, Brunellesco, Via Volturno, 19, 56100 Pisa.

Devoto, Gianluca, Cespi, Via Della Vite, 13, 00187 Roma.

Di Bari, Vito, Istituto di Fisica, Via Amendola, 173, 70126 Bari.

Di Paolantonio, Michele, Via L. Da Vinci, 64029 Silvi Marina (TE).

Dupre', Franco, Dipartimento di Fisica, Facolta' di Scienze M.F.N., P. le A. Moro, 5, 00185 Roma.

Eccher, Fausto, Uspid, c/o Fac. Scienze, 38050 Povo (TN).
Elena, Mirco, Irst, Facolta' di Scienze, 38050 Povo (TN).
Falchetta, Massimo, Dip. Fare – Enea, Casaccia, Sacco N.99, 00060
S.M. Di Galeria (Roma).
Farinella, Paolo, Dip. di Matematica, Via Buonarroti, 2, 56100 Pisa.
Ferrante, Gaetano, Dip. Fisica Teorica, Universita', 98100 Messina.
Ferrario, Massimo, Instituto di Fisica, Via Celoria, 16, 20133 Milano.
Fieschi, Roberto, Istituto di Fisica, Via M. D'Azeglio, 85, 43100
Parma.
Fiorentini, Enzo, Rosignano (LI).
Foresta Martin, Franco, Corriere Della Sera, Via del Parlamento,
00186 Roma.
Foroni, Francesco, Viale Podgora, 1, 46100 Mantova.
Frassati, Filippo, Cattedra di Storia Militare, Piazza Torricelli, 3A,
56100 Pisa.
Frediani, Carlo, CNR-Istituto di Biofisica, Via S. Lorenzo, 26, 56100
Pisa.
Fusco, Giuseppe, Via Pardi, 18, 56100 Pisa.
Giara, Alessandro, Via Roma, 4, 57023 Cecina (LI).
Gregori, Luca, Via Patriciano, 36, 34012 Trieste.
Grigani, Gianluca, Ufficio Pace E Cultura, Via Fiorenzo di Lorenzo,
06100 Perugia.
Guidi, Rossana, Via Tagliamento, 1, 57012 Castiglioncello (LI).
Gusmaroli, Franca, Centro Studi Documentazione, Storica Econo-
mica Imprese, Via G. Ferrari, 35, 00195 Roma.
Ilari, Virgilio, Univ. Macerata, 62100 Macerata.
Jacchia, Enrico, L.U.I.S.S., Via Pola, 12, 00198 Roma.
Lanzani, Guglielmo, Dip. Fisica Nucleare, Via Bassi 6, 27100 Pavia.
Lanzavecchia, Giuseppe, Enea, Viale Regina Margherita, 125, 00198
Roma.
Lazzizzera, Ignazio, Dip. di Fisica, Univ. di Trento, 38050 Povo
(TN).
Lenci, Francesco, CNR-Istituto di Biofisica, Via San Lorenzo, 26,
56100 Pisa.
Leonardi, Stefano, Dip. di Fisica, P. le A. Moro, 2, 00185 Roma.
Lepori, Vincenzo, Via Ginestre, 3, 20089 Rozzano (MI).
Longo, Giuseppe, Dip. di Fisica, Via Irnerio, 46, 40126 Bologna.
Lubrini, Patrizio, Istituto di Fisica, Via Celoria, 16, 20133 Milano.
Manfredini, Augusta, Dip. di Fisica, P. le A. Moro, 5, 00186 Roma.
Mantovani, Fabio, Via Magellano, 3, 37130 Verona.
Manzi, Sergio, Via Traversa Livornese, 90, 57011 Rosignano (LI).

Marinelli, Marino, Lega Per I Diritti Dei Popoli, Casella Postale, 87, 57100 Livorno.

Marini Bettolo, Giovan Battista, Accademia Dei XL, Palazzo Civilta' del Lavoro, Quadrato Della Concordia, 00144 Roma.

Marrazzi, Marina, Sapere, C.So Trieste, 95, 00198 Roma.

Mattiangeli, Paolo, Dip. Fisica, P. le A. Moro, 2, 00186 Roma.

Merlini, Cesare, I.A.I, Via Mazzini, 88, 00195 Roma.

Michelini, Maurizio, Enea – Dip. Fare, Cre Casella Postale, 2400 Roma.

Miggiano, Paolo, Irdisp, via Chiana 48, 00100 Roma.

Miliffi, Riccardo, Via Valdinievole, 12, 56031 Bientina (PI).

Minerva, Daniela, Sapere, Corso Trieste, 95, 00198 Roma.

Monterisi, Giancarlo, Rai Tg2 Servizi Speciali, Via Teulada, 66, 00195 Roma.

Morandi, Giovanni, La Nazione, Via F. Paolieri, 2, 50100 Firenze.

Navarra, Alfonso, Lega Per Il Disarmo Unilaterale, Firenze.

Nicolini, Claudio, Ist. Biofisica – Fac. Medicina, Viale Benedetto XV, 2, 16132 Genova.

Oppici, Silvano, Via Emilia Parmense, 8/B, 29100 Piacenza.

Ottolenghi, Andrea, Dip. di Fisica, Via Celoria, 16, 20133 Milano.

Palazzi, Antonio, Fac. Chimica Industriale, Viale Risorgimento, 4, 40136 Bologna.

Pascolini, Alessandro, Dipartimento di Fisica, Via Marzolo, 8, 35100 Padova.

Pattini, Corinna, Direttrice Didattica Del Primo Circolo, 57013 Rosignano (LI).

Petazzoni, Enrico, Via Ognibene, 2, 40135 Bologna.

Pianta, Mario, Via Siracusa, 19a, 00161 Roma.

Piazzoli, Adalberto, Dip. di Fisica Nucl. E Teorica, Via Bassi, 6, 27100 Pavia.

Piombanti, Pier Paolo, Presidente Distretto Scolastico, Via Della Cava, 109, 57013 Rosignano Solvay (LI).

Pivetti, Massimo, Via dei Serpenti, 00184 Roma.

Primicerio, Mario, Forum-Problemi Della Pace E Della Guerra, Piazza Alberti, 1, 50136 Firenze.

Ragionieri, Rodolfo, Forum-Problemi Della Pace E Della Guerra, Piazza Alberti, 1, 50136 Firenze.

Ravizza, Vittorio, La Stampa, Via Marenco, 32, 10126 Torino.

Rea, Pino, Ansa, Via Della Dataria, 95, 00187 Roma.

Reali, Enzo, Assessore allo Sviluppo Economico Comune di Castelfiorentino, 50051 Castlefiorentino (FI).

Ricci, Renato A., SIF, Via Degli I Andalo', 40124 Bologna.
Romoli, Francesco, Via Dei Cipressi, 13, 50019 Sesto Fiorentino (FI).
Romoli, Leda, Via dei Cipressi, 13, 50019 Sesto Fiorentino (FI).
Rossi, Fernanda, Via Tagliamento, 1, 57012 Castiglioncello (LI).
Rossi, Sergio, Il Sole 24 Ore, Via Grassi, 5, 10138 Torino.
Rotelli, Carlo, Capogruppo D.C., Comune di Rosignano, 57011 Rosignano M.MO (LI).
Ruffo, Stefano, Dip. di Fisica Univ., Largo Fermi, 2, 50125 Firenze.
Russo, Vito, Dip. Chimica, Via Del Risorgimento, 4, 40136 Bologna.
Saija, Antonino, Segretario Comunale, Comune Rosignano, 57011 Rosignano M.MO (LI)
Salvadori, Enrico, La Nazione, Via Paolieri, 2, 50100 Firenze.
Salvini, Giorgio, Dip. Fisica, P. le Aldo Moro, 00185 Roma.
Santoianni, Francesco, Via de Gasperi, 7, 80059 Torre Del Greco (NA).
Scaletti, Mario, Via Tagliamento, 1, 57012 Castiglioncello (LI).
Schaerf, Carlo, Dipartimento di Fisica, Fac. Scienze M.F.N., P. le A. Moro, 5, 00185 Roma.
Sensalari, Giancarlo, Via Appennini, 105, Milano.
Simoncini, Luca, I.E.I., Via S. Maria, 56, 56100 Pisa.
Spreafico, Maria Clelia, Via Corridoni, 109, 56100 Pisa.
Stabile, Giuseppe, Consigliere Comunale, Comune Rosignano, 57011 Rosignano M.MO (LI).
Tartaglia, Angelo, Dip.Di Fisica—Politecnico, C.So Duca Degli Abruzzi 24, 10129 Torino.
Tessarotto Fulvio, Dip. Di Fisica, Via Valerio 2, 34127 Trieste.
Todeschini-Lalli, Mario, Il Messaggero, Via Del Tritone, 152, 00187 Roma.
Tonello, Fabrizio, "Il Mondo", Via Rizzoli, 4, 20132 Milano.
Toninelli, Augusto, Via Della Torre, 8, 57012 Castiglioncello (LI).
Tornetta, Vincenzo, DC-Relazioni Internazionali, Via Del Plebiscito, 107, 00186 Roma.
Torre, Rocco, Via S. Andrea, 21, 56100 Pisa.
Trevi, Emanuele, Archivio Disarmo, Via Torre Argentina, 18, 00187 Roma.
Vadacchino, Mario, Dipartimento di Fisica, Politecnico, Corso Duca Degli Abruzzi, 24, 10129 Torino.
Valentini, Antonio, Il Tirreno, Via Alfieri, 9, 57100 Livorno.
Vanni, Claudio, Segretario Comitato Zona, P.C.I., Rosignano (LI).
Venerosi, Paola, CNR-IEI, Via S. Maria, 46, 56100 Pisa.

Vicidomini, Bianca, Preide Scuola Media Fattori, 57013 Rosignano Solvay (LI).
Volpe, Pietro, Ist. Medicina Sper.-Cnr, c/o Univ. La Sapienza, V. Le R. Elena, 324, 00198 Roma.

GREAT BRITAIN

Carlton, David, Dept. of International Studies, University of Warwick, Coventry, UK.
Hutchinson, George, 19 Westbourn Crescent, Southampton SO2 1EA, UK.
Leggett, Jeremy, 122 Sinclair Road, London W14 0NL, UK.
Prins, Gwyn, Emmanuel College, Cambridge CB2 3AP, UK.
Reid, Brian Holden, Dept. of War Studies, King's College London, Strand, London WC2 UK.
Rotblat, Joseph, Flat A, 63A Great Russell Street, London, UK.

WEST GERMANY

Linkohr, Rolf, Asangstrasse 219 A, D-7000 Stuttgart 61, FRG.
Wiess, Bettina, HSFK, Leimenrode 29, 6000 Frankfurt 1, FRG.

THE NETHERLANDS

Altes, Edy J. Korthals, Marinus Naefflaan 46, 7241 Ge Lochem, The Netherlands.

SWITZERLAND

Stroot, Jean Pierre, Cern, EP-Division, 1211 Geneve 23, Switzerland.

SWEDEN

Theorin, Maj Britt, Ministry for Foreign Affairs, Box 16121, S-103 23, Stockholm, Sweden.

FINLAND

Miettinen, Jorma K., Dept. of Radiochemistry, Unioninkatu 35, SF 00170 Helsinki–17, Finland.

DENMARK

Din, Allan, SIPRI, Berg Shamra, S-17173 Solna, Sweden.

POLAND

Wiejacz, Jozef, Ambasciata di Polonia in Italia, Via Rubens, 20, 00197 Roma.

ISRAEL

Freier, Shalheveth, Weizmann Institute of Science, Rehovot 76100, Israel.

LEBANON

Osseiran, Sanaa, 80 Bis Rue de Sevres, Paris 75007, France.

IRAN

Ziai, Iradj, 12 Voiei L. Terray, 95000 Creteil, France.

UNITED STATES

Bracken, Paul J., School of Organization and Management, Yale Univ., New Haven, Connecticut 06520, USA.
Brown, Paul S., LLNL, PO Box 808, Livermore, CA 94550, USA.
Chapman, Gary, CPSR, PO Box 717, Palo Alto, CA 94301, USA.
Cochran, Thomas B., Natural Resources Defense Council, 1350 New York Ave. NW, Washington DC, 20005, USA.

De Volpi, A., Reactor Analysis and Safety Div., Argonne National
  Laboratory, 9700 South Cass Av., Argonne, IL 60439, USA.
Drell, Sidney D., SLAC, PO Box 4349, Stanford, CA 94305, USA.
Dumas, Lloyd J., 2506 Chariot Lane, Garland, TX 75042, USA.
Garwin, Richard, IBM Thomas J. Watson Res. Center, PO Box 218,
  Yorktown Heights, NY 10598, USA.
Holdren, John, Energy and Resources Program, Bldg. T-4 Room 100,
  Berkeley, CA 94720, USA.
Long, Franklin A., University of Calfornia School of Social Sciences,
  Irvine, CA 92717, USA.
Paldy, Lester G., State University of New York, Center Sci., Math.,
  Tech. Education, Stony Brook, NY 11794-3733, USA.
Pike, John, FAS, 307 Massachusetts Ave. NE, Washington DC,
  20002, USA.
Postol, Ted, Center Inter. Security-Stanford Un., 320 Galvez Street,
  Stanford, CA 94305, USA.
Rathjens, George, MIT Center Int. Studies, E-38, 292 Main Str.,
  Cambridge, MA 02142, USA.
Taylor, Theodore B., 10325 Bethesda Church Rd., Damascus, MD
  20872, USA.

CANADA

Malcolmson, Robert, Dept. of History, Queen's University,
  Kingston, Ontario, Canada K7L 3N6.

AUSTRALIA

Ball, Desmond, Australian National University, GPO Box 4,
  Canberra ACT 2601, Australia.

SOVIET UNION

Blagovolin, S. E., Institute of World Economy and International
  Relations, Moscow, USSR.
Ershov, A. P., Committee of Soviet Scientists for Peace, Moscow,
  USSR.

Komzin, B. I., Institute of World Economy and International Relations, Moscow, USSR.
Kozlov, V. D., Committee of Soviet Scientists for Peace, Moscow, USSR.
Pozhela, Juras K., Committee of Soviet Scientists for Peace, Moscow, USSR.
Rassadin, A. V., Institute of World Economy and International Relations, Moscow, USSR.
Sergeev, V. N., Committee of Soviet Scientists for Peace, Moscow, USSR.
Vladycenko, Alexandr, Ambasciata Dell'URSS in Italia, Via Gaeta 5, 00185 Roma.

Note: Two co-authors of chapters, Umberto Colombo and Benoit Morel, were unable to attend the Conference.

Part I
Civilian versus Military Technology

1 Do Civilian Spin-offs Justify Investments in Military Technology?

Umberto Colombo and
Giuseppe Lanzavecchia

Though reams have been written in an attempt to justify investments in military technology on the grounds of civilian spin-offs, a really convincing apologia has yet to appear. Useful civilian spin-offs are certainly the justification put forward by the military and it is quite true, of course, that there have been many such instances of some importance. However, specific cases for or against the thesis are not sufficient to lend force to what must be a generally applicable response which is neither arrogant nor ideologically based. We shall therefore endeavour to give a strategic reply founded on technical and economic rather than ethical evaluations, although in the final part of the chapter some ethical problems necessarily come to the fore.

If the question posed in the title of the chaper is understood *sensu stricto* – namely whether investments in military technologies can be justified by the civilian benefits obtainable – it is not difficult to prove that such investments are generally too expensive when judged solely on the basis of the civilian spin-offs obtained, so it is not worthwhile making them. Direct action to attain given objectives is doubtless more effective because it permits the concentration of forces and a reduction in time. From this aspect there can be no doubt that for civilian purposes it is better to make direct investments and not to await spin-offs from military technology. There is also the fact that much secrecy attaches itself to military research, so spin-offs are only partial and there are delays in transfer to the civilian sector.

Couched in such terms the whole argument would end before it had started. But things are actually much more complex. In the first place it is impossible to be unaware of the validity or otherwise of military technology as such and the efficiency of the efforts with which it is pursued, and then consider such matters as the quantity and quality of any eventual spin-offs, the fact of whether and when certain efforts would have been made anyway by the non-military system under free-market conditions, or even whether there would have been an organised effort directed and supported by public means, and finally the actual amount of time the civilian system needs to implement programmes, especially those of a strategic nature, which the defence system can instead tackle quickly and decisively.

It must also be made clear, of course, that the question to which a reply is required recognises not only the existence of a military technology, but also that the effort to obtain the technology could be repaid purely by the acquisition thereof, without any casual spin-offs. However, if such basically gratuitous civilian spin-offs are also forthcoming without the need to justify the effort, then the larger and more timely they are the more attractive they become.

The problem therefore takes on much wider and complex connotations. Here, however, we shall touch only on some direct aspects, such as the effectiveness of the technological effort for defence proper, the spin-offs deriving therefrom, and the way these compare with the results of civilian technological development programmes. In all this, as far as possible, account is taken of the various geopolitical areas, namely the Western World and the United States particularly, the Warsaw Pact Countries, especially the Soviet Union, and the Developing Countries.

The great military technologies, involving nuclear arsenals, 'Star Wars' and advanced sophisticated systems which revolutionise even the most conventional weapons, are developed essentially by the superpowers and to some extent also by other countries both in the West and in Eastern Europe. The solutions adopted often differ more as regards the specific type of technological options considered rather than their intrinsic nature, their capacity to perform the required functions and their efficiency. It could be said that the United States makes greater use of advanced microelectronics, informatics and sensor technologies, while the Soviet Union leans towards very intelligent adaptive solutions which exploit its areas of strength in science and technology, such as mathematics, mechanics and ballistics; there can be little doubt in this regard that the Soviet Union is certainly in the forefront as regards missiles and space delivery systems in general.

It must be made quite clear that, although free-market rules and competition do not really hold good for the military sector, but rather the capacity to pursue certain objectives and to equal or outstrip the adversary, the sector is highly competitive, precisely for these reasons.

A balance must always be maintained with the adversary, who thus becomes a point of reference. This explains why the Soviet Union is certainly on a par with the United States in the military technology sector, while it lags far behind in civilian technologies, because the socio-economic system does not favour competition or risk-taking, and even if original and valid technologies are developed they are often left on the shelf or sometimes actually sold to the West. We have already discussed these aspects very thoroughly in a text for the United Nations University.[1] Then again, because of the very clear-cut separation in the Soviet Union between the military and civilian sectors – as well as the characteristics of the latter – there is substantially no spin-off at all there, while there is a very great amount in the United States; indeed this precise aspect also forms part of the US Administration's policy.

The problem is posed in quite different terms for those countries – particularly those in the Third World – which neither develop military technologies nor make weapons, but try to acquire these from the countries that do produce and sell them. Leaving aside any considerations on development strategy – namely, whether resources spent on armaments would not be better directed to stimulating production and creating infrastructure – and those of an ethical nature, it is clear that, as with all transfers of products having a technological content, there is a diffusion of 'messages', knowledge and know-how which not only modifies the country's military structure, but also its people and society itself. Furthermore certain technological and organisational solutions inherent in weapons and their use certainly make themselves felt in the long run and become the input for similar solutions in the civilian field. Of course they also foster the creation of associated local industries such as those concerned with repairs and the production of spares. Spin-off thus occurs even in the Developing Countries, though it can be readily demonstrated that it is relatively small and quite costly. But, on the other hand, there are even fewer chances of development and innovatory stimulus resulting from the import of industrial technologies, especially in the case of large technological undertakings such as refineries, petrochemicals, steelmaking, aluminium production, and hydro-electric and thermo-electric power stations. Such projects often remain isolated from the real fabric of the country's economy and its culture, and so they tend not to have any induced effects.

The Emerging Countries like India and Brazil (to name but two examples), which are capable of developing their own military technology and of producing arms, warrant separate consideration. All the reasoning advanced so far is completely applicable for these countries, but right in the forefront is the question of what precisely is the most effective development strategy. In fact these countries are not in the vanguard of either civilian or military technologies, and they do not engage in large long-term strategic themes which would bring about civilian spin-offs. A local defence industry is thus justified only by the desire to be independent of arms imports, supplies of which could become somewhat difficult in certain circumstances.

Though the military technological effort may well be effective as regards defensive and offensive capabilities, it is often not very effective in cost–benefit terms. However, it is not easy to make an evaluation and – at the limit – it may well be impossible because of the lack of the basic reference elements, for example, the existence of a non-oligopolistic market. Nevertheless there are approaches designed to try to broach the problem. For instance, various possible solutions can be compared more effectively by taking account not only of the cost of their development but especially that of the subsequent production phase. In fact the former represents a mere 15 per cent of the total cost on average and the latter a good 85 per cent. Yet the production cost is very often completely ignored because it forms part of another set of accounts and because it would change the development risk assessment criteria, broadening these and thus increasing the difficulty of evaluating the risk factor.

Evaluation of the efficacy of military technology spin-off in the civilian field in every instance calls for concrete examination of the enormous case history in this regard and, especially, of the most significant instances of spin-off. Of course, this approach tends to highlight the positive results, and it is far more difficult to take account of cases that have produced little or no spin-off. The most emblematic case is perhaps the development of microelectronics, Large-Scale Integration, and the chip, with the tremendous and increasing miniaturisation ensuing, as well as the computer and information science boom, all of which stem directly from military and space projects, and especially Project Apollo for conquest of the Moon. While the needs of these projects led to spectacular technological developments which enabled the enormous costs to be absorbed, the US Government also had a precise policy for the massive introduction of informatics in the public sector. Procurement for this policy permitted payment of the cost of implementing the technology, thus accelerating the natural times of the market response, while enabling experience to be acquired very rapidly.

The case of microelectronics and informatics warrants examination in some depth. In fact it may be said that, after the initial phase, the boom in the development of civilian microelectronics was such that there were exceptional spin-offs for defence too. Nevertheless it must be borne in mind that the initial phase which led to the breakthrough in large-scale integration was propelled forward and supported by the defence and space programmes. It was only subsequently that development was aimed especially at informatics in general and the computer in particular. This development was fostered substantially by defence and space interests in the military sector and by the public administration in the civilian sector, in the concerted programme just referred to. In any case, microelectronics is not generally an end in itself; it serves for other actitivies and developments which, in their turn, provide room and scope for microelectronics development. Hence there is a very strong feedback which is decidedly beneficial for the sectors directly involved, while it is also indirectly advantageous for the economy as a whole, with the exploitation of the ensuing new machines and fresh knowledge. Where microelectronics is concerned, the crucial point for judging the effects of military research in the civilian sector is that it sparked off extremely innovatory activity which otherwise – in free-market conditions, or even with Research and Development (R and D) projects of the conventional type – would have occurred very much later. This activity is framed in a long-term strategic panorama generally adopted for military programmes, much more frequently and much more efficiently than is usually the case in the non-military sector. In the specific instance referred to, moreover, there was a government-promoted development policy for the civilian sector which resulted in a positive cybernetics circuit that permits an enormous reduction in the development times of the relevant technologies. Finally, there is the fact that the microelectronics industry resulting from this process has turned out to be such a lively driving force that it has outclassed even the innovatory capabilities of the defence technicians. This is the most convincing demonstration of the extraordinarily positive nature of the spin-off generated by an initial technological investment. It will be appreciated, however, that, especially in a country like the United States, microelectronics defence research or a microelectronics industry without civilian ties makes no sense; neither in general does a defence technology that is isolated from the rest of the country.

The foregoing is just one example of the innumerable technological innovations and relative spin-offs that occurred in the 1960s as a result of military and space projects, especially Apollo. The latter, in fact, is particularly worthy of attention, being an enormous, complex systems project which called for the development of materials, components, instrumentation, machinery and software in a single whole in which each of the many elements had to be absolutely reliable to ensure the proper functioning of the entire system. It is perhaps the acquisition of experience with such complex systems, in which all the various pieces fit together properly, which is the most important spin-off bonus of such military and space projects. Experience of this kind brings enormous benefits and gives the country concerned a very definite advantage over its rivals.

Though the examples just mentioned are undoubtedly the most meaningful, there have been very many other instances of spin-offs worthy of consideration. Prior to the severe loss associated with the Challenger disaster, the National Aeronautics and Space Administration (NASA) was a particularly generous source of spin-off for the United States. NASA has recently focused attention on this fact with a small book.[2] This recalls the spin-offs for aviation, as well as space developments and the commercial use of space. However the accent is placed particularly on 'technologies twice used' for safety, health and medicine, transport, energy, the environment, manufacturing industry and the increase in productivity, agriculture, and the development and rational use of mineral resources. A particularly interesting chapter is that concerning spin-offs for the final consumer, the home, leisure-activities, sports shoes designed to reduce fatigue because they are better able to absorb shocks, new reflecting fabrics made to provide protection from heat or cold, pens that can operate at low and high temperatures, eyeglasses which eliminate 99 per cent of the ultraviolet radiation and technologies for the control of humidity in closed environments or for the elimination of fumes, dust and so on.

Another sector of great interest, which owes much to spin-off from military and space-research, is that involving new materials: special metals, alloys, amorphous metals, ceramics, materials with ion implants, new fibres and composites. To an ever-increasing extent, the latter are being designed for the most diverse applications, including very high temperatures, utilising metal or ceramic matrixes and carbon or ceramic fibres for the purpose. It can be said that the recent revolution in materials – both functional and structural – occurred precisely as the result of military and space spin-off.

All these examples provide no more than a few quick glimpses at what is a veritable ocean of the most disparate technological transfers, on which we shall not dwell further at this stage, though we should

like to draw attention to the fact that the US General Accounting
Office publishes sheets, technology by technology, on Military Research and Technology Transfer, while information on technological
transfer from the defence to the civilian sectors can be found in many
Data Banks. A fact that must be highlighted, however, is one which
differentiates the US situation not only from that in the Soviet Union
but also from that in European countries strong in military technology such as Great Britain, France and Sweden, namely the
organised, global, far-reaching efforts made by the United States to
exploit the results of military technology in the best possible way for
civilian use.

When speaking of military technology and spin-off, mention must,
of course, be made of the most controversial programme in recent
times, the Strategic Defense Initiative (SDI). For the purpose of our
theme, this programme must be examined not so much from the
aspect of its intrinsic soundness, of whether or not it can be implemented in whole or part, or of the political advisability of so doing,
but rather from the aspect of the transferability of the various
technologies to the civilian sector (which could be of direct interest to
the European firms participating in the programme) and the global
experience that can be acquired from the development of the whole
system (which is of importance especially for the United States).

Many of the technologies needed for the programme are closely
bound up with the space industry in the broadest sense of the term.
This industry, of course, is one of increasing importance and interest
also from the commercial standpoint, but it is one involving costly,
high-risk developments which call for collaboration between companies and countries. The most obvious spin-offs in the non-space
civilian sectors come from the development of sensors, supercom-
puters, artificial intelligence, advanced software, lasers, and new
semiconductors such as gallium arsenide, as well as from arms
development in the narrow sense. Thus the spin-offs from research on
sensors and the pertinent support technologies (such as data processing) are fundamental for remote sensing which will also benefit from
the use of rotating mirrors that permit stereoscopic detection and
redundancy. A further aspect concerns the development of new
communications systems, with pulsed lasers and ultra-rapid switches.
Yet another is the automatic production of software that will have
extraordinary consequences in the non-military sector.

Where arms are concerned, the development of those based on
kinetic energy and also those utilising superconduction, for which an

exceptional future is envisaged, opens up futuristic hopes regarding terrestrial and space transport, as well as a great number of other applications. There are, of course, many other SDI technologies that can find striking application in the civilian field. Here it is sufficient to mention what is perhaps the most praiseworthy aspect of the programme in terms of novelty and quantum leaps in the concept of systems. We have also discussed this aspect amply elsewhere.[3]

While a project such as Apollo represented a gigantic and complex but closed system – in the sense that each of its elements (component, machine, software) was pre-designed and indispensable – SDI is an open system, namely one which accepts any solution, component, machine or software that can be useful to make it work. This is in fact an approach which views a system in such a way that it can perform a certain function. The system must thus be open, perfectable, adjustable, extremely flexible, and hence be characterised by a great many degrees of freedom. To say at the outset, therefore, that it cannot operate because one or even all the proposed technological solutions are inadequate is to demonstrate ignorance regarding its most salient aspect, namely the system's self-adjusting capacity, of which it has already provided proof.

Few military programmes have given rise to such advserse reactions as SDI. Many of the criticisms stem initially from an ideological stance but then move on to conclusions that are not only political in nature but also focus on such aspects as technological feasibility and economic costs. While criticism of a military programme on ideological grounds is quite acceptable, this is hardly the case where other aspects are concerned, especially if the investments in SDI can be justified in terms of spin-off. There are, however, numerous technical reports that are critical of the programme, but even to try to review the few which are most important would involve an analysis that is too lengthy for a chapter such as this. We shall limit ourselves, therefore, to recalling the analysis by Jennifer Sims, and the book by Walter Zegveld and Christien Enzing.[4] The former indicates how participation in the programme can be advantageous for European industrial enterprises. The latter, after a broad analysis, concludes that '... from a standpoint of industrial technology policy *per se* SDI is not a cost-effective approach and could be defined as counterproductive'. With but few exceptions, this anyway is the key with which one could read any military programme in terms of thrust towards a nation's socio-economic growth, where the development of individual technologies is concerned. Yet they can make some very interesting contributions whenever they incorporate innovatory approaches to problems which the civilian system, by its very nature or because of other restraints, would not be capable of tackling or would tackle only with great delay.

The present crisis in the United States stems from the weakening of its industry, its balance of payments deficit, and some backwardness in the high-tech sectors which may be attributed in part to the fact that an excessively large share of the R and D expenditure is for military purposes, leaving too small a slice for innovatory civilian developments. However one must be careful in judging the US high-tech crisis by reference to such indicators as international trade, which was in the black to the tune of 27 billion in 1980, but $3 billion in the red in 1986, since American multinationals pursued a very precise foreign investment strategy during the period concerned. This explains the apparent deficit, because, although there was a reduction in exports, the United States was, at the same time, penetrating world markets, especially in Europe and Japan.

Even though evaluation of spin-off may remain controversial, it is clear that there are a great many pointers indicating that it is not very efficient on average and that a direct effort in the civilian sectors could be much more profitable, as shown quite clearly by the example of Japan and also by the recent upturn in Western Europe. Yet there is the fact to consider that technological programmes in the defence sectors are often concerned with very advanced themes. Moreover they take a long-term strategic view, while also ensuring an enormous concentration of means, a global systems approach and often the shortest possible development times. This situation is not generally found in corresponding civilian projects. It suffices to examine even the most valid major civilian projects for technological developments to become all too aware of this. The kind or projects concerned are those which call for a concentration of means and involve an element of risk exceeding the capabilities of even the largest individual companies or countries.

There are numerous such examples in Western Europe, Japan and more recently also the United States, for instance, Ariane and European space research, the Airbus, Community projects such as Esprit and bottom-up programmes such as Eureka, where European firms join forces to develop what are very important technologies, but within quite specific limits. The same holds good for such ambitious projects as that involving the fifth-generation computer in Japan (which, after the difficulties encountered in pursuing its initial objectives, has now been renewed in a most convincing manner). Then there is the extremely important case of the construction of Sematech, the advanced factory for chip production, promoted by the Semiconductor Industry Association and US computer producers, to which, moreover, the US Department of Defense has also been asked to contribute. These are demanding well-run projects with quite serious and often well-defined objectives. Hence they tend to be much more effective than the spin-offs – sometimes quite casual – stemming from military efforts to develop specific technologies. But such projects cannot meet all civilian needs for technology to resolve economic and social problems, because they lack an in-depth, long-term strategic outlook. Such an outlook is, of course, essential for the pursuit of the large, complex objectives of society worldwide – not just that in the advanced countries – on which a number of intrinsically different technologies must be concentrated and rendered symbiotic and synergetic by utilising a global systems approach.

It should be pointed out that, although the approach which involves the pursuit of specific individual technologies such as the chip, the supercomputer, the laser, the robot, artificial intelligence and advanced software is important and sometimes indispensable, it is now somewhat outdated. This is an approach which permits the creation of instruments, considered useful, with which it is certainly possible to do something, once they are available. However a more modern approach is that which involves studying the problems and understanding the needs, after which the objectives can be fixed and the specific technological instruments to attain them can be developed. In this regard the military system is much more modern, advanced and hence efficient. It can therefore guarantee the attainment of its own aims and also in some cases produce important spin-offs much more quickly than can civilian programmes. It must be appreciated, however, that a very profound change is occurring in the economic–industrial system. Firms which were once suppliers of products are becoming suppliers of solutions, offering systems that include products, services and knowledge. The process regarding the transformation of the function of industry is certainly one of the most important of the many changes taking place today. It involves both company organisation and the structure of production, research and services. But perhaps the most outstanding aspect is the overturning of the production–market relationship: the practice of producing goods first and then trying to sell them no longer holds good; nowadays it is the market (in the broadest sense) which to an increasing extent guides research, design and hence production, thus becoming the starting-point of a company's operations.

The independent development of technologies – leaving aside consideration of the problems they should help resolve – followed by the stage of finding uses for them, is like producing goods and then trying to sell them. This, of course, is a somewhat schematic representation of a much more complicated reality: many of the technologies now in the development stage will certainly be used to resolve a host of tomorrow's problems. However, the schematisation provides a reasonable picture of the general trend. Civilian society, too, must therefore learn to be more strategically oriented and also to set course for large global solutions without having to wait for spin-offs from the development of military technologies aimed at defining wide-ranging strategic lines that, anyway, were not originally conceived for civilian purposes.

To name but one of the problems that will certainly have to be tackled, it is sufficient to think of the environment, within the context of growth for the whole world – but especially for the Developing Countries – with a population that in the coming century will probably reach ten billion and require much greater quantities of raw materials, energy, factories, infrastructural facilities and services than today. Then there are the problems of climate, health, safety and reduction in imbalances not only at world level but also within our own countries; while last but not least is the problem of providing better, more universal education, not designed for a static, hierarchical and mechanistic society, but one that is extremely dynamic and innovatory, where the individual will have much more space for creativity in his or her work and leisure-time activities.

One of the biggest problems centres around agriculture in relation to food, the environment and resources. The situation is very different today from what it was even quite recently. Nowadays, in fact, it is not so much a matter of global scarcity of resources – with land productivity growing faster than population, although there do exist some very serious specific situations such as those in Africa – but of avoiding the excesses that are occurring, with the increasing mass of products and ensuing waste and its accumulation, the eutrophication of waters, and the continuous rise in agricultural production leading to massive surpluses. Therefore, unless the agricultural production system is rationalised, especially in Western Europe, the situation will continue to deteriorate further in the future. The problem is not that of utilising the abundance of resources anyway, since the ensuing costs could probably be borne in the next few years, through the incentive of price policies introduced to resolve real socio-economic problems. Instead, what is required is an innovatory effort involving a global systems approach, to find all-embracing technological and organisational solutions that are sound from the economic, environmental, and social aspects.

There are a number of key points, of course, such as the fact that food production cannot be further increased; indeed, in many areas it must be restricted or reduced. At the same time it is not possible to accept solutions calling for reductions in land productivity – an antihistoric, tremendously costly measure – or reductions in the amount of land cultivated. There are several reasons for rejecting such a reductionist approach, including the need for jobs in the rural areas, but more importantly the fact that, in densely populated areas, such as Europe, the land cannot be abandoned and left to deteriorate either naturally or under the pressure of human activities. Land must be defended via an active presence which guarantees protection of the soil, vegetation, and irrigation and drainage systems, while ensuring support for mountain areas, the disposal and reuse of wastes and so on. Agricultural production must thus continue to grow, though it must be channelled in different directions, where it can be useful and profitable, such as in the production of raw materials and derivatives for industry, energy, and the creation of woodlands near towns. Land-use must be planned differently. New, efficient, economic crops must be invented, and all the spin-offs of this new activity must be developed and supported, from the transformation of the biomass, to agritourism and recreation.

A particularly interesting aspect of this policy is represented by the production of energy biomass. Indeed this forms the basis of the European project known as IDEA (Innovative Dimension in Energy and Agriculture) which could lead to the large-scale production of energy biomass in Europe. Biomass energy would also help alleviate the problem which is tending to get worse and could, in the long run, become catastrophic for the planet, namely the increase in atmospheric carbon dioxide with an accentuation of the greenhouse effect and the ensuing rise in temperature. With the help of biotechnologies, new species must be identified and created or old species modified. New cultivation techniques must be found along with new processes for biomass transformation and utilisation for recovery of the most valuable by-products. And last but not least agriculture and the territory as a whole must be restructured to create, for instance, a new, sounder relationship between town and country. The scope of Project IDEA, however, extends beyond mere consideration of European problems – important though these may be – to include especially the general aspects of third-world agriculture and some specific aspects of crucial areas in the countries concerned.

Another big civilian systems programme – 'Human Frontier' – was presented at the end of 1986 by the Japanese Agency of Industrial Science and Technology. The aim is to understand the organisation of animal and plant tissues at molecular and submolecular level and to improve knowledge of the major human functions concerning the senses and movement. The final goal of these investigations is to reproduce biological systems with the numerous technologies now available or under development: biotechnologies, chemistry, new materials and informatics. The 'Human Frontier' programme is designed first and foremost to provide an example of a new scientific and technological concept, in harmony both with nature and with human societies.

These programmes and more generally the systems approach to the problems of the economy and society can take account of a great many aspects – even those that are not strictly technical – within a strategic panorama of development. Though they certainly contain Utopian elements, we consider them essential to the creation of a better future for all. After all, the search for peace and disarmament is also part of a long history of human Utopian aspirations, yet it remains an objective we cannot but pursue.

We should like to recall, in concluding, that at least in Western Europe – among citizens and countries alike – the lesson has finally been learnt that war does not pay and that there are other ways of handling differences of opinion and interests: a war among them is no longer conceivable. But, while the objective of military power manages to mobilise gigantic forces over long periods in a great strategic vision, we have yet to learn to concentrate similar forces on the great problems of civilian society in a global vision of world development. If we succeed in so doing, it will be a sign of true civilisation. To this end it is necessary first and foremost for the moral and intellectual forces which oppose armaments policies and the risk of war to unite positively so that the objectives of human society become those of development and solidarity.

Notes

1. U. Colombo and G. Lanzavecchia, 'The Situation and Future of Technology in Europe', in E. Laszlo (ed.), *Europe in the Contemporary World* (London, 1986).
2. National Aeronautics and Space Administration (NASA), *Spinoff* (Washington, DC, 1986).
3. U. Colombo and G. Lanzavecchia, 'The Great New Technological Research Projects: SDI, Eureka, GIF' (Italian text presented to the Communist Group in the Italian Senate, 17 March 1986).
4. Jennifer Sims, 'SDI and Spin-off Technologies' (Aspen Seminar, Aspen Institute Italia, Rome, 24–5 February 1986); and Walter Zegveld and Christien Enzing, *SDI and Industrial Technology Policy: Threat or Opportunity?* (London, 1987).

2 Non-Military Justification of Military R and D: Rationality or Rationalisation?

Lloyd J. Dumas

EXTERNALITIES AND MILITARY RESEARCH AND DEVELOPMENT

Those who advocate high levels of military research and development (R and D) have frequently argued that these efforts further a variety of non-military goals. It is neither unusual nor inappropriate for advocates of any public policy to point out the side-benefits of the policy they support. Such 'positive externalities', as economists call them, have long been accepted as relevant to social policy decisions. For example, supporting public education of children by taxing the general population (including those who have no children in the public school system) has been justified by the general social benefits of a more educated citizenry, not just by the value that directly accrues to the students or their families. But such arguments must be viewed with care.

For one thing, it is possible, even likely, that any activity with substantial beneficial side-effects may impose substantial indirect penalties as well. Rational decision-making requires that all significant externalities be considered, those that are negative offsetting those that are positive. Furthermore, when indirect effects become a major selling point for a policy, it is wise to ask whether this is because the alleged direct benefits of the policy are too questionable to generate much support. If so, it is worth considering whether the desirable indirect effects of this policy could be achieved more directly. If the indirect effects are really the main reason why the policy seems attractive, a more direct approach is almost always certain to be more efficient.

There is no question that military R and D produces results that

17

sometimes 'spin off' to civilian application. Yet is seems unwise to accept this as an open-ended justification either for specific military technology programmes or general levels of military R and D effort. Each programme of military R and D should be able to stand on its own merits. If it cannot, it should not be supported. Setting appropriate levels for military R and D as a whole must also contend with the negative indirect effects created when large numbers of engineers and scientists are drawn out of the pool of technologists available to pursue the non-military, technological goals of society. More will be said about this later.

The more ardent advocates of spin-off as a non-military justification for military R and D argue that societies chronically under-invest in technology in the absence of large military R and D programmes. Put simply, their argument is that people have always made the greatest technological progress during war or preparation for war. The public simply will not support extensive R and D efforts unless they feel that such efforts are critical to their security. Large-scale R and D is just too expensive, too uncertain and too long-term to garner popular support in the absence of a perceived threat. Therefore, even if it is true that those advances in civilian technology that spin off from military R and D could have been achieved faster and cheaper by a massive civilian R and D effort, such a civilian effort would never have been undertaken. Therefore large-scale military R and D is critical to the rapid progress of civilian technology.

Implicit in this argument is the assumption that people cannot be motivated to support any large, expensive and long-term investment except by fear, especially when the results of that effort are uncertain. But this is ludicrous. People do this all the time. Although the level of uncertainty is low, national programmes of road-building, power plant construction and other major infrastructure projects are certainly large, expensive and long-term, yet they are routinely undertaken with wide public support. The outcome of major investments in public education is to a degree uncertain, but neither that fact, the long-term nature of the projects, nor the considerable expense involved prevent public support. The national commitment to put a man on the moon by John F. Kennedy early in his Presidency was clearly an uncertain, expensive and long-term project, yet it was very popular, engendering not merely support but enthusiasm. Furthermore there have been and continue to be nations that support extensive civilian R and D efforts without substantial programmes of military R and D, perhaps the best present example being Japan.

Interestingly it is common, though curiously inconsistent, for this argument to be made by those who also strongly advocate the benefits of unfettered capitalism. Central to the operation of the market system is the primacy of economic self-interest, not fear, as a motivator of the human behaviour. Particularly critical is the entrepreneurial or 'risk-bearing' function, in which uncertain projects, including large-scale, long-term investments, are deliberately undertaken in pursuit of profit. Since technological advance can unquestionably open up vast new profit-making opportunities, investment in non-military R and D certainly fits into this pattern. Why then should it be impossible or even improbable that civilian R and D would advance rapidly in the absence of spin-off from massive military R and D efforts?

THE CRUCIAL ECONOMIC ROLE OF TECHNOLOGY

If the net effect of military R and D programmes was to give rise to spin-off that generated rapid advance in civilian technology, this would provide a powerful non-military justification for supporting them. The issue of whether large-scale investments in military R and D enhance or retard the progress of civilian technology takes on special importance because of the critical impact such technology has on the performance of the economy.

Civilian technological progress has clearly given rise to all sorts of improvements in existing products, and created whole generations of new products aimed at advancing the central purpose of the economy – improving the material well-being of the population. This has certainly been important. But technological advance has also played a deeper, more structural role, crucial to creating the conditions under which the vast majority of the population and not just a privileged few could experience sustained improvement in their standard of living.

Most people earn the largest part of their income in return for their labour, that is, in the form of wages and salaries; relatively few live primarily off rent, interest or profit. Consequently when wages and salaries are rising, most people's primary incomes are rising. With more money to spend, they can achieve a higher material standard of living – provided prices are not rising as fast as their money income. Rising prices erode the purchasing power of the incomes people receive. Thus increasing real income requires that wages and salaries rise faster than prices. But rising wages and salaries mean rising labour costs, and for most producers labour cost is the largest part of the cost of production. How, then, can wages and salaries rise without creating cost-push pressures that force producers to raise prices?

Producers can offset the rising cost of labour (or, for that matter, the rising cost of any resource they buy) if they can find and introduce more efficient techniques of production. With greater efficiency, that is, more output from the same amount of labour (or fuel, or materials), higher cost per unit of input need not translate into higher price per unit of product. For example a 10 per cent increase in hourly wage can be completely offset by a new production technique that generates 10 per cent more output per labour-hour.

There are many ways to increase the efficiency of production. But the advance of civilian product and process technology is unquestionably the most powerful. The discovery of improved processes, better designs for products and the machinery and equipment used to produce them, new materials, and the like, have led to the spectacular increases in productive efficiency that have so greatly enhanced productive capacity over the centuries.

A surging stream of advancing civilian technology allows wages and salaries to rise while prices remain relatively stable or even decline. Higher money incomes then mean greater purchasing power for the mass of the population, and a sustained widespread improvement in living standards becomes not only possible but highly likely. The advance of civilian technology does not therefore play the marginal role of providing a continuous flow of new gadgets. It is a key force driving economic efficiency; it is central to the creation of real economic growth.

TECHNOLOGICAL MISCONCEPTIONS

Many popular misconceptions about technology arise from the failure to see it as the result of a social as well as a scientific process. Perhaps the most common is the idea that the development of technology proceeds along a single path whose direction is prescribed by the laws of nature and the nature of scientific inquiry. Society may be able to influence, though probably not to determine, the speed at which this path is travelled. But it can neither influence nor determine its direction. Technology has its own imperative. Nothing could be farther from the truth. Rather than a single path of possibilities, technology is more like a complex, interconnected network of roads leading off into the as yet unknown. Which of the many possible paths are taken and which forsaken is a matter not of scientific necessity but of social choice.

Scientists and engineers are the repositories of the vast store of accumulated scientific knowledge and the techniques for applying that knowledge. As such they are specially capable of performing the research and developing the applications crucial to technological advance. But there is nothing in the training or technical experience of engineers or scientists that makes them any more capable than the lay public of choosing which of the available general directions for future research should be explored and which ignored. Nor does their knowledge make them more capable of deciding to what purpose the existing body of scientific and technical knowledge should be applied.

Another commonly held belief is that when one particular branch of technology is abandoned and another pursued, it is because the new technology is scientifically superior to the old. But many times a line of technological development is abandoned because its characteristics are judged not to fit prevailing or projected economic, social or political conditions or requirements as well as the new technology – not because it is somehow technically inferior. For example, earlier in this century the pursuit of lighter-than-air aircraft technology was largely abandoned in favour of pursuing emerging heavier-than-air technology. The technology of heavier-than-air craft was superior in that it offered the potential for greater speed and less sensitivity to weather conditions, but it was also far less energy-efficient and required larger and more elaborate ground support facilities. Had speed been less of a social (and military) requirement and energy much more costly, technological development of aircraft would have taken a very different direction.

Another common misconception is that technology proceeds as the result of scientific breakthroughs that spring out of the air; they are cut of whole cloth, the result of genius, of a single flash of insight – a misconception greatly encouraged by film-makers and playwrights. In fact, the progress of technology depends heavily on the integrated efforts of many different individuals. Much of it is a methodical, systematic process of discovering small pieces of new knowledge and fitting them together to create incremental improvement of existing products and processes. Even when breakthroughs occur, as they sometimes do, propelling technology spectacularly forward, they are built on a vast store of previous discovery which sets the stage for this new leap forward.

For example, the invention of the incandescent light bulb by Thomas Edison, one of the most prolific inventors of the twentieth century, depended directly on a great number of prior discoveries relating to properties of electricity, vacuums and materials (among others) developed as a result of the efforts of many earlier researchers and inventors. Nor is it just the successful pieces of discovery that support such breakthroughs – knowing what does not work, what is not true is important as well. Furthermore Edison did not simply sit down one day and invent the light bulb. Countless experiments were slowly and carefully carried out with the aid of his assistants, trying out, for example, hundreds – perhaps thousands – of materials as filaments before finding that tungsten worked best.

Advances in basic research tend to be of interest primarily to the scientific and engineering community. Very often they are barely noticed and but dimly understood by the wider society. Advances in applied research, however, tend to catch the public's attention and imagination to a much greater extent. The laboratory development of a potential AIDS vaccine, the first creation of a fully three-dimensional visual image by holography, the initial use of fibre optic material to transmit information – these are much more widely reported and discussed.

On the other hand, the final development phase of technological advance, the 'commercialisation' part of the 'D' of 'R and D' is often regarded as far more pedestrian and far less interesting by both the technical community and the lay public. It is as if all the excitement, all the fire, all the achievement attends to the first technically successful innovation. Yet it is development which turns a technically interesting discovery into an economically viable, practical product or process.

The main point is that the game of technological progress is not won when the laboratory prototype of applied research succeeds. There is a great deal of engineering and science yet to be done and until that succeeds the innovation remains a technically intriguing phenomenon, not a potent force that can penetrate society, with all the social, economic and political effects of new technology about which so much has been written. The mere existence of a device in a high-priced, unreliable, difficult-to-operate form that may be appropriate for some specialised use does not guarantee that it will successfully spin off into more general use. The efforts of large numbers of engineers and scientists may still be required before that crucial transition can be made.

Both breakthroughs and the steady, unspectacular progress in science and technology that cumulates to major advance are thus social processes, which crucially depend on the interaction of the efforts of many individuals. The wider the dissemination of previously developed and newly discovered scientific and technical information, the greater the opportunities for interaction among contemporary researchers, the more technological progress tends to flourish. The breakthrough, flash-of-insight, solitary-genius view of technological advance is misconceived.

Yet another common misconception is that it is relatively easy to reshape particular technological R and D results to fit alternative social application. This may be due to a lack of appreciation of the extent to which particular operational goals established for R and D programmes shape the technology that emerges. While it is certainly true that those engaged in R and D are never really sure exactly what they will find, it is also true that what they find is very strongly conditioned by what they are looking for.

Operational R and D goals are established by social and political decision, as was discussed earlier. They shape the nature of experiments done and the way results are interpreted. They also shape the decision as to which lines of inquiry seem to be most effective in pursuing the established goals and should therefore be followed, and which should be abandoned. Different goals will substantially alter the process by affecting all of these decisions, and consequently affecting the R and D results. It may therefore be very difficult to apply the results of R and D directed to one set of operational goals to a very different purpose.

Much of the discussion surrounding the Strategic Defense Initiative (SDI) or 'Star Wars' proposal illustrates this particular misconception very well. It has been argued that laser research carried out under SDI offers great non-military benefits, since lasers have a wide range of application to medical, industrial and consumer-oriented purposes. Because of the publicity given such innovations as laser eye surgery, laser printers and music played back by laser on compact discs, these arguments strike something of a responsive chord in the general public. But R and D aimed at finding technology appropriate to the development of very-high-powered lasers that can follow a rapidly-moving target at a great distance and burn through its metallic skin or knock it out of the sky is very different from the kind of R and D appropriate to the development of very-low-powered, extremely precise lasers that can gently repair a detached retina. There may be some results that transfer easily between these two types of research, but it is very likely that they will be extremely rare.

SPIN-OFF AND DIVERSION

Large-scale continuing programmes of military R and D divert significant numbers of engineers and scientists from civilian-oriented technological development. It has been roughly estimated that something on the order of 25 per cent of the world's technologists are pursuing military R and D objectives. In the United States, it seems that 30 per cent is a fairly conservative estimate.[1] In all likelihood, it is higher still in the Soviet Union.

There can be little doubt that, if the talents of these individuals were instead devoted to the civilian-oriented R and D, considerable direct contribution to the advance of non-military technology would result. Since this contribution is currently foregone, it represents a social-opportunity cost, a penalty attached to their present employment. On the other hand, the results of the military R and D they perform sometimes spin off to civilian application. Both the negative 'diversion' effect and the positive 'spin-off' effect of military R and D on civilian technological progress do exist. The real question concerns the relative strengths of these two effects. It is this which determines whether the net effect of military R and D is to stimulate or to retard the progress of civilian technology.

If spin-off were the more powerful effect, nations heavily engaged in military R and D would have a considerable advantage in developing civilian technology over nations than do comparatively little military R and D. We would therefore see nations like Great Britain and the Soviet Union at the forefront of civilian technology, while nations like Japan and West Germany lagged far behind. But, of course, just the opposite situation obtains.

Furthermore, if spin-off were the larger effect, the huge amount of resources poured into military R and D in the United States over the past few decades would have made this decade a golden age of American technology. Instead it has been clear for more than ten years that the relative rate of civilian technological progress in the United States has been seriously retarded. The sorry state of American civilian technological advance has been widely chronicled in such popular journals as *Business Week*, with stories bearing titles such as

'The Breakdown of US Innovation' and 'Vanishing Innovation'.[2] It has been documented in reports of the National Science Board, the governing board of the National Science Foundation.[3] Even the 1985 Report of the Commission on Industrial Competitiveness, specially appointed by President Ronald Reagan and heavily weighted with executives from military industry, strongly emphasised the problems of American commercial innovation.[4] In short, it is clear that there is not nearly enough spin-off to compensate for effects of the sustained diversion of such large numbers of engineers and scientists as have been drawn into military R and D. Spin-off is a singularly poor non-military justification for military R and D.

Those who advocate the spin-off argument typically make reference to examples of spin-off such as radar or early computer technology that originated during the Second World War or in the early postwar years. In that era military and civilian technologies were much more similar than they are today. There has been a very striking divergence. The gap between a Second World War vintage bomber and its contemporaries in civilian aviation was minuscule as compared to the gap between a B1 Bomber and a Boeing 747. That divergence has sharply reduced the extent to which military R and D generates usable spin-off to civilian technology.

Further, in the spirit of the earlier discussion, the capability for high performance under extreme conditions demanded by present military forces, combined with very little attention to cost, generates technology that requires considerable additional development to spin off effectively to economically viable civilian application. In other words, it is not possible to capitalise on what limited spin-off potential does exist without the efforts of a considerable pool of engineers and scientists. A nation diverting a large fraction of its technologists to military R and D may simply not have enough engineers and scientists working in civilian R and D to make even the limited spin-off available happen. For example, the spin-off from American military technology that does occur comes, increasingly, out of Japanese industry.

In addition, the secrecy that nearly always surrounds military R and D programmes inhibits the free flow of information and collegial interaction so vital to technological advance. A civilian application cannot spin off from a military technology to which non-military researchers have no access. And the paranoia that flows from obsessive militarism also attempts to close off channels of information and interaction involving non-military technologies deemed likely to have military application. This fear of 'reverse' spin-off further retards the progress of civilian technology.

The implications of the net drain on civilian technological progress created by the persistence of large-scale programmes of military R and D is well illustrated by the case of the United States. The nation that was once a fountain-head of so much industrial innovation has dramatically slowed the rate at which it generates new commercial technology relative to many of its economic rivals. As the stream of civilian technology slowed, American industry progressively lost its ability to offset costs. As a result, higher wages, higher fuel costs and the like were rapidly translated into higher prices. During the past twenty years, the United States experienced considerable cost-push inflation that not only wore away the purchasing power of American consumers, but priced US products out of the market. The consequent loss of export markets and domestic markets as well to foreign producers caused the loss of many high-paying industrial jobs, generating high unemployment and much higher under-employment. It also threw the US balance of trade, which had been positive for 77 years in succession, into deeper and deeper deficit. It took only two years to double the record-breaking trade deficit the United States experienced in 1983.

The much celebrated victory over inflation in the United States in the past few years is a hollow one indeed. It was achieved as a result of widespread wage concessions, the shifting of considerable employment to low-wage occupations, a fortuitous external fall in oil prices, and an avalanche of low-priced imported goods. But it makes very little difference if the standard of living slips because prices are rising faster than wages or because wages themselves are falling while prices are more stable. There has been no improvement in the cost offsetting capability of American firms in the past few years, which remains the key to achieving stable prices with *rising* wages, the path to an improving rather than deteriorating standard of living. The standard of living of Americans did in fact decline between the mid-1970s and the mid-1980s. A man and a woman, each earning the median income for their respective gender, saw their combined purchasing power fall by more than 10 per cent between 1973 and 1985.[5] There is nothing special about the United States that produced this result. It is the predictable outcome of sustaining an excessively large military, industrial and scientific establishment over an extended period of time. Whether capitalist or socialist, large or small, more developed or less developed, any nation that follows such a path can expect eventually to bear the consequences of economic deterioration. The shape of the decay may differ, but the reality of decay will not.

CONCLUSIONS

R and D directed at the improvement of product and process technology for civilian-oriented industry plays a crucial role in creating the gains in productive efficiency that make sustained improvements in the general standard of living possible. The justification of military R and D on the grounds that it greatly stimulates the economically powerful process of civilian technological progress is based on a series of misconceptions about the nature of technology and the forces that shape its advance.

Technological development is a scientific process that is socially directed. It is not unidirectional, but is capable of following whichever of the many technically feasible paths are chosen by the interplay of various social forces. Dramatic advances in technology do not develop mainly through the work of the solitary genius, but through a combination of incremental advance and breakthrough, both of which depend on the work of many individuals. This is true of all stages of the R and D process, including the often denigrated but socially vital 'commercialisation' phase of development. R and D results are very much shaped by the goals and conditions under which the R and D process goes on, and are often not nearly as malleable as is commonly believed.

The net result of all of this is that the spin-off of civilian application from military R and D is nowhere near as potent a stimulus to civilian technological progress as is commonly argued. Instead, the persistent diversion of large numbers of engineers and scientists to military R and D is a powerful counterthrust to the engine of civilian technological advance. As such it retards the improvement of industrial efficiency, undermining the ability of producers to offset rising wages and generating pressures towards higher prices and/or lower wages. Under such conditions, continued, widespread improvements in the material standard of living become exceedingly difficult to achieve.

It is a grand delusion to believe that large-scale programmes of military R and D offer us the prospect of a better, more abundant life. For the concern of military R and D is after all with finding more efficient ways of destroying goods and killing people, not with finding more efficient ways of creating goods and fostering life. If we allow ourselves to continue in our obsessive pursuit of security through R and D aimed at generating more and more sophisticated weaponry, we may not even have a future – let alone a secure or prosperous one. Surely we are capable of doing better. We must find the political will to redirect the energies of those we have entrusted with the accumulation of humanity's scientific and technical brilliance. Finally it is up to us how the great power of that knowledge is wielded. In the end, we will get what we have paid for.

Notes

1.   L. J. Dumas, *The Overburdened Economy* (Berkeley, California, 1986) pp. 208–11.
2.   'The Breakdown of US Innovation', *Business Week*, 26 February 1976; and 'Vanishing Innovation', ibid., 3 July 1978.
3.   National Science Foundation, National Science Board, *Science Indicators, 1974* (Washington, DC, 1975); and *Science Indicators, 1978* (Washington, DC, 1979).
4.   President's Commission on Industrial Competitiveness, *Global Competition: The New Reality* (Washington, DC, 1985).
5.   US Bureau of the Census, *Money Income of Households, Families and Persons in the United States: 1982*, Current Population Reports, Consumer Income, Series P60, no. 142; and *Money Income and Poverty Status of Families and Persons in the U.S.: 1985*, and Current Population Reports, Consumer Income Series, P60, March 1986.

3 Non-Military Justifications for Investment in Military R and D

Roberto Fieschi

INTRODUCTION

The connection between military-oriented and civilian technology (and scientific) research dates back a long time, but only recently has world-wide attention been paid to it. There are two reasons for this. The first is the large proportion of public funds invested in military research, 70–90 billion dollars a year on a world-wide scale (of which 85 per cent in the United States and Soviet Union alone) equivalent to about 40 per cent of all the funds invested throughout the world for research. Figures 3.1, 3.2 and 3.3 show the situation in capitalist nations. Probably there are comparable investments in the Soviet Union.[1] It has been estimated that on a world-wide scale one-quarter to one-third of the engineers and scientists are working on military research. The second reason is the need of governments to justify these investments to their citizens, when traditionally accepted reasons, such as national defence, or more generally, the political–military national interest, are not convincing because of the size of the financial effort. The most recent example of this is the Strategic Defense Initiative (SDI) Project which, more than any other case to date, necessitates the provision of large amounts of funds for the research phase. Even in Italy, when the decision to support industrial participation had to be made, authoritative figures commented that the research spin-off from advanced technological sectors would benefit competitive developments in advanced, scientific and technological, non-military sectors.

During periods of crisis, choices as to what should be done are simple and clear. On 13 May 1940 when the German offensive in the West was moving quickly towards its full development, Winston

29

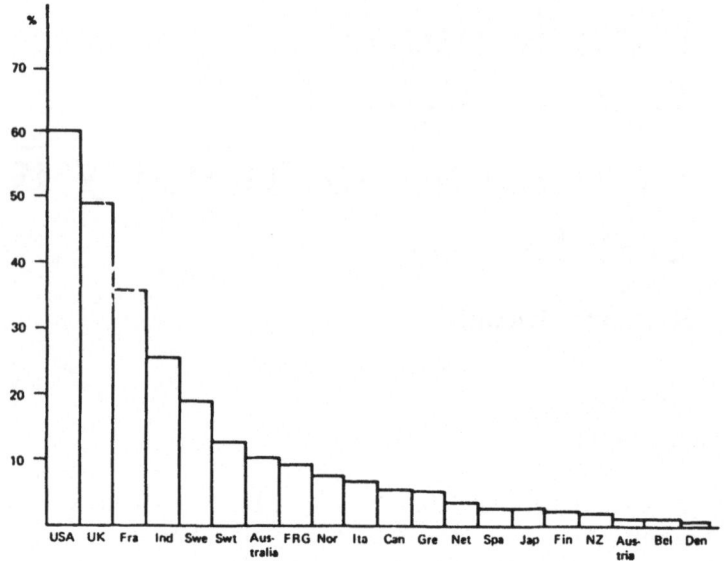

Figure 3.1   Military R and D as a percentage of government R and D, 1981–4 averages
Source: Stockholm International Peace Research Institute.

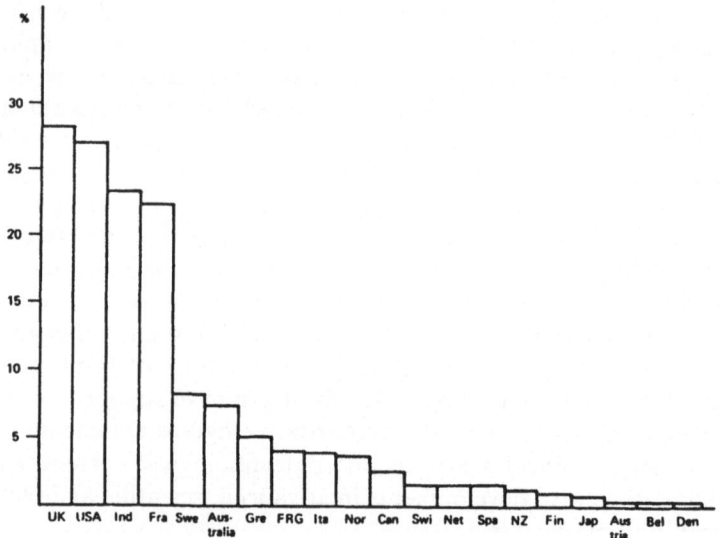

Figure 3.2   Military R and D as a percentage of gross domestic expenditure on R and D, 1981–4 averages
Source: Stockholm International Peace Research Institute.

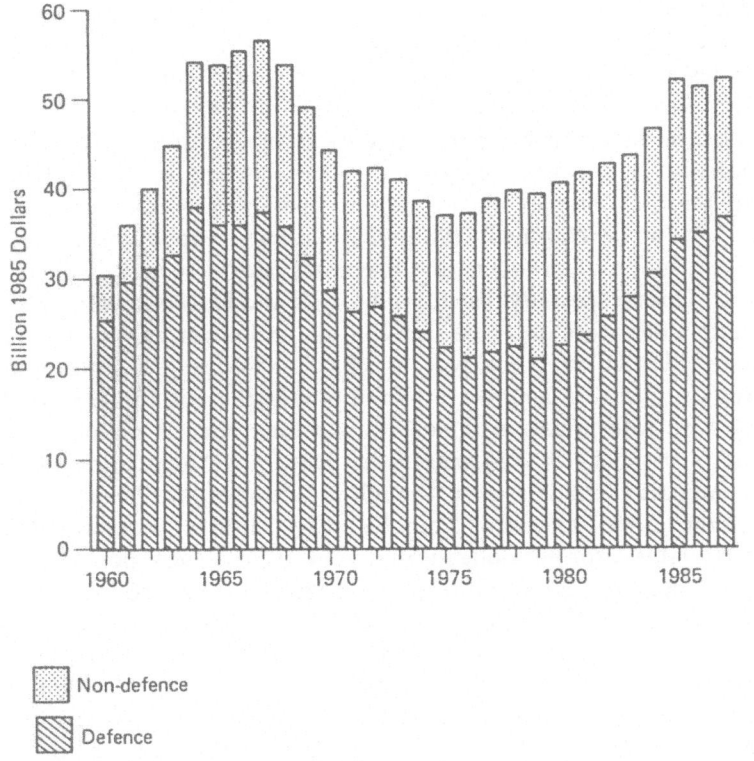

Figure 3.3   Federal R and D outlays, FY1960–87
Source: *FAS Public Interest Report: Journal of the Federation of American Scientists* vol. xxxix, no. 7 (1986).

Churchill made his famous speech to the House of Commons: 'I have nothing to offer but blood, toil, tears and sweat'. In 1941, after the Germans had invaded the Soviet Union, Stalin decreed 'All for the battle, all for the Victory' and from that moment, for a period of four years, nothing which was not directly connected with the enormous national defence effort was produced. During that same year the United States was putting all of its energy into the construction of the atomic bomb. The leaders of the Alliance did not have to justify military investments and thus it was not necessary for them to claim that war technology would lead to civilian benefits in these fields, even if this did actually occur (radar, jet aircraft, nuclear reactors and electronic computers). The reason for the rearmament was the defence of the nation against aggression and this was understood and accepted by their citizens.

Independently of the uses of research, the problem of the non-military consequences of military research does exist and it has to be tackled; in fact, this has been done by several authoritative experts, especially in the postwar period. The problem of technological innovation is complex and in general it is difficult to evaluate how research and development (R and D) investments have contributed to the innovation of products and processes and to economic development.[2] This is especially true for military R and D. If one examines the available literature it can be seen that an evaluation cannot be easily quantified in an analytical way; it is hard to find solid intersubjective data. Thus, in this field more than in others, there is ample room for subjective interpretation which, in the long run, is biased by the writer's or speaker's *Weltanschauung*. Naturally, the present writer is not immune from this, even though information has been collected from the 'hawks' as well as from the 'doves'. It must be admitted that the available information leans in favour of the latters' point of view. A great deal of clear and interesting information can be found in the books edited by F. A. Long and J. Reppy[3] and by J. Tirman.[4]

The consequences of these large military investments should now be considered. Probably the belief of some people that military research greatly influences civilian applications can, in large part, be traced to the fact that large investments in the military sector have had important and positive bearings on the economy, especially with respect to jobs, in some nations and at a specific time – in Germany after the rise of the Nazis, for instance, or in the United States after Pearl Harbor, and then again during the postwar period. However, this problem is more general than that of the R and D influence which needs to be discussed here, since it is primarily connected with the massive production of war *matériel* and full use of the productive manpower. Counterexamples, such as the difficulties which the economies of Eastern European countries have experienced as a result of rearmament, or the success of capitalist countries which invest little in the armament sector, can be cited. In addition, W. Leontiev and F. Duchin have pointed out that if a part of the world military expenditure had been invested in the development of poor countries there would also have been a positive spill-over into technologically-advanced countries; international commerce, production and development would generally have benefited from it.[5] At any rate, in the event that it were shown that what might be termed military Keynesianism is indispensable for the development of a given economic system, two alternative conclusions can be drawn: first, that it is necessary to favour massive military investments; and secondly that there is something very wrong with such a system.

A more specific element supporting the view that military needs are at the basis of qualitative jumps in technological development,which are then inevitably reflected in technological progress and civilian production, can be found in the fact that some important innovations emerged or became important during the two World Wars; for example, propeller airplanes during the First World War; jet aircraft, radar, nuclear energy and electronic computers during the Second World War. It cannot be denied that war needs gave a substantial impetus to these new technologies and to their diffusion. Nevertheless a closer analysis shows that, in various instances, the essential elements for the innovation had already been introduced into civilian technology. During the mid 1800s the Krupp steel industry was already highly advanced before it began supplying artillery to the British, Prussian, and Russian armies. Similar statements can also be made about the Armstrong Industry, whose frontloading cannons were not accepted by the British Armed Forces, as well as Vickers in England and Carnegie in the United States.[6] In Germany and the United States, even the inventors of the jet engine did not obtain government support, which arrived only after the new technique had shown its worth. It is a general consensus that civilian R and D is responsible for the major part of the scientific basis and much of the fundamental technology upon which military R and D is built; thus advanced products of civilian technology are normally incorporated into military systems. Innumerable examples of this can be given.

The foregoing, as well as the following comments, should not be seen as efforts to underrate the value of the impulse which the massive military technological development has passed on to technology and production development in the civilian field; quite the contrary: when large amounts of money are funnelled into co-ordinated programmes, and channeled towards military R and D *it is inevitable* that some of the innovations and results can be used in the civilian sector (the spin-off). In addition those supporters of militarily sponsored R and D programmes add that the advantages derived from this type of undertaking in high-risk areas and in long-term programmes as well as the advantages gained from the lushly-funded advanced pro-grammes, which private industry could not afford, all lead to qualita-tive progress, and push forward substantially the cognitive and technological frontier. In this debate a first problem is to see whether a comparable investment in non-military R and D would give better results, minimise the amount of natural resource waste, correct the distortion of the objectives and that of the limitations imposed by secrecy requirements and so forth. A second and more serious problem, namely that of military influence on the arms race, on strategic evolution and on national security has been amply dealt with elsewhere, and need not be discussed here, save for citing an example of technological pull which has created insecurity, namely the sequence of developments: early Anti-Ballistic Missile (ABM) defence systems led to the development of Multiple Independently-targetable Re-entry Vehicle (MIRV) offensive systems, which resulted in a vulnerability window (presumed), and an increase in the number of strategic arms, culminating in SDI, which, in turn, resulted in a disarmament negotiations stalemate.

ECONOMIC IMPACT OF THE BIG PROGRAMME

Some writers claim that large military programmes positively influence many sectors of social development, an influence which outlasts technological breakthrough. Analogies may somehow be made with the great scientific endeavours in the fields of space exploration, elementary particle physics and astrophysics. The 'spin-offs', in addition to promoting new products and processes, also involve organisational and managerial methods, quality control, and commercial benefits.[7] The National Aeronautics and Space Administration (NASA) has made quantitative global evaluations. An early 1971 study showed that the $29 billion invested during the 1959–69 period would create an increase in the Gross National Product (GNP) of the order of $200 billion during the next two decades. A more analytical study, carried out in 1976, demonstrated that the NASA programmes resulted in a benefit of seven million dollars in four innovative sectors (cryogenic insulation, integrated circuits, gas turbines and computer science); there were similar advantages in the fields of pacemaker technology and of surface treatments of zinc. These studies have been a target of criticism from a number of different quarters. Figure 3.4 shows the results of a study of the indirect benefits of the European Space Agency (ESA) programmes. Similar studies of the large R and D programmes could give the same type of results. The ESA is, however, at present very worried about its

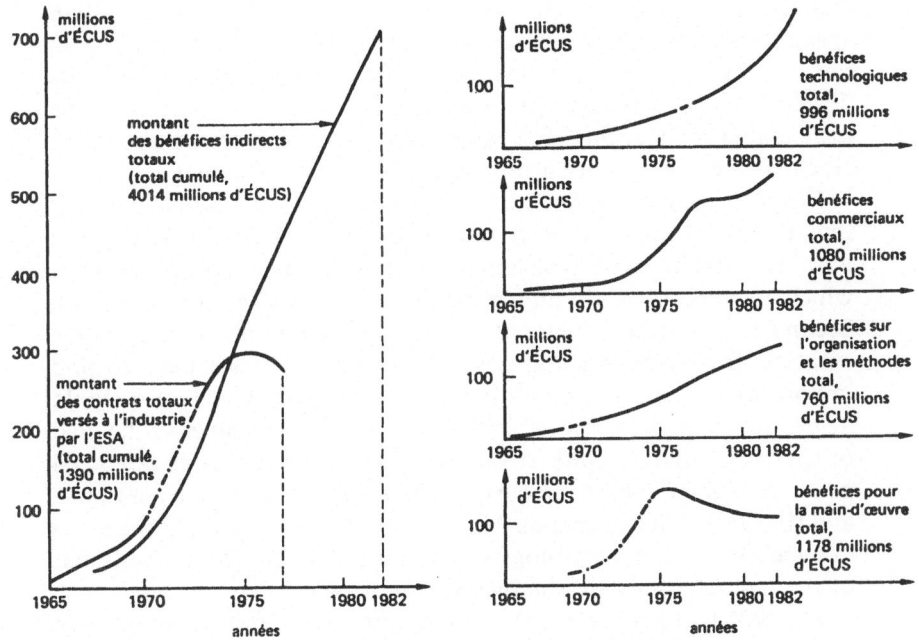

Figure 3.4　Les courbes de la figure A indiquent respectivement les montants des contrats versés à l'industrie européenne par l'ESA pour réalités les grands projects spatiaux européens, et les montants des bénéfices indirects générés par ces dépenses de l'ESA. A partir de 1975, les "bénéfices" dépassent pour la première fois les "crouts" et ces bénéfices poursuivent par la suite une croissance très forte, bien que les dépenses de l'ESA se ralentissent assez sensiblement. Les courbes de la figure B résultent d'une "ventilation" de la courbe des bénéfices indirects de la figure A en quatre grandes composantes: bénéfices technologiques, bénéfices commerciaux, bénéfices pour l'organisation et bénéfices pour la main d'oeuvre. Cette ventilation permet de mesurer la compleité du processus de diffusion de l'innovation: les bénéfices technologiques tardent à se faire sentir mais prennent ensuite une évolution exponentielle, les avantages commerciaux dépendent du succés des réalisations spatiales, les avantanges sur l'organisation croissent régulièrement mais semblent tendre vers une limite, quant aux avantages pour la main d'oeuvre, ils dépendent conjoncturellement de l'importance des dépenses realisées.
Source: *La Récherche*, vol. 17, no. 183 (1986) 1542.

Columbus programme. This programme involves European civilian space research in collaboration with the American space station; in fact, the SDI programmes have created problems for the ESA in its development of the space module which is to be connected to the principal space station.[8] The Americans are also worried about the risk that the SDI might curtail space science development.[9]

From the above analysis we can reach two conclusions. First, it remains to be shown that greater advantages could not have been obtained, at the same investment level, without huge programmes which, in any case, could only be justified by their scientific objectives. Secondly, even if the superiority of the big programmes as innovative catalysts for much research could be shown, it is not hard to find important civilian projects of great scope and which also could aggregate wide international collaboration. For example, in the field of space research, in spite of the fact that the ESA spends only 0·04 per cent of the GNP of the participating nations for the programme, and the United States spends 0·2 per cent in space projects, 'the areas in which European technology is commercially competitive with that of the United States are significant and growing'.[10]

The SDI is the most impressive military research programme ever launched. One of the most ambitious projects deals with the development of several types of very powerful lasers. Nevertheless it seems that the laser-producing firms do not consider the SDI as being commercially viable in this field.[11] Paradoxically, it seems that the most attractive prospects for the R and D are offered by those firms which can build the offensive systems which will be necessary to break through an eventual Soviet defensive system, if this new chapter in the arms race is ever 'written'. Some even believe that the SDI instead of favouring the development of lasers and of optical and detection devices will distort the research mainstream and will slow down its development.[12] It is no mere coincidence that the best industrial laboratories, such as IBM and Bell, stay clear of SDI.

AN EVIDENT LIMITATION OF MILITARY R AND D

It is rather obvious that in some technological sectors there is a big overlap between military and civilian needs; a typical example of this can be found in the field of microelectronics, computers, material technology, machinery and communications. In other sectors the military needs are specific and are unrelated or even in contrast with civilian ones; thus the spin-off would be occasional and very limited. The military is searching for very manoeuvrable fighter aircraft, silent submarines, compact nuclear reactors, stealth cruise missiles and bombers, and systems capable of overcoming very exceptional conditions such as high levels of radiation. All of these have little in common with consumer products.[13]

Consider the following examples. It is a well-known fact that a nuclear explosion under certain specific conditions results in a strong electromagnetic pulse (EMP) which can damage normal electronic circuits; because of this huge amounts of funds have been invested in developing circuits and weapons and communication systems shielded against EMP. No civilian market would ever be interested in this development since nobody would ever dream of producing items for consumers worried about the survival of their goods after a nuclear war. Secondly, the laser is widely used for civilian purposes: it is extensively employed in science, industry and medicine. But no need is felt for a laser as powerful as tens of megawatts, capable of driving a hole through metal or ceramics from thousands of kilometres away – as would be required by the SDI. An X-ray laser is the dream of many researchers and if it could be produced it would have many important applications. There is, however, little connection between these expectations and the Livermore project (the Edward Teller toy), since in this case the laser pulse would be triggered by a nuclear explosion which in a fraction of a second would destroy the metal wire carrying the coherent pulse, together with the laboratory and part of the city in which the instrument is located.

Electromagnetic rail guns capable of firing normal bullets weighing a few kilograms at a speed of 100 kilometres a second are at present on the drawing-board; the heart of the gun is a compact generator, similar to the one used for high power pulse welders. Here the main difficulties lie in the electronic switching system which has to produce a rapid firing rate. Even if the single components can be used for civilian technology, the total effort necessary for producing the weapon is not proportional to the limited 'fall-out'. Finally, for the past decades military aviation development has been directed towards supersonic aircraft with ever-increasing velocities, whereas civilian aviation in most cases uses subsonic models; the commercial flop of the Concorde is well-known.

THE ROLE OF MILITARY CONTRACTS

Companies which are awarded important military contracts have a sure market and profits for many years. This could stimulate the growth of an industry which needs large investments paid in advance, thus reducing the risks involved in entering a new market. This general economic mechanism could favour the innovation of products and processes more than direct R and D support could do.[14] In the United States military demands have led to the development of civilian production in the fields of electronic computers, jet aircraft, nuclear energy and space communications, to such an extent that a civilian market has been established. Undoubtedly these markets would have been created anyway, but their connection with the military gave them a boost, thus placing American firms in the lead in the international market.

The development of microelectronic industry is a good example of this. During the 1950s the Department of Defense actively supported R and D in the semi-conductor sector, but integrated circuits were invented independently of military programmes. Between 1949 and 1959 only five of the 112 Texas Instruments patents were taken out under the auspices of government contracts; and of these only two were marketable. This occurred even though the government was financing two-fifths of the firm's R and D. At the end of the 1950s the two principal innovations in the field of semi-conductor devices, the integrated circuit (Texas Instruments) and planar technology (Fairchild) had not been financed by the military. The military's direct interest in micro-miniaturisation was a guarantee of a market for the first integrated circuits, which were relatively costly, and it boosted this sector's commercial development. In 1963 100 per cent of all of the monolithic integrated circuits was sold to the Pentagon for use in the guidance system of Minutemen II missiles; ten years later, however, this had been reduced to 15 per cent. During this same interval the cost of these circuits fell from 50 dollars to one dollar. Thus we are dealing more with an alternative mechanism to the normal 'market pull', which implements an inefficient 'technology push': the military, writes R. De Grasse, has acted as 'creative first users' of the industrial products by acquiring the innovations before they have been commercialised in a civilian market and by financing the efforts to improve their quality and reliability.[15] At the end of the 1960s and the beginning of the 1970s, the inventions of the MOS

(Mostek) and of the Microprocessor (Intel) did not receive Pentagon support. Subsequent Pentagon financing played a very insignificant role compared with the dominant role of the civilian market.

At any rate, even if it is not essential, the military's role in electronics will be very important also in the near future, because the weapons of the next generation and the Command, Control, Communications and Intelligence (C^3I) systems will always rely heavily on electronics. The American integrated circuit market budget should increase from 900 to 1900 million dollars between 1983 and 1988. It can be predicted that there will be a strong influence on high-velocity integrated circuits and on gallium arsenide integrated circuits.[16] With respect to the former, D. C. Mowery doubts that direct intervention by the Pentagon would lead to more marketable products.[17]

The military market is very different from the traditional one – both under free enterprise as well as under oligopolistic conditions, since in the former the seller and buyer have converging interests: thus costs and prices are pushed into the background. Table 3.1 lists various examples of defects within the defence market.[18] If in the initial stages of the evolution of a new technology it is possible that military investments either directly or indirectly act as a stimulus, thus minimising risks, in later stages the possibility of a spin-off is reduced since civilian commercial demands split away from the military ones: 'the needs of modern weapon systems, which are increasingly more mysterious and exotic', minimally influence the non-military situation. More generally, it can be stated that, rather than producing for a market which is constrained by minimum cost, the war industry tends to satisfy performance requirements for advanced and complex products; therefore this leads to very high production costs and, in the case of a capitalistic economy, enormous profits. The Maddox Report (1979), prepared for the British Government, states in fact that military technological products are too complex and costly to be easily transferred to the civilian market.

In Great Britain, during the 1950s and 1960s, the Ministry of Defence supplied almost all of the public funds for the R and D in the field of semi-conductor research. The end-product of this was some very sophisticated devices which, however, because of their specificity, could not be easily adapted to the civilian market. The commercial weakness of the British electronic industry can probably be attributed to this excessive influence of the military bias.[19] The British Electronic EDC has clearly expressed its concern about the fact that

Table 3.1   Some examples of 'market imperfections and failures' in defence

Free-Market Theory	Defence Market
Many small buyers	One buyer (DoD)
Many small suppliers	Very few, large suppliers of a given item
All items small, perfectly divisible and in large quantities	One ship built every few years, for hundreds of millions of dollars each
Market sets prices	Monopoly or oligopoly pricing–or 'buy in' to 'available' dollars
Free movement in and out of market	Extensive barriers to entry and exit
Prices set by marginal costs	Prices proportional to total costs
Prices set by marginal utility	Any price paid for the desired military performance
Prices fall with reduced demand	Prices rise with reduced demand
Supply adjusts to demand	Large excess capacity
Labour highly mobile	Greatly diminishing labour mobility
Decreasing or constant returns to scale	Increasing returns to scale in region of interest
Market shifts rapidly to changes in supply and demand	7–10 years to develop a new system, then 3–5 years to produce it
Market smoothly reaches equilibrium	Erratic behaviour from year to year
General equilibrium–assumes prices will return to their equilibrium value	Costs have been rising at approximately 5 per cent (excluding inflation)
Profits equalised across the economy	Wide and consistent profit variations between sectors; even wider between firms
Perfect mobility of capital (money)	Heavy debt, difficulty in borrowing
Mobility of capital (equipment) to changing demand	Large and old capital equipment 'locks in' companies
No government involvement	Government is regulator, specifier, banker, judge of claims, etc.
Selection based on price	Selection often based on politics, or sole source, or 'negotiation'; only 8 per cent of dollars awarded on price competition

Table 3.1 Some examples of 'market imperfections and failures' in defence

Free-Market Theory	Defence Market
No externalities	All businesses working for DoD must satisfy requirements of OSHA, EEO, awards to areas of high unemployment, small business set-aides, etc.
Prices fixed by market	Most business, with any risk, is for 'cost plus fee'
All products of a given type are the same	Essentially, each producer's products are different
Competition is for share of market	Competition is frequently for all or none of a given market
Production is for inventory	Production occurs after sale is made
Size of market established by the buyers and sellers	Size of market established by 'third party' (Congress) through annual budget
Demand sensitive to price	Demand 'threat'-sensitive, or responds to availability of new technology; almost never price-sensitive
Equal technology throughout industry	Competitive technologies
Relatively stable, multi-year commitments	Annual commitments, with frequent changes
Benefits of the purchase go to the buyer	A 'public good'
Buyer has the choice of spending now or saving for a later purchase	DoD must spend its annual congressional authorisation

Source: J. S. Gansler, *The Defense Industry* (Cambridge, Massachusetts, 1980).

the electronic industry reaps benefits which are too small considering the quantity of money the Ministry of Defence invests in the R and D.[20]

The Science Policy Research Unit states that this overwhelming concentration on defence problems within Great Britain has weakened British industry's chances of penetrating the consumer market, especially that of electronics.[21] In addition to this, according to a recent trade union report, the failure of Britain's electronics industry to exploit this technology commercially can be traced to excessive dependence on the Ministry of Defence's demands.[22] During the past ten years this has led to the closing down of an alarmingly large number of firms and the loss of 100 000 jobs per year. There are others who, on the contrary, claim that the British electronics industry owes its existence to the stimulus given to it by military research which aims at the creation of newer and more advanced ideas, processes, materials and devices.

The available data mainly concern the United States. There is no evidence to support the view that within the Soviet Union there is an interest in developing military technology for an eventual spin-off. Soviet military technology is slightly ahead of the civilian technology, even though it is far behind that of the United States. The Soviet Union has rarely been ahead in innovative military technology. The Soviet system is rigid and a horizontal transfer of technological know-how across various departmental lines is discouraged even though the need to exploit military developments better within civilian produc-tive systems has occasionally been discussed. The obsession with secrecy increases the difficulties of technological transfer even within the military sector.[23] In any case, in the Soviet Union, civilian aviation has developed partially on the basis of military aviation prototypes. The military's principal advantage seems to be that, in addition to abundant financing, there is less red tape and more efficiency in production management.

NEGATIVE SPIN-OFF

Also within the fields of a given technology the demands of the military and consumer markets differ; thus it is possible that, when the latter market evolves, building on products already developed for the former, initial flaws will remain as the project proceeds. For instance, the development of Pressurised-water Reactors (PWRs) for electro-nuclear plants was strongly influenced by uranium-enrichment capa-bilities (used for making nuclear weapons) and by the development of military reactors, specifically those used in propelling submarines and for the nuclear propulsion aircraft project, terminated in 1953. Compactness and high power density are essential prerequisites for military reactors; these characteristics, however, imply more complex and less secure facilities. The chosen development direction has led to a drastic reduction in the initial development and engineering costs, but, increasingly, more stringent safety regulations for civilian applications have caused production costs to soar while a satisfactory safety level has still to be achieved. The electronuclear industry might now be better off it is had started out without the conditioning of the military technology spin-off.[24]

Another example is aviation. Many civilian aircraft are adaptations of military aircraft designs (the Boeing 707 being the civilian version of the B47 bomber, for instance). However, according to a Boeing engineer, the more a commercial aircraft resembles a bomber, the less likely it is to be successful.[25] The recent commercial success of the Airbus in the field of low-consumption, large civilian jets, seems to confirm this type of criticism. Because of this, Boeing is beleaguered with problems whilst almost all of the other American firms are out of business. More generally, according to Seymour Melman,[26] in order correctly to evaluate the role played by this governmental orientation of university research, one must examine not only the accomplished work but also the work that was *not* done. What is the cost of the control by defence ministries on a university in terms of lost opportunities or sacrificed potential? In 1962 the Joint Council of Engineers published a report on 'National Needs in the Technological Research Field 1965–1985', in which mention is made of the opportunities in the field of civilian technology which have been passed over because of the priority given to state-oriented research. Other information regarding mistaken paths taken by science and technology owing to the military investments is available in the United Nations 1972 report, 'Disarmament and Development'. The criticism reported so far is plausible but not cogent; it is possible that it has been biased by the opinion of authors who are inclined to highlight the negative aspects of large military expenditures.

OTHER DATA AND OTHER LIMITS

An analysis of three research projects financed at the Massachussetts Institute of Technology (MIT) by the Office of Naval Research at the end of the 1940s and the beginning of the 1950s has shown that a substantial flow of knowledge, which would otherwise have been delayed for many years, resulted from these projects. Let us recall that in the late 1950s, a great deal of West European fundamental research was financed by the US Armed Forces and that about 20 Nobel Prize winners have had the benefit of Pentagon support.

On the other hand, a study of fifteeen industrial economies has shown that in the period between 1960 and 1970 there was a negative correlation between military spending and GNP growth. The United States, Great Britain and particularly the Soviet Union, which heads the list for military spending and investments in military R and D, are at the bottom of the list as far as production growth and foreign market competition are concerned. The Japanese government policy of controlling commerce, directing investments and supporting research successfully influence production development efficiency and competitiveness. For example, in 1974 the sale of digitally controlled machinery in the American market was 4 per cent, while it jumped to 50 per cent in 1983.[27]

According to *Business Week*, the United States is losing its dominant position in the high technology sector because it is spending too little on civilian-oriented research.[28] A recent report of the Senate Armed Services Committee[29] claims that the Pentagon takes more from the technological base of the nation than it is capable of giving back. The committee was deeply impressed when a comparison was made with the activities of official Japanese support of civilian technological development in the field of computers, artificial intelligence and new super-conducting materials.

One of the reasons for the low efficiency of the defence industry is that it tends to favour a few big firms, which are less innovative and have fewer job openings with respect to the medium to small firms. Many of the more critical analyses stress two other disadvantages, the brain-drain and secrecy. The wide range of job offers, the funds for R and D and sometimes the attractiveness of the big programmes claim the best scientists and engineers, thus drawing them away from endeavours which are strictly civilian. This point of view is held by many. In some cases, however, the opposite results might occur. In Israel a considerable number of technicians who were trained within the defence industry system moved over to the civilian sector, bringing with them their skills. This knowledge was then used to promote civilian initiatives.[30]

There is a strong connection between the leaders of military industry and their defence ministers. A certain degree of legalised corruption has emerged from this relationship. In two years the principal military contractors have spent about $17 million in their Washington offices and a Political Action Committee has been set up to contribute to the campaign expenses of politicians who are members of the Armed Services Committee and of the Defense Appropriations Committees of the House of Representatives and of the Senate respectively. In addition to this one should consider that in 1973 in the United States 3233 persons who had previously held important positions in the armed forces were hired by 100 firms which had obtained two-thirds of the Pentagon grants; during the same year 275 managers of military industries went on to high-ranking positions in the Pentagon.[31] It is evident that there is the risk that an official in the Defense Department, who hopes to obtain a good job in industry, might be swayed in his decisions on projects presented and developed by the industry itself; too close a scrutiny might jeopardise his chances of having a second career after retirement. This situation favours the short-term profit agencies more than the civilian ones. N. Cousins has recently re–examined the numerous cases of bad management, excessive spending, abuse and fraud, which have been attributed to the Pentagon's principal suppliers.[32].

The rigorous secrecy which surrounds the R and D military sector, often from the moment explorative research is begun, introduces an extraneous element into normal scientific practice; this element acts as a barrier between the exchange of ideas and information, as well as discouraging constructive criticism, competition and the growth of new ideas. This obsession with secrecy is typical of the Soviet Union where, as an example, it is forbidden to visit many of the Institutions of the Academy of Sciences. More alarming is the fact that, with the advent of SDI, restrictions are becoming more frequent in the United States. During the past few years orders to limit further the circulation of scientific and technical information have been given.

THE SITUATION IN ITALY

There is little available information concerning the role of the Defence Ministry in R and D. The most accurate and recent study has been written by Paolo Miggiano.[33] The 1985 White Book indicates that personnel working on research projects can be divided into the following categories: 300 full-time and 119 part-time researchers; 218 full-time and 247 part-time technicians (with high school diplomas or college degrees).

The funds marked for R and D during the 1983–5 period on the average constitute about 2·5 per cent of the defence budget (this adds up to about 100 billion lire in 1984), and is expected to increase in the 1986–9 period. According to the Stockholm International Peace

Research Institute, during the three-year period 1981–4 government expenditures for military R and D were 6 per cent of the overall budget for R and D and 2 per cent of military expenditure. As is evident from the discrepancies among different sources, we are dealing with significant sums which are certain to increase. Correspondingly there is the risk that the military will increase its control of the funds assigned for R and D through channels such as the Ministry of Industry. In spite of the propaganda campaign carried out by the Strategic Defense Initiative Office (SDIO) and the mobilisation of almost all of the national industry in that sector, Italian participation in the SDI is a very modest one. Personnel who are employed directly by the Ministry do not play an active role in research. This research is delegated instead to industrial firms which receive grants from the government or from abroad (about 200 firms employing 80 000, or 2 per cent of all the positions in the manufacturing industry). The White Book states:

The development of the national arms industry during the 1960s and 1970s was due more to the entrepreneurs' efforts to meet the increasing foreign demand than to an overall plan of development of the specific industrial sector. Relatively small and middle-sized, but well-qualified industries also played a leading role compared to numerous other productive activities. The industries in this sector, once they have acquired the initial technico-scientific expertise (such as patents and know-how), mainly from the United States, have successfully improved on these already patented systems, and in some cases have come up with original and autonomous solutions.

According to M. Frosi, who had been Secretary of the Technical-Scientific Council of the Defence Ministry, the R and D work is actually done primarily in civilian industry, while the Defence Minstry's role has been limited to defining the characteristics of the requested product and to the testing of prototypes. According to Miggiano, the technological fall-out from military to civilian, despite the emphatic yet vague statements made by political leaders or research groups, is very limited indeed.[34] Firms such as Elettronica, Spa, Selenia and Aeritalia were involved, yet the overall result is marginal with respect to the national economic development. The end-result of these dozens of military, naval and aeronautical projects is that little of importance has been produced in corresponding civilian sectors.

Assuming that these so-called spin-offs – albeit modest – might concern composite materials, helicopters and components for aviation, control and communication systems, it is likely that the know-how acquired from a given firm will have remained firmly within the confines of one company. At any rate, even the White Book claims that 'it will be necessary to produce a comprehensive cognitive picture'.

CONCLUSIONS

● Civilian and military technologies overlap in many areas. Military technology widely uses developments which have been achieved in the civilian field; in various cases, military R and D has spin-offs into the production of goods for the civilian market.

● The spin-offs are due mainly to the fact that military grants guarantee a secure and remunerative market which facilitates technological innovations in the initial phase, when the risks and production costs are still high.

● In the majority of cases the technology developed for war purposes cannot be easily transferred because it has to meet needs which are too specific and of little interest for civilian production. However it is necessary to remember that the difficulty of transferring technology is not a problem only of military investments.

● Military investments tend to favour the large firms which are less innovative.

● The restrictions of secrecy, typical of military R and D, have a negative influence on science and technology.

● Military R and D diverts economic and human resources and leads to the neglect of important scientific priorities and civilian programmes.

● The spin-offs would be more substantial – if the amount of money invested is equal – if the R and D were used directly for civilian objectives.

● Countries with a state-run economy and the majority of the capitalist economies can invest in civilian R and D. In the United States, however, it is more difficult to obtain large government investments in civilian R and D. Thus military research is more in tune with the system.

• In some instances R and D military successes negatively influence civilian production (negative spin-off).

• In Italy the military R and D is almost exclusively carried out by those firms which have military grants. The civilian spin-offs are few. Statements to the contrary made by high-ranking politicians and industrialists are unfounded and politically motivated.

• The real justification for military R and D is to be found in the nation's political and military needs. The supporting thesis that civilian technology needs military R and D is strained and bound to fail. The correct position is the following: in the case of security needs necessitating a military R and D, a way of minimising waste and resource distortion as well as a way of capitalising on civilian applications and commercial exploitation must be found.

• Military R and D encourages the arms race (technological pull) and influences strategy. Today this effect extends beyond a nation's or an alliance's purely defensive needs; it introduces a sense of insecurity and contributes nothing to international peace. Consequently those scientists and technicians working on new arms development are faced with an enduring moral problem.

Notes

1. Marek Thee, *Military Technology, Military Strategy and the Arms Race* (Springfield, Vermont, 1986).
2. R. Fieschi, *L'invenzione tecnologica* (Milan, 1981).
3. F. A. Long and J. Reppy (eds), *The Genesis of New Weapons* (Oxford, 1980).
4. J. Tirman (ed.), *The Militarisation of High Technology* (New York, 1984).
5. W. Leontiev and F. Duchin, *Military Spending* (Oxford, 1983).
6. A. Sampson, *The Arms Bazaar* (London, 1977).
7. P. Cohendet, 'Les retombées économiques des grands programmes', *La Recherche*, vol. xvii, no. 183 (1986) 1541; and M. Bianchi-Streit, 'Economic Utility Resulting from CERN Contracts', CERN–84–14 (1984).
8. H. Gavaghan, *New Scientist*, 29 January 1987.
9. N. Heneson, *Nature*, vol. cccxxv (1987) 477.
10. Office of Technology Assessment, *Civilian Space Policy and Applications* (Washington, DC, 1983).
11. P. Speser and K. Mancuso, 'R and D Opportunities in the SDI', *Laser Focus*, August 1984.

12. D. Chaplin and J. Hassard, 'Star Wars – Money from Heaven?', *Physics Bulletin*, vol. xxxviii (1987) 19.
13. *FAS Public Interest Report: Journal of the Federation of American Scientists*, vol. xxxix, no. 7 (1986).
14. N. Rosenberg, 'Le ricadute civili della spesa militare in R and S: l'esperienza statuitense dalla seconda guerra mondiale a oggi', in G. Lunghini and S. Vacca (eds), *Cambriamento tecnologico e teoria dell' impresa* (Milan, 1987).
15. R. De Grasse, 'The Military and Semiconductors' in J. Tirman (ed.), *The Militarisation of High Technology*.
16. L. A. McClelland, 'Military Requirement Push IC Development', *Defense Electronics*, February 1985.
17. D. C. Mowery, 'Innovation, Market Structure and Government Policy in the American Semi-conductor Electronic Industry', *Research Policy*, vol. xii (1983) 183.
18. J. S. Gansler, *The Defense Industry* (Cambridge, Massachussetts, 1980).
19. K. Dickson, *Research Policy*, vol. xii (1983) 113.
20. National Economic Develoment Office, *Civil Exploitation of Defence Technology* (London, 1983).
21. C. Hilsum, *Physics Bulletin*, vol. xxxviii (1987) 17.
22. Greater London Trade Union Resource Unit, *Lost Jobs, Wasted Skills* (London, 1987).
23. M. Kaldor, *The Baroque Arsenal* (New York, 1981).
24. Rosenberg, *Civilian 'Spillovers' from Military R and D*.
25. Kaldor, *The Baroque Arsenal*.
26. Seymour Melman, *Pentagon Capitalism* (New York, 1970).
27. Tirman (ed.), *The Militarisation of High Technology*
28. *Business Week*, 11 March 1985.
29. M. Mecham, in *Aviation Week and Space Technology*, 18 May 1987, p. 21.
30. E. Berglas, *Defense and Economy: The Israeli Experience* (M. Falk Institute, 1983).
31. G. Rumble, *The Politics of Nuclear Defense* (Cambridge, Massachusetts, 1985); and *The Defense Monitor*, vol. 16, no. 3 (1987).
32. N. Cousins, *The Pathology of Power* (New York, 1987).
33. Paolo Miggiano, 'Note sulla ricerca e sviluppo militare in Italia', *IRDISP*, June 1986; and Paolo Miggiano, 'La ricerca e sviluppo militare non produce sviluppo', *Poltica ed Economica*, February 1987.
34. Ibid.

4 Non-Military Justification for Investments in Military Technologies

Shalheveth Freier

INTRODUCTION

It is sometimes assumed that non-military justifications play a part in the choice of the military technologies in which a country will invest. Such justifications, however, play no part in decisions on military technologies – or defence Research and Development (R and D) – on the part of countries which have a potential adversary, are ultimately responsible for their own defence and have only their own human and material resources to fall back on in order to develop the technologies needed to satisfy their defence requirements. There are only a few such countries. They include the United States, The Soviet Union, and the present writer would add the country he knows best, Israel (which must develop and produce what it cannot buy in defence hardware, and whose access to the defence market is contingent upon political circumstances). In these countries, defence R and D is dictated by perceived defence requirements alone. Derivatives – or spin-offs – which benefit the economy are desirable and are looked for, but they have no influence on the original choice of projects in defence R and D.

Non-military justifications as a component in decisions on defence R and D apply only in countries which are not solely responsible for their defence. This is true, for instance, of the countries allied within the North Atlantic Treaty Organisation (NATO), which rely upon the alliance and on the US commitments within NATO to ensure their security. Presumably the same is true of the Soviet Union and the states associated with the Warsaw Treaty. This latitude in choice applies equally to countries which either have reliable defence arrangements and sources of arms procurement or have no identifiable adversary. Such countries can, for example, doubt the defence value of the Strategic Defense Initiative, and yet wish to be associated with the R and D effort generated by it, fearful lest they fall behind in new technologies and hence in their competitiveness in the market. These countries can indeed decide which military hardware to buy ready-made or which to develop, and determine their total investments in R and D, funded by the public purse, with due weight being given to alternative investments in the civilian sector. It is the problem of the first set of countries, those which have no alternative but to engage in defence R and D for their proper defence, which will be discussed here in more detail, together with the interaction of such defence R and D with technologies which benefit civilian industry or the common weal.

DEFENCE R AND D SPIN-OFFS AND THEIR CHARACTER

It can easily be argued that investments in R and D for defence purposes will benefit non-military applications as well. Two premises support this argument, both of which seem reasonable. First, many technologies whose speedy development is due to military require-ments are not intrinsically of interest to the military alone. Presuma-bly no technology has only one realm of application. Nuclear energy and radar are instances of developments hastened by military neces-sity, but with the passage of time, certainly not confined to military applications alone. Secondly, endeavour in private industry (say the use of electro-optical devices) or in the public domain (say the sensing of pollutants in the atmosphere) tends to use existing technologies, to improve or adapt them. If such technologies owe their initial develop-ment to defence requirements, it is indeed *these* technologies which are at the disposal of private industry or serve as tools for diagnosis and remedy in areas of public concern.

If these two premises are true, it is readily apparent that the level of investments in R and D for defence stands in direct relation to such derivatives as benefit the civilian market or the public good. Such positive correlation between investments in defence R and D and the resulting non-military spin-offs are a troublesome proposition in two respects. The technologies which become available on the market as spin-offs are the ones which are of interest to the military. They are the ones on which defence R and D money is spent and, conversely, they have an influence on industrial R and D efforts for the products of which the military is a customer. Also it can be argued that the measure of support accorded to defence R and D is not balanced with respect to other areas of public interest which fare less well in allocations, such as alternative sources of energy, and which could produce technologies different from the ones developed under the influence of defence requirements.

THE PARTICULAR ATTRIBUTES OF DEFENCE R AND D

It would therefore be useful to enquire what it is that gives defence R and D such an advantage over other public R and D concerns. If technological progress is necessary and desirable, it would be gratifying to think that such progress does not necessarily depend on the pressures of defence needs, but that non-military R and D objectives could command an equal measure of support, and yield useful spin-offs also outside their defined scope.

Defence R and D probably has the following attributes. First, defence has a popular and parliamentary appeal, in particular when an adversary exists or is perceived. There is little to substitute for defence R and D under such circumstances. Secondly, the goals of defence R and D have a crude appeal to the imagination, such as greater power or performance per volume, greater precision, better intrusive intelligence, better control at remote distances. Thirdly, Defence R and D is presumably distinguished from all other R and D efforts in the public domain, in that its goals determine the expenditure, and cost is less of a constraint than in any other area, in which the public purse is called upon to support R and D. Finally, under these circumstances, risk money is available for exploratory research and development at levels inconceivable in private enterprise using its own funds.

Such public funds are farmed out to government research laboratories, universities and private industry. Even though they carry the defence label and have overall defence objectives, the latitude inherent in exploratory research, so funded, often does not differ from the latitude afforded by funds for basic research. It is presumably this latitude in exploratory research and development which yields the quality of military hardware and concurrently the non-military variants.

These are potent attributes, which set defence R and D apart from all other R and D paid for from public funds and determine the technologies which can be exploited for non-military applications as well. Except for the first attribute, the real or perceived defence need, which cannot be argued away, there is no reason in principle why alternative goals pertaining to the common need cannot be determined and why R and D in their service cannot be similarly guided and endowed by the public purse, and why technological progress, possibly with different emphases, cannot be assured. There is no reason in principle why this should not be so but of course there is difficulty in practice. Just as people will rally more readily to defend their country in war than they will to ensure its survival in peace, defence as such is less demanding of judgement than other priorities on which there is less of a public consensus.

The foregoing discussion has set out in some detail the rather trivial point that military technologies have non-military spin-offs and have a definite influence on areas in which technological innovation and progress are registered and applied. In principle, then, R and D in areas other than defence can equally well serve their objective and yield spin-offs. Other areas of public concern, not recognised as being as urgent as defence or as frightening as cancer, multiple sclerosis or, more recently, AIDS, have had less attention and endowment, and therefore less influence on technological progress. But this need not be so. If, for instance, alternatives to fossil and fissile nuclear energy had commanded a concerted response, their influence on technological progress would of course have been related to *their* properties and requirements. It is not only in the nature of defence R and D that spin-offs benefit the civilian industry. Any non-military objectives that are defined, presented, managed, and endowed like defence research will of course have beneficial spin-offs in areas other than the stated objective. It may also be the case that too much emphasis is placed on the direct relation of R and D spending in one area and industrial success in the area in which the money is spent. High-speed ground transportation, for instance, is as much a development of the aerospace and electrical industries as it is of the automotive and railway industries; electrostatic copying did not come from the office equipment industry; nor did instant photography have its origins in the photographic industry; and synthetic fibres were developed in the chemical industry and not in the textile industry, where they found their first application. These are not singular instances. Indeed the unexpected benefits that one sector of industry can derive from R and D conducted on behalf of another is tending to multiply.

Nor should we forget the inventions due to independent inventors. Most of them had problems with high-risk money, which even in the United States was hard to come by, except in the Boston or Palo Alto areas, until the 1960s. The availability of such high-risk money from any source, including defence, cannot but encourage such independent inventors. A few instances of such independent inventions, which play no small part in our lives are: DDT, rockets, penicillin, catalytic cracking of petroleum, the zipper, the gyrocompass, the helicopter, air conditioning, FM radio and the ball-point pen.

There are those who argue cogently the adverse influence of defence R and D on scientific and technological progress in vital non-military areas. The present writer often cannot help feeling that much of this criticism is actually directed against the defence value of such defence programmes, and that many defence R and D projects would qualify for more general approval if conducted under a different aegis.

None of the foregoing should be taken to imply that the industrial success of a country is necessarily tied up with the existence of indigenous defence R and D. This is not at all so. Take, for example, Japan or West Germany, in both of which the pressures on defence R and D are small. And it is not so even in the industrialised countries which do maintain a large defence R and D effort. It is the policies in support of basic, applied and industrial research in Japan and West Germany, financed by public funds, which explain their success.

CONCLUSION

As long as defence needs exist, there will be defence R and D. Its relevance to defence can be disputed, but not its very pursuit, as long as military confrontations are possible. Similarly, its portion of the total R and D budget funded by public or corporate sources can be argued about, but defence R and D cannot be argued away. Moreover, in assessing non-military applications, account should be taken not only of direct derivatives, but also of the contingencies mentioned – invention in one sector of industrial endeavour being of benefit to another – and the risk money available to the independent inventor.

There are, as already stated, attributes of defence R and D which, suitably applied to non-military objectives would both promote these objectives and yield spin-offs in other areas. But this is really a matter of determining priorities of social importance and imparting this importance to the public and the decision makers.

Part II
The Strategic Defense Initiative: Military and Civilian Space Technology

5 Scientists and the Strategic Defense Initiative

Sidney D. Drell

Reviewing the last 40 years, it is striking what breathtaking progress mankind has made on his voyage of discovery. We are now dealing with concepts we could hardly imagine 40 years ago. In the same way that we look back on the seventeenth century and the time of Newton as a time of wonder at the discovery of the universality of nature's laws, which applied to the solar system as well as to phenomena on earth, so will the inhabitants of the twenty-first century and beyond look to this amazing era. They will see that enormous progress was made towards achieving a unified theory of nature's forces and building-blocks. They will see the first bold strokes of a picture – of a theory rooted in data – of how our universe evolved from the Big Bang of some 15 or so billion years ago. They will also find momentous achievements of both basic and applied significance that are remaking the structure of the familiar world – high temperature superconductivity being just the most recent. But, just as we have dreamed of profound new understanding and progress – perhaps of a complete theory of everything – we have also created a nightmare with our scientific advances. The nightmare is simply this: because modern science has given us weapons of such enormous destructive potential – bacteriological as well as nuclear – the dream may end in a holocaust of almost unimaginable proportions and consequences.

Mankind faces no greater challenge than to avoid such a holocaust. Towards the end of the Second World War the United States Strategic Bombing Survey was established to conduct a study of the effects of aerial attacks on Germany and Japan. After visiting Hiroshima and Nagasaki, the Survey team wrote as the conclusion to their report issued in June 1946:

No more forceful arguments for peace and for the international machinery of peace than the sight of the devastation of Hiroshima

and Nagasaki have ever been devised. As the developer and
exploiter of this ominous weapon, our nation has a responsibility,
which no American should shirk, to lead in establishing and
implementing the international guarantees and controls which will
prevent its future use.

This remains a challenge to all Americans, but it has a special
significance for the international community of scientists, who have
created the nuclear weapons which pose so grave a danger to our
survival. Moreover this challenge presents to us an especially difficult
dilemma. We are trained to approach issues such as the nuclear
weapons danger as a technical issue. We bring important physical
insights to an understanding of the mechanisms of a nuclear explosion, the significance of its power, the reality of radioactive fall-out
and its deadly consequences for humanity. These facts are necessary –
and must be understood because they limit the range of practical
policies for facing the nuclear danger and seeking to avoid conflict.
Laws of nature cannot be coerced or ignored in setting policy goals.
But we face a very profound, fundamental moral issue in dealing with
nuclear weapons of mass, indiscriminate destruction. And *this* is a
realm to which the scientist brings no special expertise. Furthermore,
to add to the scientist's dilemma, outside the laboratory the scientist
enters a world of policy and politics where shifting and often
apparently irrational laws of political and social interactions replace
what he is expert at, namely, the disciplined study of fixed, rational,
and – when fully understood – beautiful laws of nature. In this much
more turbulent political realm the scientist may end up disillusioned,
even harmed and embittered. Just recall the sagas of Robert Oppenheimer and Andrei Sakharov.

If is not at all new, of course, for scientists to be involved in war and
weapons of death. Archimedes designed fortifications and instruments of war, including a great catapult, to help thwart the Romans
besieging Syracuse in the third century BC. Leonardo da Vinci was
renowned as one of the greatest military scientists of his time.
Michelangelo was once the engineer-in-chief of the fortifications of
Florence. So there is a very distinguished roll of honour of those who
have designed and built weapons. But in our time there is a new
element that acutely heightens the scientist's dilemma. Never before
have scientists dealt with weapons of absolute destruction, with
weapons whose use could mean the end of civilisation. Never before
have we had so little margin for error. At the same time, the gulf

between our leaders and their understanding of the technological weaponry they command has grown to dangerous proportions. The former British Prime Minister, Harold Macmillan, lamented this fact in his book *Pointing The Way*:

> In all these affairs Prime Ministers, Ministers of Defence and Cabinets are under a great handicap. The technicalities and uncertainties of the sophisticated weapons which they have to authorise are out of the range of normal experience. There is today a far greater gap between their own knowledge and the expert advice which they receive than there has ever been in the history of war.[1]

This is a dangerous gap; one we must try to close. Success in closing it will require efforts from both sides: from the scientists as well as from society's political leaders.

This effort should not be viewed as a moral obligation for every scientist. Each individual, scientist or not, must choose his or her path and goal in society. However, the scientific community as a whole has a special obligation to assist society, through its governments, to understand the implications of the products of our scientific advances and to shape the applications of those products in ways more beneficial than dangerous to the human condition. The American Physical Society's study of the science and technology of directed-energy weapons, undertaken in 1985, should be viewed in this spirit. This subject was thrust into the forefront of public interest by President Ronald Reagan's famous 'Star Wars' speech of 23 March 1983, in which he called on the scientists 'who gave us nuclear weapons to turn their great talents now to the cause of mankind and world peace, to give us the means of rendering these nuclear weapons impotent and obsolete.' He held out hope of a world freed from the nuclear threat of retaliation by advanced defensive technologies that 'could intercept and destroy strategic ballistic missiles before they reached our soil.' Scientists can and should make an important contribution to informed public debate on the desirability and dangers of such weapons systems by independently evaluating the physical basis, the technical feasibility and the related implications of such weapons. The issue raised by Reagan is one of profound importance. What he is calling for is nothing less than a fundamental change in the basic strategic relationship between the United States and the Soviet Union. Nor is he the first president to raise this issue in search of a safer world to be achieved by applying new technologies to strategic defence.

Every president since Dwight Eisenhower, upon entering the White House, has asked quite properly, 'Can we not defend ourselves? Can science and technology not provide a shield against the threat of nuclear annihilation? Can we not do better than live as mutual hostages under the condition of mutual deterrence?' Upon further analysis and after detailed study of the technical realities, each of them concluded that the answer was 'No'. This conclusion led in 1972 to the negotiation and ratification of the Anti-ballistic Missile (ABM) Treaty of unlimited duration that is in force today. It was, of course, proper and vital for Reagan to ask this question again in 1983, in view of the prodigious technical advances in the more than ten years since SALT I was negotiated. Do these advances now offer us new and better prospects for dealing with the nuclear threat as we look ahead into the future?

The present author has been very heavily involved for many years in the technical and arms control implications of strategic defence. Hence it is appropriate in this chapter to set down clear views on what the United States should be doing in strategic defence. These views are based on technical judgement and can be summarised in three points:

• A high-quality and balanced research and technology programme in strategic defence that is consistent with the ABM Treaty is in the security interests of the United States. Compliance with the Treaty will pose no harmful technological burden on a properly structured American programme for the coming decade. The gap between today's technology and an effective defensive system is so great that it is now far too early for any programme in strategic defence to involve technology demonstrations of types that could interfere with the Treaty. Indeed, the SDI programme will be superior and more likely to achieve its goals of determining what advanced technologies look promising and what systems concepts look practical if its research priorities are not distorted by premature emphasis – or by politically motivated requirements to stage early demonstrations. A considerable body of evidence has shown that early demonstrations of new technologies have two deleterious effects. First, they tend to freeze the technology being demonstrated before it is fully mature, thus guaranteeing less than full capability. Second, they tend to absorb money from the associated R and D programme (because of cost overruns), thus eliminating the possibility of better solutions.

• A decision to begin premature deployment of a partial strategic defence system would, to put it politely, make no sense. It could not be justified technically and would not serve the security interests of the United States. We must still surmount major technical obstacles before we can consider systems testing, let alone deployment. Consider, for example, the problem of boost-phase defence. This is the crucial first layer of a nationwide defence which must destroy the enemy's salvo of attacking missiles during their boost-phase, within minutes of their being launched and before they can deploy their many thousands of multiple warheads and their potentially hundreds of thousands of decoys. The only technology presently available to accomplish that critical defensive mission must rely on space-based chemical rockets guided to impact by heat-seeking sensors. These are known as 'smart rocks'. They must be based in space because there is not enough time for them to be launched from the ground. This technology of hit-to-kill rockets has been successful in recent demonstrations against single targets that were planned and, therefore, known to the experimenters in advance. However, tens of thousands of them would have to be orbited by the United States on many hundreds or even thousands of satellites to counter just the current Soviet missile threat – and the required numbers increase very rapidly if the Soviets introduce countermeasures. With the currently available technology for sensors and guidance the hit-to-kill interceptor rockets will weigh many hundreds of pounds. Thus these weapons will require that the United States launch many millions of pounds into space – tens of millions of pounds – in order to deploy the satellite fleet that must carry them. Remember that ten million pounds is more than 160 shuttle loads. And, once in space, the satellites will be extremely vulnerable to direct attack as they circle the earth like ducks in a shooting gallery. The existing technology of ground-based interceptors with nuclear warheads – already deployed – poses a particularly acute threat to such satellites. This kind of space-based defence is at a serious disadvantage – to use a popular phrase, it might even be called fatally flawed – if it tries to enter into a countermeasures, counter-countermeasures competition with a ground-based anti-satellite threat. First, the ground-based interceptors require considerably less energy to raise one nuclear warhead weighing several hundred pounds to a given altitude than it costs the defence to insert into orbit at the same altitude one interceptor weighing many hundreds of pounds. Secondly, there is the problem of the absentee ratio: the attacker need only punch a hole in a small fraction of the defensive constellation – perhaps no more than 10 per cent of it – that is over his launch sites at the time of his attack. The defence, on the other hand, must equip the entire constellation to counter the ground-based anti-satellite threat.

Thirdly, the satellites in space must protect themselves against threats from the entire 4π solid angle around them; and they are vulnerable for long periods of time. In addition, as the Soviets replace their present lumbering boosters with modern rockets whose solid-fuel engines burn faster and that can lift warheads into space in little more than a minute or two, both the time and the space above the atmosphere available for intercept will shrink virtually to zero. A technology programme would have to succeed in reducing the weight of the individual interceptor rockets from today's designs of many hundreds of pounds to no more than a few tens of pounds before this concept could conceivably be of any practical value – although serious issues of systems effectiveness and countermeasures would still remain. Such a programme will take many years to accomplish and can be, and should be, pursued within the limits laid down by the ABM Treaty. The present writer vigorously opposes the current early deployment push that seems to be largely motivated by political considerations. The element of time urgency is distorting the research effort by driving 'proof of feasibility' experiments that are expensive and time-consuming. Not only is it technically unjustified; it will almost certainly spell the demise of the ABM Treaty. We have to be realistic in assessing the enormously difficult challenge to achieve not just one individual technological objective – such as a very light, high-acceleration rocket or a very high-powered laser – but an effective nation-wide defensive system in the face of enemy countermeasures designed to defeat that system. This is not a matter of man versus nature, which is the character of a technical problem. It is much more difficult: it is man versus man; the system must remain effective against all efforts to deny it by a determined opponent.

● There is a need to relate the provisions of the ABM Treaty to the 'other physical principles' – that is to say, the new technologies – for strategic defence. The ABM Treaty has provided the basis for US–Soviet efforts to achieve stability, to seek reductions in nuclear armaments, and to avoid war. Although progress in achieving reductions in nuclear arms has been disappointing since the Treaty was ratified, we should certainly continue to enforce it unless technical and political developments make it possible to supersede it with a superior peace-preserving regime. What is needed is an effort to strengthen the Treaty in consultation with the Soviet Union by resolving outstanding issues of compliance and by clarifying its application to new technologies in strategic defence. In particular, we need to relate the provisions of the ABM Treaty, on a case-by-case basis, to the new physical principles for strategic defence that include

space-based battle stations, directed-energy beams, and sensors utilising lasers, optics and particle beams for threat acquisition, tracking
and hand-off. Unrestrained testing, including testing in space, of the
new technologies as advocated under the Reagan Administration's
so-called 'broad interpretation' of the ABM Treaty is not in the
strategic interests of the United States. The Treaty, as it was interpreted prior to 1985, contains provisions that were designed liberally
to allow the signatory nations to pursue strong research and technology programmes in strategic defences, whilst at the same time
preventing preparations for rapid deployment of nation-wide defence
following break-out from the ABM Treaty. It is in our interests to
maintain these restraints on the extensive Soviet programme in
strategic defence. They, too, have their SDI programme, including
powerful lasers. With regard to the Soviet effort, the author disagrees
with claims occasionally made that the Soviet programme threatens
to pose a military threat to the United States in the near term.
Particularly in the critical areas of battle management relying on
computers and software their technology lags seriously behind that of
the United States.

An agreement to abide by the provisions of the ABM Treaty as they
have been enforced up to now will provide a strong impetus to the
continuing negotiations aiming at deep cuts in the offensive strategic
arsenals of the United States and the Soviet Union. Achieving such
deep cuts is a stated policy goal of both countries. It is clearly a
desirable goal in its own right. But, in addition, it will be important to
the United States' strategic defence programme. Unless the offensive
threat is greatly reduced numerically and limited by appropriate
technical constraints as a result of major progress in the US–Soviet
political process and dialogue, there can be no prospect of building an
effective nation-wide defence now or of achieving one in the foreseeable future. In short, the strategic defence programme of the United
States in the coming decade must not contribute to the dismantling of
current treaty achievements. While it may be justly concluded that
SDI played a constructive role in bringing the Soviet Union back to
the arms control negotiation table, it should not stand as a barrier to
progress.

Scientists bear a special responsibility to assist our society to
manage better the technological consequences of our scientific achievements. There is another responsibility that needs to be discussed,
and that is the responsibility of scientists to speak accurately and

responsibly on the technical challenges to society. All scientists are, of course, citizens and passionate human beings too. They accept no restraints on their political involvements and actions. And may they always speak out passionately and with outrage when appropriate on issues of social injustice, prejudice and freedom, as Sakharov has done so courageously. However when scientists speak on the technical issues themselves it is in their own interest to maintain the same high standards that they set in their professional lives. It enhances their credibility, individually and, in the long run, the credibility and value of the entire scientific community.

It is to be hoped that this credibility survives the current debate over SDI because this is an issue that has sharply divided scientists into warring camps and has led to a deep schism between a large segment of the scientific community and the US Government. The fault for this schism may well be attributed to the failure of the US institutions. The debate raises an issue of fundamental strategic importance, and one that presents an extraordinarily difficult technical and operational challenge. Yet it burst upon the political agenda, based on no prior technical analysis. As John Bardeen, who served on the White House Science Council (WHSC) charged with advising the President's Science Adviser, wrote on 13 September 1985:

> President Reagan prepared his speech with no prior consultation either with technical experts in the Pentagon concerned with research in the area or with his own Science Adviser, Jay Keyworth. I was a member of the WHSC at the time. Although we met only a few days before the speech was given and had a panel looking into some of the technology, we were not consulted.

This incident shows that effective use is not being made of the scientific contribution to technical defence issues at the highest level. The US Government has a need here. It cannot dispense with scientists and it must also bear its share of the burden to close the gap between itself and scientists. Its obligation is to arrange for the best possible scientific analysis and advice before risking major decisions or raising impractical expectations. Furthermore the scientific advice must be neutral and free of political and doctrinal biases. This is not easy to achieve.

Eisenhower understood these issues well when he created, in 1957, the position of a full-time science adviser in the White House and also established the President's Science Advisory Committee (PSAC), which served the president and the nation well for more than a decade before it was disbanded in 1973. He also showed his wisdom when he dismissed the issue of their political affiliations by saying, according to first Science Adviser, Jim Killian, that he liked scientists for their science and not for their politics. Since the mid-1970s, after PSAC was abolished, numerous scientific groups and numerous individual scientists have all urged the US Federal Government to re-establish a PSAC-like entity in the executive branch. There is a clear need for a pre-eminent science council with working panels, a degree of continuity and independence of action, functioning to give the President an accurate and informed view of the technical components of his basically political decisions. The Americans have no such effective mechanisms at the present time. The effort to compensate, at least in part, for this need is being made at several universities and research centres. In particular, at Stanford University's Center for International Security and Arms Control, a programme exists to train, in mid-career, established scientists in technical issues of national security. Not only do we need their critical and independent analyses of important arms control and defence issues with a major technical component. We also need to bring new talents and energy – a new generation – to bear on these issues. This is just one modest effort to replenish a small and diminishing pool of technically-trained independent voices.[2]

Notes

1. Harold Macmillan, *Pointing the Way* (London, 1972).
2. This chapter is adapted from the author's speech when retiring as president of the American Physical Society in 1956. See 'Thoughts of a Retiring APS President', *Physics Today*, vol. xl (1987) 56–62.

6 Defensive and Offensive Weapons in Space and Civilian Space Technologies

Richard L. Garwin

INTRODUCTION

This chapter contains a discussion first of defensive weapons in space, then of offensive weapons in space, and finally of the high-technology civil use of space. It concludes that the security of the United States and its allies (and of the Soviet Union and other nations) is best served by keeping all weapons out of space, and by banning the use or testing of weapons to or from space. This is a necessity if there are to be truly deep cuts in the numbers of nuclear weapons, and should enable the reduction of the 50 000 weapons now possessed in total by the United States and the Soviet Union to around 2000 within about 10 years.

DEFENSIVE WEAPONS IN SPACE

A truly defensive weapon in space would have as its function the destruction of weapons of another side – logically either those that travel through space or those that do not. We shall first discuss defence against nuclear weapons travelling through space. In his famous speech of 23 March 1983, President Ronald Reagan asked the scientific community to give the world the means to *replace* deterrence of nuclear war by *defence* against nuclear weapons. He went on to say that deterrence of nuclear war is effective but morally repugnant, and he asked for the means to intercept strategic nuclear weapons before they could touch the soil of the United States or that of her allies, and in this way render (at least one class of) nuclear weapons impotent and obsolete, making possible their elimination.

Of course, strategic nuclear weapons carried on ballistic missiles are in the form either of Intercontinental Ballistic Missiles (ICBMs) or

Submarine-launched Ballistic Missiles (SLBMs). Each has a boost phase, lasting 100–300 seconds, while the powerful rocket engines are giving the warheads the required speed of some 7 km/s so that they can fall to their targets a quarter of the world away. If a missile has Multiple Independently-targeted Re-entry Vehicles (MIRVs), then at present there is a post-boost vehicle, which over a period of perhaps 600 seconds gives each successive RV a small amount of additional velocity, to direct it to a target within a few hundred kilometres of the other targets of that missile. For the remaining 20 minutes of the flight, the RVs fall through space freely in 'mid-course' and eventually re-enter the atmosphere in a final minute of flight before arriving at their target and exploding. In this terminal phase, the nuclear warhead is protected against the meteoric heat and forces of atmospheric re-entry by the re-entry vehicle – a long cone-shaped enclosure covered with material which ablates and thus carries away the frictional heat of re-entry. A defensive system would have to detect the missile during some phase in flight, assess the situation and assign weapons, track the target, direct the weapons, kill the target, assess the effectiveness of the weapons, and continue until the incoming nuclear warheads are all or nearly all eliminated.

In March 1984, the United States created the Strategic Defense Initiative Office (SDIO) directed by General James A. Abrahamson. The SDI programme has had as its goal the research and development of the technologies necessary and adequate for realising President Reagan's dream of an effective defence against nuclear weapons carried on strategic ballistic missiles. To this end, one would require sensors, interceptors, battle-management capabilities and the like. In the traditional ballistic missile defence – the Anti-ballistic Missile (ABM) systems which the United States deployed and operated in 1975–6, and which the Soviet Union has in operation around Moscow – the sensors are early-warning satellite detection of the infra-red emission from the flame of the missile, wherever launched, and radars to detect the RVs thousands of kilometres away, as well as additional radars to direct nuclear-armed interceptors to intercept the RVs above the atmosphere in the terminal area or within the atmosphere. By itself, so-called exo-atmospheric intercept is strategically ineffective, even with powerful nuclear weapons of five megaton (5 MT) yield because in the vacuum of space a very light object can serve as a decoy for a heavy re-entry vehicle. If radar is the only sensor, the decoy need weigh only perhaps 1 kg to mimic a 300-kg RV, but if the decoy must mimic the RV not only for a radar sensor but also for visible-light and infra-red sensors, then it might have to be as heavy as 3–5 kg. Of course a defensive system has to work in our world of some 6000 km radius – exactly 40 000 km circumference. The surface area of the globe is 500 million square kilometres.

Our world also has the 1972 ABM Treaty between the United States and the Soviet Union, which commits the two signatories not to defend the national territory against attack by strategic ballistic missiles, nor to lay the basis for such a defence. The Treaty does permit a defence limited to 100 interceptors at a single site, which in the Soviet Union has been centred on Moscow, and in the United States near Grand Forks, North Dakota. Like any other treaty, the ABM Treaty could be terminated at any time by agreement between the two parties, and it can be terminated unilaterally upon six months' notice by one side if it judges that its supreme national interest is at stake. The ABM Treaty reflects the views of the American and the Soviet leaders at the time of its inception that it was not then feasible (or foreseeable) to have a defence good enough to replace deterrence of nuclear war, and that the influence of a defence would, instead, be to stimulate the growth of offensive weapons, of means to attack the defence, and so on.

In connection with the SDI programme, Paul H. Nitze, Chief Arms Control Advisor to Reagan, formulated three criteria that an SDI system must satisfy if the United States is even to contemplate its deployment. These criteria were written into national policy by Reagan himself in NSDD 172, and they are that the defence must be militarily effective; it must be adequately survivable; and it must be cost-effective at the margin. The official document published by the Department of State discussing NSDD 172 notes that a defence which is not adequately survivable is likely to *provoke* nuclear war rather than to prevent it, while one which is not 'cost-effective at the margin' will *stimulate* the deployment of more offensive weapons rather than terminate it.[1]

The problem with the defences thus far conceived under SDI is that they are not adequately survivable; nor are they cost-effective at the margin. Very relevant to this matter is the Report of the American Physical Society Study Group on 'Science and Technology of Directed Energy Weapons', published in April 1987 and which has now appeared in *Reviews of Modern Physics*.[2] This study does not go into detail on countermeasures to the SDI, in large part because the SDIO is not happy to have information published about possible countermeasures to defensive systems, even though much of this work is already published and all that it was proposed to include in the study can be derived from simple physics and calculation. Indeed, some of the most severe survivability and cost-effectiveness problems of SDI weaponry will be demonstrated in this chapter. The 'silver bullets' which can be used against SDI are the most effective, reliable techniques available, and they are substantially simpler in technology and lower in cost than the defensive system themselves. Just to name them, they are space mines, single-warhead ICBMs, and direct-ascent nuclear-armed anti-satellite (ASAT) weapons.

In a logical framework, one can talk about *passive* counters to SDI. These can include decoys which resemble boosters (to foil boost-phase intercept), or which resemble warheads in mid-course flight. Because it may be difficult and costly to have a decoy resemble precisely a finely machined re-entry vehicle, one wants to use the tool of *anti-simulation*, by which one makes the costly element resemble the *decoy*. In this way, if one had the job of preserving against assassination the life of an unpopular king, one could find 20 beggars and provide them with limousines and jewels, or whatever a king now wears, so as to reduce the chance that an assassin could choose the correct king. But one could do this at considerably lower cost if one dressed the king to look like the beggars – this is anti-simulation. Another example of anti-simulation is to be found in the provision of a booster decoy – an old missile launched from a soft launch pad in the silo fields from which the real ICBMs will come. Rather than housing the obsolete booster in a hardened silo, and giving it a guidance system which will resemble in every way the boost-phase flight of a real ICBM, one would want to use anti-simulation, by which some small fraction of the warhead-carrying missiles are based on soft pads in the silo fields, and the entire fleet of accurate ICBMs has its guidance system programmed to provide a slight stagger during last phase, resembling the behaviour of a simple programmer rather than a guidance system. This can be done without in any way impairing the accuracy of arrival of the warheads at their targets.

'Closed-spaced objects' in space – the RV is tied by relatively rigid cord of some 50 or 100 m length to a structure of decoys – can foil the non-nuclear hit-to-kill intercept vehicles proposed for SDI. For the interceptor is unlikely even to be able to strike one of the complex of close-spaced objects, and if it does have this ability, it will be unable to determine which is the real warhead. Finally, among the passive techniques one can list large balloons; here the RV is enclosed (but not housed at the centre of) a balloon larger than the collision radius of the interceptor, so that it cannot be killed reliably with a single collision.

Among *active* countermeasures against space defences, one can have space mines, both nuclear and non-nuclear.

- A space mine accompanies its quarry satellite always within lethal radius, ready to explode and to destroy it on radio command or in case of attack on the space mine; the space mine has no need to be *covert*, although after years of emphasis that an overt space time is the real threat, Abrahamson continues in his Congressional testimony to dismiss the space mine threat because it is difficult to hide them!
- Direct-ascent nuclear-armed ASAT weapons (DANASAT) already exist in the form of the Soviet high-yield exo-atmospheric GALOSH interceptors deployed for many years with the ABM system around Moscow. With only minor modifications these would be highly effective weapons against the space-based weapons of an early SDI. The SDI of the other side would have an easier job to destroy the satellites in predictable orbits than to destroy the enemy rocket force.
- As a final example of an active countermeasure one has the x-ray laser. As emphasised in the aforementioned study published in *Reviews of Modern Physics*, the x-ray laser can shoot *up* from deep within the atmosphere, from an altitude at which it is totally shielded from the spreading beam of a space-based x-ray laser trying to destroy it.

Among *threatening* countermeasures to a space-based defence are additional nuclear-armed ICBMs and SLBMs. The single-warhead small ICBM would be a likely growth path for the existing offensive forces, because (if silo based) it would cost about as much per warhead as the existing weapons, but it would require ten times as many boost-phase interceptors to destroy its complement of warheads than if they were based on the 10-warhead SS-18 missiles. Furthermore it is easy to make a small single-warhead ICBM a 'fast-burn booster' which would protect it totally from an intercept from space-based kinetic kill vehicles, as emphasised in the study published in *Reviews of Modern Physics*.[3]

Over the last year or so there has been a lot of emphasis in the United States on 'near-term deployment' of a space defence system, which would be unable to make any use of the directed energy weapons such as lasers or neutral particle beams or x-ray lasers, but would perforce rely for any boost-phase intercept component (still called by SDI proponents 'critical' or 'essential' to the success of the defence) upon Space-based Interceptors (SBI) – the so-called space-based Kinetic Kill Vehicles (KKVs).

In a very interesting paper by Chris Cunningham and colleagues of the Lawrence Livermore Laboratory, it is demonstrated that, even if the SBI do not have to descend from the altitude at which their garages travel, the minimum mass launched to orbit to do the boost-phase job in a boost-time T is obtained when the interceptors leave their garages with a speed of 6 km/s (more generally, twice the exhaust velocity, which in this case is about 3 km/s).[4]

Now it is time to do some simple calculations which go far towards showing the inadequacy of a space-defence in the real world. The mass ratio (that is the ratio of the mass reaching final velocity ΔV) to the initial rocket mass (final payload, plus engines of various stages, plus rocket fuel) is given by the conservation of momentum by the so-called 'rocket equation' which is reproduced as Equation 6.1.

Mass ratio: $m = M_f/M_o = e^{-\Delta V/V_e}$

If we take $V_e = 2\cdot5$ km/s $(= I_{sp}g)$, $\qquad\qquad$ (6.1)

then for $\Delta V = 8$ km/s, $m = 0\cdot04$

A rocket is propelled by the reaction to the momentum given to its jet of rocket fuel, in turn maximised by burning the propellant in a *chamber*, and having the hot gases flow through a convergent–divergent nozzle, so as to expand them, uni-directionally, to as low a temperature as possible. In this way, all of the potential thermal energy of the fuel is available as kinetic energy of the jet, although the payload of a rocket reaching a high speed is substantially limited by the weight of engines and rocket chamber. It has long been known that one *stages* rockets for high speed, so that one can approach the mass ratio of Equation 6.1. If, for purposes of calculation, the actual exhaust velocity of the fuel of some 3 km/s is reduced (rather than to take explicitly into account the existence of engines and tanks) to about 2·5 km/s, for a velocity gain of 8 km/s, $m = 0\cdot04$. Thus, about 4 per cent of the initial mass of a rocket can be launched into Earth orbit at low altitude, which requires a velocity of some 8 km/s. If one wanted to go on an extensive tour of the solar system, perhaps 16 km/s would be required (in comparison with 11 km/s to leave the Earth completely). The mass ratio for reaching 16 km/s would be the *square* of the mass ratio to reach 8 km/s, or a payload of about 1·6 kg for every ton of initial rocket mass.

Surprisingly (and we will return to this later) the *efficiency* of conversion of thermal energy of the rocket fuel to kinetic energy of the final payload is not at all small:

Energy efficiency:

$$\varepsilon = \tfrac{1}{2} M_f (\Delta V)^2 / \tfrac{1}{2} (M_o - M_f) V_e = (m/1 - m(V_f/V_e)^2 \qquad (6.2)$$

For $\Delta V = 8$ km/s, $\varepsilon = 31\%$.

This shows precisely this ratio, the numerator being the kinetic energy of the payload, while the denominator is the thermal energy of the fuel of all the stages of the rocket. In propulsion to Low Earth Orbit (LEO), for which the mass ratio is only 4 per cent, the energy efficiency from Equation 6.2 is still 31 per cent. And to reach 16 km/s, for which the mass ratio is only 0.16 per cent, the efficiency is still 9 per cent.

The same equation is important in determining the mass of the defensive system required to counter a current offence. For instance, 'near-term deployment' would achieve the essential boost-phase inter-cept within the boost-time T of the ICBMs by basing the SBI on space *garages*. Those garages in their 90-minute orbits of the Earth which are in the vicinity of the Soviet Union could contribute an SBI out to a distance from the garage of T multiplied by ΔV, and although by appropriate choice of orbit the density of garages over the Soviet Union could be increased by about a factor 3 from that of uniform coverage of the Earth's surface, that is about the best that could be done. Since the *area* which can be reached by an SBI from a garage is proportional to the *square* of this outreach distance, the number of garages goes inversely as the square of the SBI velocity-gain.

If there is a homing warhead mass M_f required (at present 15 kg for the small homing warhead of the US aircraft-based ASAT, after the two-stage rocket falls away which brings it to orbital altitude), the initial *mass* of the SBI housed in the garage is given again by Equation 6.1, in terms of the homing warhead mass and the required velocity gain from the garage. One can reduce the mass of a given SBI by having only a small velocity gain, but then a lot of SBI will have to be placed on orbit in many garages in order to ensure that the required areal *density* of SBI will be available to handle the boosters. Because the boosters in flight are not spread uniformly, one must provide an SBI areal density which can at every point equal this maximum required. The number of SBIs on orbit thus increases as the reciprocal of the *area* to whch an SBI can be dispatched from its space garage, so that the total mass of SBI on orbit has an exponential factor from Equation 6.1, and a square factor in velocity gain as just indicated.

SBI area coverage:

If $A_o = \pi n (T\Delta V)^2$, and $n \simeq (1/\Delta V)^2$ (6.3)

mass on orbit $= M_f n \, e^{\Delta V/V_e}$ and is
proportional to $(e^{\Delta V/V_e})/(\Delta V)^2$;

The calculations by Cunningham *et al.* show that only 13 per cent of the orbiting SBI can in any way attack either the SS-18 booster *or* its post-boost vehicle (2·5 per cent in boost phase, 10 per cent against the MIRV bus) while only 2·3 per cent could attack the Soviet SS-25.[5] Furthermore, if the Soviets change to a 6-warhead 'minibus' only 1·9 per cent of the KKVs could attack the SS-24, compared with 9·5 per cent against the current version SS-24.

In summary, a 15-kg homing kill vehicle would require about 300 kg per SBI (including the SBI itself). This is *more* than the weight of a nuclear warhead (some 300 kg, including re-entry vehicle), and if 10 per cent of the SBIs in flight can reach a booster or RV, this is the minimum required against a 10-MIRVed missile to be effective in destroying boosters carrying 10 MIRVs instead of warheads. With only 2 per cent of the SBIs capable of boost-phase intercept, even a 10-warhead booster cannot be destroyed cost-effectively in this way, and a single-warhead offensive missile *certainly* cannot be.

But if NSDD 172 says that we will not even consider an SDI for deployment unless it is both cost-effective at the margin and surviv-able, why are the Soviets worried about it? Probably for the same reason that Caspar Weinberger, US Secretary of Defense, says that a Soviet SDI system would be the worst strategic nightmare he could imagine. For, although it could not be used as a defence against a first strike, it could be popped into orbit (in principle), so that it would be in place over the Soviet offensive fields at the time that a *US* first strike arrives, giving the Soviets the choice between allowing their ICBMs to remain in their silos to be destroyed, or to fire them and have them destroyed by the space defensive system.

OFFENSIVE WEAPONS IN SPACE

In fact, the most threatening offensive weapon system in space would be just exactly this hypothetical space-based defence which would not be sufficiently effective against a first strike but could do a pretty good job (in principle) in suppressing a retaliatory strike. But one can consider more strictly offensive weapons. These are nuclear weapons in orbit, which could be brought down on their desired targets. This is not a very useful deployment for weapons, since they are far more vulnerable than if kept under the oceans in submarines or in silos. Furthermore a single weapon in orbit would require *more* propulsion to strike an arbitrary target on the ground within the 30-minute ballistic missile flight time, than if the weapon were launched by ICBM on the ground, and would thus be a far more costly way to destroy that target. However, a vast *system* of bombs in orbit could have the closest ones de-orbited in just a few minutes onto a target complex.

Lasers in orbit have been described as capable of setting a million fires in a few minutes, and that is true. However, thermite bombs (a mixture of aluminium powder and iron oxide) in orbit could do the same job as quickly and at lower cost, and without hindrance by clouds. In either case, the fires can readily be extinguished if there is no high-explosive attack at the same time, and this is a difficult way to start the Third World War. One could use powerful lasers in orbit (or ground-based lasers with their beams directed by orbiting relay mirrors) to destroy aircraft in flight, or especially to destroy satellites. However, there are alternatives available sooner and more cheaply for these tasks. Then there is the space-based anti-satellite weapon. Even a weapon with far less capability than needed to provide a defence against strategic ballistic missiles will be a highly effective ASAT system. However, one does not at present have nearly enough satellite *targets* to justify such a system.

Finally, there is considerable discussion of 'non-nuclear strategic weapons' and those have even been described in some cases as 'long-rods' which would be based in orbit and brought down with precision of a few tens of centimetres to destroy in a highly specific manner various strategic targets. This could probably be done,[6] but it is easy to *defend* a set of strategic targets against such non-nuclear attack – by mobility, by hiding the precise location (even with an umbrella), or by explosive-backed armour a few metres above the target. And, although a long-rod penetrator can penetrate many metres into the ground, by the same token it would pass right through a command headquarters without doing much damage.

In short, space is not a very effective or efficient place for the *offence* to deploy its weapons, nor are space-based offensive weapons a threat of a new kind which could not be overcome at considerably lower cost than would be required to mount the threat.

CIVIL ACTIVITIES IN SPACE

There has been great benefit to the world from satellites for the relay of communications and broadcasts, for weather observation satellites, imaging for civil and military purposes, for high-precision navigation by the TRANSIT satellites and now by the US Global Positioning System (Navstar-GPS) and the similar Soviet system. As for more futuristic systems requiring enormous launch capability or mass in orbit, space manufacturing may have some merit at some future time and will probably be carried out effectively by unmanned recoverable vehicles. Satellites to capture the light of the sun, converting it into laser light or to microwave beams and relaying it to Earth, are not economical in comparison with ground-based systems. For some decades it has been proposed to deploy very large mirrors of very thin metallic foil, which could be used to light up portions of the Earth at night or on moonless nights, and this is a feasible task. However such environmental modification of national territories by another nation hardly seems acceptable in the modern world. The deployment of such a capability, even if its actual *use* is restricted to the illumination of the territory of one's own nation or that of nations which request it, surely deserves a better economic analysis and legal basis.[7]

There is, however, one 'civil' use of space which greatly overlaps space weapons. This is the desire on the part of some to use ground-based lasers to heat material carried by vehicles, thus providing a ground-based energy source for a rocket engine. The material would be heated and expand into space, providing exactly the kind of propulsion that a rocket engine does, but (one hopes) with a higher exhaust velocity. Indeed, if one could only match the exhaust velocity of a chemical rocket, then it would be irrelevant that the energy came from Earth – the mass ratio of the rocket on launch to the mass of its payload after a certain ΔV gain would be the same as that for a chemical rocket which needs no external supply of energy! But the enthusiasm should be tempered by some absolute limits on laser launch requirements. For instance, to launch 10 tons into low-Earth orbit (8 km/s velocity gain) in 320 seconds (during which time the vehicle would have moved under constant acceleration some 1500 km) would require the transfer *to* the payload of some 320 gigajoules of kinetic energy, and to do that in 320 seconds would require the payload to gain kinetic energy at the rate of 1 gigawatt. This is the power output of a large civil nuclear power station; but even lasers proposed for space defence, discussed above, have power outputs of the order of 0·1 GW, and not 1·0 GW. Furthermore one can hardly conceive of a rocket efficiency much better than 50 per cent, and even that would require a 2 GW laser source, if all the laser light traversed the atmosphere without scattering or loss. If one assumes a factor 10 for loss of laser light in traversing the atmosphere and for spill over the relay mirror, that would now require a laser of 20 GW as a source. Similarly, it would be remarkable to obtain a laser efficiency of 30 per cent, which would require a prime power input of some 70 GW during launch, and would demand a laser which would be enormously overpowered even for the space defence role.

Finally, the *benefit* to be achieved by such an enormous programme is minimal, since we have already emphasised (and calculated) that the efficiency of conversion of rocket fuel energy to satellite kinetic energy is 31 per cent for an ordinary rocket.

CONCLUSION

Given the fact that no space-defence system thus far proposed looks as if it will satisfy the Nitze criteria (and thus its deployment would be a threat to our *own* security), and since offensive weapons are not really very attractive, and since none of the space weapon technologies has significant economic or scientific promise justifying its cost, it is time to put this expensive education to use. The United States and the Soviet Union should adhere strictly to the traditional form of the 1972 ABM Treaty.

Of course, 'strict adherence' to the 1972 ABM Treaty cannot be accomplished while the Soviet Union continues to construct a violation of the Treaty – the early-warning radar at Krasnoyarsk. This is a pair of buildings housing transmitter and receiver, apparently identical to the early-warning radar at Pechora. The latter is permitted by the ABM Treaty, because it is a large phased-array radar located on the periphery of the Soviet Union and looking out.

The Krasnoyarsk radar looks over thousands of kilometres of Soviet territory before reaching the border. The radar at Krasnoyarsk cannot play an effective *military* role as part of a Soviet ABM system (and its early-warning role is prejudiced by its being located so far into the interior), but it *is* a violation of the Treaty. It should either be destroyed or converted into what the Soviets have at times said it is – a space-track radar. However it is of the essence of arms control agreements that there be 'timely warning', and to limit Krasnoyarsk in actual operation to space track (by the computer programme which determines in which direction the pulses are directed) would provide no protection against its immediate conversion to another role in time of nuclear war. An alternative to destruction of the radar would be to build in front of the transmitter at a distance of some 50 m a hill of earth 50 m high. Such a hill in an arc of some 150 m length would prevent significant detection and elevation angles of 22 degrees or lower, thus preventing the radar from seeing a re-entry vehicle which was about to destroy it.

The construction of such a hill would cost only a few million dollars and would provide visible evidence that the radar could not be used for any purpose other than space track. There would then be no doubt about its compliance with the ABM Treaty. That would, of course, raise the question of *US* compliance with the ABM Treaty in regard to the radar construction at Thule, Greenland, and Fylingdales, Great Britain. However, the ABM Treaty did not ban development or even use of anti-satellite weapons. It would add to the security of both the United States and the Soviet Union (and it would be highly desirable for the rest of the world's participation in space as well), to have a multilateral ban on ASAT tests, deployment and use. Furthermore, because many nations active in space do not have nuclear weapons, it is not at all true that the only threats to the peaceful use of space come from the United States and the Soviet Union. There should be an *international* ban on weapons to and from space, as well as on ASAT.

Beyond the strict adherence to the 1972 ABM Treaty, these other aspects of ensuring a peaceful and productive exploitation of space are covered in a brief 1983 draft treaty prepared with the support of the Union of Concerned Scientists and presented by the present writer and two colleagues to the Senate Committee on Foreign Relations on 18 May 1983. Much of the substance of this draft was incorporated in a Soviet draft treaty presented to the UN General Assembly in August 1983, and either of these could be the basis for an early bilateral treaty between the United States and the Soviet Union, which (like the

Biological Weapons Convention) could be opened for accession by the other nations of the world.[8]

Unfortunately, technology enthusiasts in the United States (and I have heard from some in the Soviet Union and other nations as well) continue to demand no restrictions at all on their activities. They say, 'Give us the money and enough time, and we will think of something'. But their enthusiasm is self-serving in many cases, as shown by the disaster to the US civil and military space programmes represented by the space shuttle. The space shuttle programme initially promised launch to low Earth orbit at about $5 per pound, and was committed with a goal of $50 per pound. Now Abrahamson, who directed the space shuttle programme for several years before his accession to the Directorship of the Strategic Defense Initiative programme, says that the shuttle provides launch to orbit at a cost of $1500 per pound, and that this is at least ten times as much as can be accepted by the SDI programme. Not only did the shuttle programme and its supporters request money for totally unrealistic goals, they used their bureaucratic power to suppress the evolution or even the continued acquisition of expendable boosters for launch to LEO, which would have shown the shuttle programme to be an uneconomic approach, but which would have provided the United States with a durable, economic launch system such as we had in the 1960s before the shuttle distortion entered the US picture.

To conclude, consider the following three hypothetical cases:

- If with my revolver in my pocket I accost a prosperous-looking person on the street and demand, 'Give me $100 or I'll kill you' and I am caught, I will be sent to jail for armed robbery.
- If without the revolver in my pocket, I accost the same person and say, 'Give me $100 or my brother will kill you' and I am caught, I will go to jail for extortion.
- But if I go on television and demand from the American public, 'Give me $300 billion for our military activities, or the Russians will kill you', I will be deemed a great patriot and perhaps elected to high office.

Keeping weapons out of space is the key to deep cuts in nuclear weapons, which could reduce the world's present 50 000 weapons to some 3000 within ten years. This cannot happen if the nations of the world create a new arms race in space, whether or not the space weapons are nuclear-armed.

Notes

1. US Department of State, 'The Strategic Defense Initiative', Special Report no. 129, June 1985, p. 4.
2. Nico Bloembergen, C. Kumar, N. Patel *et al.*, 'APS Study: Science and Technology of Directed Energy Weapons', *Reviews of Modern Physics*, (July 1987) pp. S1–S202.
3. Ibid, p. S30.
4. Chris Cunningham, Tom Morgan, and Phil Duffy, 'Near-Term Ballistic Missile Defenses', Strategic Defensive Systems Studies Group of the Lawrence Livermore National Laboratory (1987).
5. Ibid.
6. The projectile would have great problems in seeing through the sheath of superhot air which would surround it if it preserved its initial speed to low altitude, but it could slow to about 2 km/s and use radar or optical sensors. Alternatively, it could be command-guided, or use a 'differential GPS' navigation scheme.
7. Without giving a full analysis, one can consider a foil at 40 000 km altitude (geosynchronous altitude). The light could not be restricted to a region on the ground smaller than about 400 km diameter, and to equal sunlight intensity the foil would need to be of some 400 km diameter or more. But to provide light level 100 times that of moonlight over the same 400 km diameter circle, only some 13 km diameter of foil is needed. At a foil thickness of 0·1 milligram per square centimeter (1 g/sq m), this would be some 100 tons of reflective foil. Properly distorted, this 'mirror' could provide moonlit levels over a 4000 km diameter region on the ground.
8. Both the Union of Concerned Scientists' draft and the Soviet draft are reproduced in conjunction with articles: Richard L. Garwin and John Pike, 'History and Current Debate' and Yevgeny P. Velikhov, 'Effect on Strategic Stability', *Bulletin of the Atomic Scientists* (May 1984) pp. 10S–11S.

7 Clarifying ABM Treaty Ambiguities: Threshold Limits

John E. Pike

INTRODUCTION

New developments in ballistic missile defence (BMD) technology pose major challenges to the Anti-Ballistic Missile (ABM) Treaty of 1972. There is a growing need to find a way to keep the ABM Treaty current with the evolution of BMD technology. While the ABM Treaty is of indefinite duration, it needs and deserves periodic updating. Over the past two years discussion has focused on the debate over the Reagan Administration's broad interpretation of the ABM Treaty, which holds that the Treaty does not limit the testing of exotic BMD technologies, such as lasers. But this is now a false issue. It is increasingly clear that this interpretation of the Treaty is without legal or factual merit. Congressional opposition seems likely to ensure that it will not be implemented. In any case, the Administration has had considerable difficulty explaining what additional tests would be conducted should the broad interpretation be implemented. Resolution of the broad interpretation debate does not resolve the inevitable conflict between the permissive and restrictive readings of terms of the traditional interpretation of the ABM Treaty.

The Reagan Administration espoused a permissive reading of the traditional interpretation of the ABM Treaty in its first annual Report to the Congress on the SDI in early 1985, and in each subsequent edition of this report. The Administration asserted that the SDI programme was consistent with this permissive reading of the Treaty, arguing that the SDI was not developing ABM 'components', and that SDI tests would not be conducted 'in an ABM mode', or demonstrate 'ABM capabilities', and thus the technologies tested under the SDI would not be 'capable of substituting for' ABM components. But a more restrictive reading of the Treaty's terms leads to the conclusion that many of the tests under the SDI do involve components with ABM capabilities, and thus are inconsistent with the Treaty. Unfortunately, the Treaty provides inadequate guidance for choosing the proper reading of these critical terms.

New definitions of what constitutes ABM 'capabilities', and focusing on thresholds rather than categorical bans, could resolve this problem. Devices with capabilities above a certain 'threshold' would be subject to the testing and deployment limits of the Treaty, while those with inferior capabilities would not. Similarly, there are questions about what is an ABM 'component' or what constitutes 'development', terms that are central to the ABM Treaty, but which lack sufficiently precise definition. But it should be possible to determine whether or not the mirror of an ABM component such as a laser or sensor telescope is larger than two metres in diameter, within an acceptable margin of error. Thus threshold limits would provide a less ambiguous operational definition for the 'development' of an 'ABM component' which has 'ABM capabilities' or has been 'tested in an ABM mode'. This approach was first proposed in a meeting sponsored by the Federation of American Scientists (FAS) with Soviet scientists in early 1984. Over the past year a series of private talks have confirmed the usefulness of threshold limits. At the February scientist's forum in Moscow three days of intense discussions led to agreement on a communiqué by Soviet and American technical experts endorsing the threshold limits approach.

Formulation of the American stance at the Geneva negotiations has previously led to consideration of this approach. In September, 1985, press reports indicated that 'agreement on common definitions of precisely what is permissible research' would be one of the priorities for the American delegation. There have been a number of indications of Soviet interest in this approach. Press reports prior to the 1985 Geneva summit suggested that the Soviets wanted to ban space-based testing of kinetic kill mechanisms, set ceilings on the power levels of laser kill mechanisms to be tested in space, and regulate testing of high-energy power sources in space. Soviet General Secretary Gorbachev has endorsed the idea of technical discussions along these lines, suggesting in April 1987 'let experts of the two countries take their time, ponder on the subject, and agree on the list of devices that would not be allowed to be put into space in the course of this research'. While the senior US arms control adviser, Paul Nitze, has publicly expressed interest in reaching an agreement of this kind, the American government remains committed officially to the broad interpretation of the Treaty.

Although the two sides did agree in January 1987 to establish a special working group at the Geneva negotiations to discuss what activities are permitted under the ABM Treaty, thus far this effort has not borne fruit, because the American side is under instructions only to reiterate US support for the broad interpretation of the ABM Treaty, and is prohibited from exploring Soviet proposals for a compromise. Discussion of threshold limits to clarify the existing commitments under the ABM Treaty should be the focus for discussions in the special working group at the Geneva negotiations that was formed for this purpose early this year and in the Standing Consultative Commission (SCC). It will not be possible to resolve all of these issues at once, as this will be a continuing process as new technologies are identified. But such negotiations limiting grey areas was precisely the role originally intended for the SCC, and it ought to be entirely feasible, if the will to agree exists.

THE GOALS OF THE ABM TREATY

The ABM Treaty serves two mutually supporting and yet contradictory roles in banning the deployment of nation-wide anti-missile systems. First, the Treaty allows sufficient research and testing by both sides to permit the design of countermeasures needed to ensure that any anti-missile system that might be deployed would be ineffective. But the second role is to keep such deployments so far in the future that offensive forces are not built to discourage break-out from the restrictions imposed by the Treaty. The first role of the Treaty is to permit research in support of the Treaty regime. Research on BMD technology provides reassurance that BMD systems would not be effective, because both sides understand the technology well enough to design countermeasures. Designing a decoy requires a practical familiarity with the performance characteristics of the sensor that it is designed to fool. And when technical experts have argued that ballistic missile defence systems would be ineffective, their critique was not simply the result of idle speculation on a blackboard. It was the result of concrete, actual work that provided a practical understanding of these technologies. The second purpose of the ABM Treaty is to establish a long lead-time for deploying an anti-missile system. Sufficient time should elapse between the point at which the treaty regime was exceeded and the time an anti-missile system was actually deployed so that there would be no requirement actually to implement countermeasures in existing force structures.

One factor in the current high levels of strategic offensive forces is the efforts to offset the other side's actual or potential strategic defences. Deep offensive reductions may prove very difficult to achieve in the absence of greater constraints on strategic defences. However strategic defences are more difficult to constrain than strategic offences, and this may impose limits on offensive reductions. Possibly two-thirds of the current US inventory of strategic offensive weapons were initially rationalised at least in part as a response to current or potential Soviet strategic defences. This includes multiple warheads (MRVs and MIRVs) which were deployed on SLBMs in the 1960s; these were justified in part by the need to respond to Soviet ABM systems, and later to discourage Soviet break-out from the ABM Treaty. If one regards the 656 SLBM warheads initially deployed on the Polaris A-1 as the level required for basic target coverage, over 5500 additional SLBM MIRVs are a response to Soviet ABM systems. More multiple warheads (MIRVs) were also deployed on ICBMs in the 1970s, and justified in part by the need to discourage Soviet break-out from the ABM Treaty. If the roughly 1000 weapons on single warhead ICBMs in the 1960s were required to cover Soviet targets, about half of the 2100 ICBM warheads currently deployed are a response to strategic defence. Finally, more bomber payloads, which today include about 2000 gravity bombs, are needed to attack primary targets. But US bombers also carry 1400 Short-Range Attack Missiles (SRAMs), which are used to suppress Soviet air defence, and over 1700 Air-Launched Cruise Missiles (ALCMs), which were initially justified primarily as a response to Soviet air defence.

The essential problem is finding a balance between the low level of research that is necessary to develop effective countermeasures and thus uphold the ABM Treaty regime, and the high level of activity that would permit a system to be deployed quickly – provoking the deployment of those countermeasures, particularly the proliferation of offensive forces, and thereby impeding negotiated limits on offensive systems. The ABM Treaty of 1972 is the only major bilateral arms limitation agreement in effect between the United States and the Soviet Union. The Treaty reflects their shared judgement that limitations on strategic defences and offences are interrelated. Since 1972, technology has evolved considerably, as might be expected over any

15-year period. In addition, differences in national policies relating to strategic defensive systems and the scope of the ABM Treaty have emerged, as reflected most recently at the Reykjavik Summit. However, quite apart from these differences, the issue of the threshold between permitted and prohibited development and testing activities would have arisen in any event, and should now be dealt with by the parties to the Treaty. The Treaty prohibits the development, testing and deployment of ABM components that are space-based, air-based, sea-based or mobile land-based.

The Treaty provides several criteria for establishing what devices are subject to these limits:

- The components of ABM systems at the time of the signing of the Treaty, namely interceptors, launchers and radars.
- Devices that have been 'tested in an ABM mode' (that is, tested against strategic ballistic missiles or their components in flight trajectory).
- Devices that have 'ABM capabilities' or are 'capable of substituting for' ABM components.

The Core Problem

The central problem is that the march of technology has complicated the interpretation of the terms of the Treaty. In 1972, verification of testing in an ABM mode was a fairly straightforward process. The operation of a radar could be monitored by electronic intelligence satellites, and the launching of an interceptor, and the flight of a target re-entry vehicle could be monitored by various means. These activities provided a rather unambiguous basis for defining 'tested in an ABM mode'. But the new BMD technologies pose a greater challenge for determining whether a device has been 'tested in an ABM mode'. Passive sensors such as telescopes which can be used to track targets do not emit signals, and thus their association with an anti-missile test can be difficult to determine. Long-range interceptors can be tested against satellite targets which mimic the characteristics of a strategic ballistic missile. Unfortunately, the determination of whether a device is capable of substituting for an ABM component or whether it has ABM capabilities is also very difficult, particularly if the device is based on other physical principles (such as lasers). The ABM Treaty does contain a precise threshold definition of what constitutes a radar that has ABM capabilities, but the Treaty provides no guidance on the point at which a tracking telescope is capable of substituting for an ABM radar.

Threshold definitions of ABM capabilities agreed to by the United States and the Soviet Union could resolve this problem. There may be questions about what is an ABM 'component' or what constitutes 'development', but it should be possible to determine whether or not a mirror is larger than two metres in diameter, with an acceptable margin of error. These threshold limits would provide a less ambiguous operational definition for the 'development' of an 'ABM component' which has 'ABM capabilities' or has been 'tested in an ABM mode'.

Alternative Regimes

A number of possible regimes limiting anti-missile systems have been discussed in recent years, and the attitudes of the United States and Soviet Union have evolved over time. Both countries originally agreed to the traditional interpretation of the ABM Treaty in 1972. But in recent years the Reagan Administration has modified its adherence to the Treaty, moving first to the permissive reading of the Treaty in early 1985, and subsequently moving to the broad interpretation. The Soviets called initially for a ban on all purposeful research and subsequently moved to a proposal that would ban all testing outside laboratories. In 1985 the Soviets moved to a ban on testing that would either ban all testing in space or all testing of space-based elements of BMD systems, and in 1987 the Soviets proposed an approach apparently based on threshold limits.

The initial Soviet reaction to the Strategic Defense Initiative was to call for a complete ban on research of this type. And this continued to be their position through early 1985. Senator Gary Hart reported on 17 April 1985 that when he met Andrei Gromyko in Moscow in January 1985: 'Gromyko responded that a moratorium on space weapons could not be based on testing alone, but would have to include research as well, given that 'research is 90 per cent' of the process of weapon development.' However the Soviet chief negotiator, Victor Karpov, appeared to move away from this position in October 1985 when he stated that the Soviet government had never opposed basic scientific research, although they did seek a ban on development and testing. Another Soviet negotiator, Yuli Kvitsinsky, elaborated this approach in late October 1985, noting that 'what cannot be observed does not exist'. Subsequent discussions in the spring of 1986 at the Geneva negotiations confirmed this position.

A less restrictive regime would prohibit all field testing of anti-missile or anti-satellite components or elements (however that might be defined), while permitting laboratory testing. In a meeting with several American Senators in Moscow in September 1985, General Secretary Gorbachev stated that 'you can't verify what's going on in the brain, and that's what we refer to as fundamental or basic research. But as soon as you go beyond the laboratory, go to mock-ups, models, contracts with defence contractors, here surely verification can be done. We want a ban on that phase of research that approaches design and manufacture'.

Although the American side gained the impression at the Iceland summit that the Soviets were still insisting on a ban on all activities outside the laboratory, following the summit the Soviets insisted that this was not their position. Unfortunately, Soviet statements on this matter are somewhat confusing. In particular, it is unclear whether their position calls for a ban on *testing in space*, which would ban testing above the atmosphere, or whether the ban would be on *space-based testing*, which would permit testing in space of devices that were on ballistic rather than orbital trajectories. The draft agreement submitted by Gorbachev at the Reykjavik summit on 11 October 1986, which is the clearest formulation of the Soviet positions, stated that:

All testing of *space-based* elements of a ballistic missile defence *in outer space* will be prohibited except research and testing in laboratories.

That will not require a ban on tests allowed by the ABM Treaty – of fixed land-based systems and their components. The sides must find mutually acceptable solutions in this area during negotiations in the next several years.

Both sides agree to make additional efforts to reach mutually acceptable agreements to ban ASAT's (anti-satellite weapons).

While it is unclear what the Soviets mean to include when they use the term 'elements', another interpretation of their approach is that it would include a ban on space-based testing prohibiting testing of elements that complete one full revolution of the earth. This regime would permit unlimited activities above the atmosphere, as long as the element was on a ballistic trajectory – orbital trajectories would be prohibited. There would be considerable difficulty in distinguishing between the two previous regimes since in most cases elements or devices or components that could be tested in a space-based mode could also be tested in space flying on a sub-orbital ballistic trajectory in which the test elements fall to Earth at the conclusion of the test.

The first version of the ABM Treaty is a restrictive reading of the narrow interpretation of the Treaty. The narrow interpretation of the Treaty applies the Article V prohibition on testing space-based and air-based components to all ABM *capable* components, regardless of whether they are based on traditional or exotic technologies, and applies these limits to devices with even rudimentary capabilities. This regime would establish a variety of threshold limits on the performance of devices with potential anti-missile capabilities. This would clarify some of the ambiguities posed by the restrictive and permissive reading of the ABM Treaty through agreed thresholds that would define permitted and prohibited activities. Under the ABM Treaty today categorical limits are implicitly defined by threshold limits unilaterally determined by both sides. This regime would seek bilateral agreement on these threshold limits in explicit numerical terms.

The ABM Treaty provides precedent for issues of interpretation through threshold limits. At the time the Treaty was signed, conventional rocket and radar technologies were fairly well understood, and the Treaty contains a variety of specific threshold limits on such systems; namely:

100   Interceptors at permitted deployment areas (Article III-a/b).

150   Kilometre radius of permitted deployment areas (Article III-a/b).

6   ABM radar complexes at the national capital deployment area (Article III-a). 3-Kilometre radius for each ABM complex at the national capital deployment area (Article III-a).

2   Large phased-array radars at the ICBM deployment area (Article III-b).

18   Small radars at the ICBM deployment area (Article III-b).

15   ABM launchers at test ranges (Article IV). 3 000 000 Power aperture product defining large phased-array radars (Agreed Statement B).

1300   Kilometre separation of the two permitted deployment areas (Agreed Statement C).

1   Maximum permitted number of independently guided war-
        heads per interceptor (Agreed Statement E).

The permissive reading of the narrow interpretation of the Treaty
recognises that the Treaty does apply to all types of anti-missile
components (including exotic systems). But it holds that Article V's
restrictions on mobile components do not constrain the SDI since
none of the devices tested under the programme have *all* of the
characteristics of an ABM component, and that Article VI's restrictions do not constrain SDI testing since these tests would either not be
conducted in an ABM mode, or would not actually and totally
demonstrate ABM capabilities.

The American delegation at the Geneva negotiations has previously
considered this approach. In September 1985 press reports indicated
that 'agreement on common definitions of precisely what is permissible research' would be one of the priorities for the American
delegation. There have been a number of indications of Soviet interest
in this approach. In October 1986 the Soviet Academician, Roald
Sagdeev, stated: 'If a powerful laser is able to produce effects needed
for SDI, a demonstration of these types of devices would be quite
destabilising. But if tests are with modest instruments, they could be
considered permissible under the ABM Treaty.' In April 1987 General
Secretary Gorbachev suggested: 'Let experts of the two countries take
their time, ponder on the subject, and agree on the list of devices that
would not be allowed to be put into space in the course of this
research.'

The difference between the restrictive and permissive readings of
the Treaty is best illustrated by the case of the Airborne Optical
System (AOS), also known as the Airborne Optical Adjunct. AOS is a
modified Boeing 767 aircraft that carries an infra-red telescope for
tracking and identifying re-entry vehicles while they are still above the
atmosphere for interception by mid-course and terminal defences.
The Reagan Administration offers three lines of reasoning under its
permissive reading of the ABM Treaty to support its contention that
AOS is Treaty-compliant.

The first rationale is that the Boeing 767 cannot stay aloft for a
sufficient period of time to be an effective ABM component. This is
the least compelling part of the permissive case for AOS, and resort to
such a tenuous line of reasoning suggests the weakness of the
permissive case as a whole. The Boeing 767 currently has a maximum
airborne endurance of about ten hours. This is comparable to the

endurance of the E-3 AWACS which performs an air defence function analogous to the BMD function performed by AOS. Contractor studies have suggested that, even with its current endurance, a fleet of less than 40 767 aircraft would be adequate for an operational system. And if needed, the endurance of the 767 could be extended to several days through the use of aerial refuelling. The second part of the permissive case for AOA is the assertion that AOS is compliant because of its limited signal and processing capability. But the prohibitions in Article V apply to the development of components that can be monitored by national technical means of verification. And the permissive interpretation would require a detailed understanding of the computer software and communications capabilities of AOA which is clearly beyond the capabilities of national technical means.

The third argument under the permissive interpretation assumes that a device would not be a Treaty-accountable ABM component unless it could perform the complete function of or substitute for a 'stand alone' basis for an ABM component as defined in Article II of the Treaty. Although there are some missile defence systems with a single sensor (such as the proposed Site Defence system that was under development in the United States in the 1970s) they are the exception, rather than the rule. In practice, most missile defence systems have more than one sensor component, each of which plays some role in the management of the battle. For example, the early Nike-Zeus system had not one or two, but four separate types of radars, for target acquisition, decoy discrimination, target tracking and interceptor tracking. Under the permissive interpretation of the difference between a 'component' and an 'adjunct', all of these radars would be considered to be adjuncts to one another, and none of them would be considered to be a component.

The Airborne Optical System performs a role similar to that of the Perimeter Acquisition Radar (PAR) in the Sentinel/Safeguard system. Radars such as the PAR were clearly considered to be ABM components, and subjected to strict limitations in the Treaty. The initial configuration of AOS has a passive infra-red telescope sensor which can provide some target tracking data, but not the range from the sensor to the target. This limits the utility of this system, and on this basis some would argue that AOS is not a component. However, as early as 1990 AOS will be upgraded under the Airborne Laser Experiment effort with a laser range-finder to provide target range information. This would improve the performance of the sensor, and raise more serious questions about the systems compliance with even the permissive interpretation of the Treaty. Although the permissive interpretation case for conducting the initial tests of AOS is not compelling, it must be conceded that the plain text of the Treaty, as well as what is publicly known of the negotiating record, does not provide a clear-cut basis for choosing between the permissive and restrictive readings of the Treaty.

The third version of the ABM Treaty is the Reagan Administration's broad interpretation of the Treaty, which holds that the testing limits of Article V do not apply to exotic systems based on other physical principles, and that the only provisions of the Treaty relevant to devices other than conventional rocket and radar systems is the Agreed Statement D limit on deployment of exotic systems. A variant of the broad interpretation of the Treaty holds that even kinetic energy weapons are not subject to the Treaty's testing limits, since such devices have on-board guidance systems, and thus are not totally dependent on external guidance from ABM radars or other sensors.

The recent Senate debate over this issue has made it clear that the broad interpretation is very unlikely to become the basis for the SDI programme. It is increasingly clear that this interpretation of the Treaty is without legal or factual merit.

NEW THRESHOLD LIMITS TO CLARIFY THE ABM TREATY

A ballistic missile defence system is composed of four elements – weapons, weapon launchers, sensors and battle management. It is generally recognised that battle management poses the greatest technical challenge for perfecting an anti-missile system, and that sensors pose a greater technical challenge than do weapons and weapon launchers. It is an unfortunate paradox that the most technically challenging aspect of a BMD system (battle management) also poses the greatest problem for verification, while the least technically challenging part of the problem (weapons) poses the least problem for verification. The ABM Treaty places no constraints on battle management systems, since it was recognised by both parties that such limitations would be difficult if not impossible to verify. At the time the ABM Treaty was signed, BMD sensors were very large radars that required years of construction, and thus were easy to verify, so the

Treaty provided a strict regime of limitations on the deployment of such radars. But future systems using passive sensors may be much more difficult to verify. This does not mean that they would be impossible to verify, or that they should be exempted from constraint. But more stringent constraints on weapons testing may be needed to compensate for the difficulties of limiting sensors.

First, the threshold should ideally apply to a wide range of technologies. One of the challenges posed by a new and emerging anti-missile technologies is their dazzling variety. The search for threshold limits should focus on a small set of parameters that cover a wide range of weapons, sensors, or both. A common limit of five square metres on the aperture of laser beam director mirrors, satellite sensor mirrors, and the windows on airborne sensor aircraft would constrain a wide variety of weapon and sensor technologies. Second, the limits should apply to technologies that are of interest for ballistic missile defence. While limits on some systems, such a rail-guns, might be imagined, the low priority currently assigned to such devices suggests that more immediate issues, such as lasers (particularly ground-based), rocket interceptors and passive infra-red telescope sensors should be addressed as a matter of priority. Third, the threshold limit should be related as directly as possible to the actual performance of the device in question. The power aperture product, the radar threshold limit that was agreed to in 1972, does this very well. The brightness of a laser is similarly a very good measure of the laser's military performance.

Fourth, it must be possible to distinguish permitted and prohibited activities. Ten or 20 years ago the volume of an interceptor was a fair indicator of its anti-missile potential. But the recent advent of very small terminal homing sensors has reduced the size of interceptor warheads, and thus of interceptor rockets. In the future, very capable anti-missile interceptors may be much smaller than today's anti-aircraft rockets. Fifth, the threshold should provide adequate insurance against break-out from the Treaty limit. A typical weapon system might take about five years from the point of conception to initial field testing, another five years from initial testing to an initial operational capability, and an additional two to five or more years to reach a fully operational status. Limits on deployment provide at most five years lead-time, and may provide much less, while development and testing constraints may provide a ten year lead-time. Major reductions in offensive forces will increase the utility of limits on development and testing. And sixth, the threshold limit clearly must be verifiable. This means that data related to the limit can be collected using national technical and other means.

There are in turn several criteria that should be applied to verification. First, the required technical collection systems and other means of verification should be available during the time frame in which the parties to the Treaty are likely to encounter the thresholds they are intended to monitor. The development and deployment of entirely new dedicated space-based sensors for monitoring limits such as laser brightness might require as much as ten years. Setting a brightness threshold limit at the level anticipated in the late 1990s, by which time a new satellite monitoring system might be in place, would permit such extensive testing of lasers at potentially very high brightness levels as to call into question the utility of the limit. In such cases, co-operative measures such as in-country monitoring stations should be considered, since they could be deployed much sooner. The second criterion is the cost of monitoring. Since one of the canonical goals of arms control is saving money, the cost of the monitoring system should be less than the cost that would be incurred by not placing a limit on the activity in question. And third, technical collection systems should not be so capable that they reproduce the anti-missile systems that they are intended to limit. Large space-based infra-red telescope sensors used for verification may be difficult to distinguish from sensors that would form the basis for an ABM battle management system. It would be perilously paradoxical if it were necessary to develop or deploy reasonable facsimiles of an anti-missile system to verify limits on the development or deployment of such as system.

In some cases verification may require the use of non-intrusive co-operative measures. While national technical means may be adequate to measure the aperture of the beam-director mirror of a ground-based laser, when not in use such mirrors are normally screened from the environment by a moveable cab or dome, and thus out of sight. Agreement would have to be reached that such screens would have to be temporarily removed on a periodic basis to permit monitoring by national technical means. Some threshold limits might require more intrusive means of verification. In-country monitoring stations may be needed for the verification of limits on threshold limits on the brightness of lasers. During its passage through the atmosphere a fraction of the energy of a laser beam will interact with the atmosphere, through such mechanisms as aerosol scattering. An automated collection device, stationed a few kilometres from the laser beam director, could observe this scattering, and determine the laser's wavelength. With the addition of some small low-power lasers, and other devices, this station could also assess the scattering properties of the atmosphere in the vicinity of the laser, and thus provide the basis for determining the fraction of the laser's power that would be scattered, and thus the brightness of the laser. Pre-launch inspection of all satellite payloads could determine the presence of a reactor core. This would require placing a radiation monitoring device next to the exterior of each launcher's payload shroud shortly prior to launch. This would not require actual viewing of the satellite, and thus would not compromise the design characteristics of the payload.

And finally, some limits may require creative approaches to verification. It may be difficult to distinguish prohibited anti-missile technologies from benign scientific endeavours. The participation of the international scientific community in the development and execution of projects such as large space-based astronomical telescopes or nuclear powered planetary probes could provide reassurance that they were not being used as a cover for military developments.

FUTURE TREATY CONSTRAINTS

In conclusion, continuing treaty constraints would include the following:

- Directed energy systems such as lasers should have a limit of ten to the 19th power watts and joules per steradian on their peak and average brightness, which is a function of the laser's power and energy, as well as the laser's wavelength and the diameter of the primary beam director mirror. Brightness is the most useful measure of a laser's performance. Brightness levels needed for effective ABM systems would probably be hundreds or thousands of times higher than the proposed threshold. Even though the maximum brightness of American military lasers has increased at the rate of a factor of 100 every five years since the early 1970s, the proposed threshold would provide a five to ten year lead-time protection. The minimum brightness level required for ABM purposes is about 10 to the 16th power watts per steradian, and no other applications require lasers of this brightness, except for anti-satellite weapons. However, lasers of such brightness may be relatively small and difficult to identify.

Verification of such a limit would probably require the use of non-intrusive in-country monitoring stations located near identified or suspected laser facilities.

- A ban on the testing of ABM interceptors (defined as the approach within ten kilometres at a relative velocity in excess of ten metres per second) above an altitude of 40 kilometres would preclude the further development of exo-atmospheric interceptors for area defence. This would also effectively ban anti-satellite weapons. Systems tested below 40 kilometres with a relative velocity in excess of 4 kilometres per second would be subject to the deployment limits of the Treaty, thus reducing concerns about the strategic implications of anti-tactical ballistic missile systems, while permitting testing of short-range endo-atmospheric ABM interceptors. The 10 metres per second threshold would permit the rendezvous and docking of manned spacecraft, since such vehicles have very low closing velocities during the final several hundred kilometres of the rendezvous. Such a threshold could be monitored by national technical means.

- A limit of 5 square metres (a diameter of about 2·5 metres) on the aperture of ground- and space-based laser beam director mirrors, space-based sensor satellite mirrors, and the windows on airborne telescope systems, would constrain the ABM potential of all these systems, and could be monitored by national technical means. Although the area of beam director mirrors could vary widely, calculations of the performance of space-based systems typically use areas substantially in excess of 5 square metres. This threshold would usefully constrain space-based lasers, which might pose problems for monitoring a brightness limit. Anti-missile sensor satellites require much larger optical systems than simple early warning satellites. Airborne telescopes need much larger windows on the aircraft than are required for astronomy and intelligence collection.

- A limit of five kilogrammes on the mass of Plutonium 239 or Uranium 235 launched into orbit on a satellite would preclude the use of reactors to power space-based ABM sensors. This could be verified by pre-launch inspection of satellites by radiation monitors. Exceptions could be made for scientific spacecraft, verified through international participation in the project.

- A limit on the total number of permitted large-phased array radar transmitter faces (perhaps the 15 that both Parties appear to plan) as well as specification of the distance from the national border (for instance 350 kilometres) that construction is permitted and specification of what constitutes a space-track radar would resolve the Krasnoyarsk and Fylingdales issues. Lowering the Treaty's power/aperture product threshold definition of an ABM radar by a factor of ten from 3 000 000 to 300 000 would lessen concerns about anti-tactical ballistic missiles.

- Agreement not to place more than 300 tons of payload into orbit each year would permit both parties to conduct current and projected space projects, while providing reassurance that a space-based defence was not being covertly deployed.

8 Cosmic Space and the Role of Europe

Rolf Linkohr

INTRODUCTION

The evolution of technology, which has taken place at an almost exponential rate, has allowed mankind to reach and use outer space. The number of objects orbiting in space is very high. A reasonable estimate is in the tens of thousands – manned or unmanned – of artificial satellites, most of which will orbit our planet for a very long time. The history of the conquest of space has started only recently. During the Second World War the V2s, launched by the Germans, could only reach a height of about 215 km. On 4 October 1957 the Soviet Union launched Sputnik I, which was the world's first artificial satellite. On 12 April 1961 the Soviets launched the first man into outer space, orbiting several times around the Earth. By July 1969 American astronauts had set foot upon the Moon. In May 1983 the Pioneer 10, launched by the United States, escaped from our solar system and entered infinite cosmic space. Man in space is thus already a reality. A French publication describes the appearance of a new human species, the 'Spacepithecus', successor of the famous 'Australopithecus', who represented an important step in human evolution. This may be considered 'science fiction', but very often fantasy is based on rational analysis and romantic 'feelings' become a part of the process of the analysis of reality.

There are advanced spatial projects in the industrialised countries, primarily in the Soviet Union and the United States. But in Europe and Japan as well ideas are circulating concerning the possibility of space colonisation, which will be discussed shortly. In this chapter it is further intended to consider three questions, with particular attention being given to the European contribution. These are:

- What is the driving force behind the conquest of space?
- How is it possible to avoid the militarisation of space?
- What kind of advantages may we get from space technology?

NEW HORIZONS – OLD MOTIVATIONS?

It appears that the main argument for the development of space technology is not only advances in science or the result of human curiosity, but the close links that develop between civilian and military technological potential. It is no exaggeration to say that peace borders on war in space. The fantastic amount of money that the two superpowers spend on space research is not only based on their desire to analyse chemically a piece of the Moon, but on its military potential. To be present in space means power. Even if the use of space was reserved exclusively for civilian purposes, a presence in space could ultimately promise world domination. Francis Bacon was right when he affirmed in his *Novum Organum Scientiarum* that 'knowledge is power in itself'.

Let us examine an historical example. In 1714 the English naval captains asked the House of Commons to consider the problem of the determination of the longitude at sea through the knowledge of the position of the sun and with the help of a calendar. They could, in fact, measure latitude rather easily, but had problems with longitude. Knowledge of their exact position at sea was crucial for both the Royal Navy and the merchant navy. Parliament, with the advice of famous scientists such as Sir Isaac Newton and Edmund Halley, announced a public competition with a substantial prize: £10 000 for anybody who was able to determine longitude within a degree, £15 000 within two-thirds of a degree and £20 000 within one-half a degree. This corresponds to about 50–100 km at the Equator, and proportionally less towards the Poles.

Parliament repeated its offer in 1765, 1770, 1780 and 1781, since most of the proposals were not good enough. Forty years after the first announcement, a German scientist from Göttingen, Tobias Meyer, presented his famous lunar tables. After some tests, they were found to be precise within a degree and a quarter. The English mechanic, John Harrison, was able to determine the longitude at sea within half a degree. In 1765 the Longitude Commission suggested that Parliament award £3000 to Tobias Meyer, who had recently died, and £10 000 to John Harrison. In 1770 the British Admiralty published the first tables.

This consideration of the past can help us understand present problems. Superiority in space can be obtained only by having the best capabilities and technology for navigation and determination of position. The US Strategic Defense Initiative Office (SDIO) has shown a great interest in the Instrument Pointing System (IPS), a very sophisticated device developed in West Germany for the Spacelab. IPS is, in fact, one of the few SDI projects with West German participation. The United States is also developing the GPS (Global Positioning System), a navigation system based on 18 satellites, which allows a position to be determined within three metres. This device has had one very important side-effect. Intercontinental ballistic missiles have currently an accuracy of targeting within a range of 300–500 metres. If this accuracy were increased to 200 metres, it would be possible to destroy all Soviet missile silos.

This example shows the close link between civilian and military technology. A system, the West German IPS, which has been developed for purely scientific reasons, can be utilised as a basis for a substantial increase in the accuracy of American missiles. Bacon was certainly right. If knowledge means power, it ought to be shared by all nations; but shared knowledge makes it impossible to acquire too much power. It is therefore vital to internationalise space technology, and not only to save money.

EUROPEAN PROGRESS IN SPACE

West European expenditure on space technology is relatively modest, if it is compared to that of the United States. Western Europe spends about 2 billion ECU – something more than $2 billion. France has the highest share, followed by West Germany, Italy and Great Britain. The reason for this relatively modest European expenditure in space research is that its goal is purely civilian. The European Space Agency (ESA), which is the European equivalent of the US National Aeronautics and Space Administration (NASA), pursues a well defined scientific goal. There is a general European consensus about this: space programmes have and will continue to have purely civilian goals. This is causing disagreement with the United States. As is well known, the Americans, West Europeans, Japanese and Canadians would like to build a space station in 1992, on the 500th anniversary of the discovery of the Americas by Columbus. The West Europeans, like the Japanese and Canadians, do not want this space station to be utilised for military experiments within the SDI programme, whereas the Americans, under pressure from the Pentagon, are pushing for its possible utilisation for military experimentation.

In Western Europe there are two prevailing points of view. The first is concerned with the peaceful nature of West European space agencies. It will not be possible to involve ESA in any pseudo-military project. The second argument is economic in nature. The military utilisation of the space station would render technology transfer very difficult, as it would be classified by the Americans. In Western Europe and, if the present writer's information is right, in Japan as well, there is a growing scepticism about the 'Columbus' project. Experience has shown that it is very difficult to co-operate with the Americans in such an ambitious project.

Besides the debate with the Americans, the West Europeans are concerned at the great cost of any future space programme. It is not yet certain that all the projects proposed by ESA will be completed, but the appropriate ministers have already signed a common declaration which requests a considerable increase in expenditure. At the beginning of August 1987, the Director-General of the British Space Centre, Roy Gibson, resigned in protest at the refusal of the British Government to triple space expenditure to £300 million. Even if his proposal had been accepted, British expenditure would still only be one-half of the French or West German outlay. Important decisions on the future of the space policy within ESA should have been taken in Autumn 1987. It is an ambitious project, which would allow West Europeans to build a space station of their own, independently of the United States. The password is autonomy. The amount of money under discussion is no longer modest. It is about 50 billion ECU, over a period of about 20 years. After the completion of other technological projects over the last few years, such as the self-breeding reactor or the supersonic Concorde, it is easy to imagine that such a decision will provoke a lively debate in the various European Parliamentary commissions, especially in the budget commissions, mainly due to the likely inflation of expenditure.

On 31 January 1985, eleven Ministers representing members of ESA, reached an agreement on a long-term programme, concluding in the year 2000. Its political purpose was the autonomy of Western Europe and co-operation with the United States. The programme is ambitious and its costs are going to be much higher than those of previous programmes. Its scientific aim is the exploration of space in the vicinity of the Earth and the Sun, of the planetary system and of the cosmos. Telecommunications and remote detection, sectors which have already seen great advances, will constitute other important areas of development. A great part of this activity will be devoted to building up transport capabilities, of both men and equipment, in space.

The long-term plan includes the participation in the project of the international orbiting station under the supervision of the United States – the Columbus project. It is an autonomous and independent module within the framework of the international station with four elements: a manned laboratory with two astronauts (Attached Pressurized Module), connected to the station, whose launch with the American space shuttle is scheduled for 1996; a polar station, whose launch is scheduled for 1997 using Ariane 5, a new European rocket not yet developed; a laboratory module with two segments free-flying but served by the space station (Man-Tended Free-Flyer), whose launch with the Ariane 5 is scheduled for 1998; and the platform Eureka B (European Retrievable Carrier) with unlimited mission, but unmanned, whose launch is scheduled for 1994 with the space shuttle. Two transport systems will be used. One is that already mentioned, the Ariane 5 rocket, a development of the Ariane 1 to 4 family, with a useful load of 15 metric tons. It should be available in 1995. The second system is the space vector Hermes, a shuttle launched by Ariane 5 with some astronauts aboard.

Hermes and Columbus are part of the long-term strategy called EMSI (European Manned Space Infrastructure). Columbus will thus serve as an intermediate step towards a manned West European space platform, a project which should be carried out in the next century. These projects have yet to be approved by the various European national parliaments. As it is not possible to do anything without money, we are forced to wait for the decision of these bodies. The European Community is not involved in the space programme, as ESA is an interstate programme; there are, however, some European research projects on cosmic space, such as remote detection, which is being carried out with great success in the Common Research Centre in Ispra.

People are already beginning to talk of the next generation of space technology. There are projects for developing horizontal landing or take-off to reduce the expense of transportation. As is well known, the United States has approved an ambitious programme under the direction of DARPA (Defense Advanced Research Project Agency); this programme is aimed at the operational development of a supersonic aeroplane, the so-called X-30. In Europe there are two projects, a British one, called HOTOL (Horizontal Take-Off and Landing), and a West German one, called Sanger. These two projects are based on different approaches; HOTOL, for instance, has a single engine which can use alternatively oxygen from the atmosphere or fuel carried on board. There still remain very difficult technical problems to solve. Beyond Mach 3, the engine temperature calls for completely new concepts and materials. On the other hand, Sanger uses techniques which are already proven. It consists of two parts, one of which burns oxygen from the atmosphere in a classical engine, thus carrying the second part – a shuttle – up to a certain height, where it fires a solid-fuel rocket jet. Both parts can land at the end of their mission

Summing up the European space policy we can conclude that:

• European nations have succeeded in building up an impressive and very sophisticated technical potential, both in terms of scientific application and transport capabilities. Arianespace, the private enterprise which was created to make the European rocket commercially available, now has a growing share in the world market for satellite launching.
• Co-operation in space technology is part of several projects, which are aimed at protecting European technological auton-omy; included among these projects are the Airbus, Eureka or the Research Programme of the European Community. Through industrial co-operation the nations of Western Europe are inte-grating in an unexpected way; this process ought to be acceler-ated by 1991 with the final integration of the Community internal market.
• Space technology pursues a civilian purpose.

THE SDI TEMPTATION

The general West European orientation is towards civilian develop-ments but, since in space civilian and military applications are rather close, civil industry has been involved in military projects. As an Italian saying goes, *'chi va al mulino si infarina'* (you cannot touch pitch without being defiled). In Europe too, civilian and military sectors have common points of reference. It has already happened with the aeronautical industry, a classical case, where civilian and military applications are closely connected. The same companies often build fighters and commercial aircraft, so that there is no longer a psychological threshold between the two sectors. Space industry, unlike the aeronautical one, does not cater for a military demand from West European states. These are not military projects, except perhaps the project for the development of a remote-detection military satellite, which is being developed within NATO. But if there is a demand, the offer follows. The secret contracts, which are not so secret after all, would not have been signed if Europe had not enjoyed a highly developed technical potential which engages the interest of the SDIO.

According to the information available, Great Britain has so far received $40 million from the SDI budget, and out of this amount $15·5 millions go to nine private companies. The Ministry of Defence has received $10 million for a so-called study of European architecture. The basic idea is to apply in Europe the American experience in the field of space defence. Ten million dollars more will be invested in this study of European SDI. Laser research has received $10 million. It is carried out by public research centres. Ion sources for star weapons are being studied, while Heriot-Watt University is undertaking research on high-speed optical computers. The West Germans too are exploiting this American cornucopia. Dornier has acquired $21 million for an infra-red system. Schott and Zeiss too are playing the game, Schott with a high-precision mirror and Zeiss with a laser. Italy, too, has reached a secret agreement with the United States. Even some French companies are involved, even if France has no official agreement with the United States. Estimates exist that suggest that the European share in the SDI project will finally amount to $300 million. This participation in SDI sheds a new light on Eureka, by definition a project for civilian co-operation. The same companies often participate in Eureka and SDI. The research fields are also similar: computers, lasers, software, new materials, robots and telecommunication. After the meeting between François Mitterrand and Helmuth Kohl in December 1985, a project group was created to study a European defence initiative against short-range missiles. Some parts of SDI, such as ADA (Airborne Optical Adjunct Program) could be utilised. As the Anti-Ballistic Missile (ABM) Treaty does not involve short-range missiles, there is the possibility that the system could be developed and tested. Kinetic energy weapons could well be equipped with ADA. The formal borderline between SDI and Eureka is not always what we perhaps think it is. This situation is serious, largely because of the lack of parliamentary control over Eureka. The European Parliament is informed about single projects, but cannot influence procedures or its general orientation. Moreover,

Eureka is basically an extra-parliamentary project. This lack of democratic accountability has often been criticised, but without success.

Moreover, the secret agreements involving some states which are members of the European Community hinder the development of a common European research programme, since there are severe restrictions on knowledge transmission in projects connected to SDI. British firms cannot co-operate with a Spanish laboratory in a project involving SDI. While Europe is trying to create a common research programme, some countries have established privileged links with the United States. Recently, however, many people have been disappointed. The commercial and industrial value of SDI has probably been exaggerated with undue optimism about West European shares in contracts. There are also many serious studies which show that the SDI research programme has not had much value in increasing the competitiveness of individual nations. On the other hand, this is also true of civilian and military space technology. A recent study, carried out by the 'Kieler Institut for Weltwirtschaft', by comparing the competitiveness of several countries, concludes that there is no link between the economic power of a country and its expenditure in military and space research: this result deserves careful consideration.

THE LIMITS OF RATIONALITY?

This chapter will conclude with some thoughts on the motivation behind space technology and draw three conclusions. First, it appears to be difficult to demonstrate the need for huge expenditure in space research in terms of science or industrial competitiveness. The most competitive countries in the world are certainly not those which spend more money on space technology. There are other reasons. One derives from the fact that to be present in space means increased power. Even if we do not share this view, we have to recognise that it is possible to justify in this way the military polarisation which exists on our planet. But there is a second reason, pertaining to Western Europe, where a military presence in space is not considered so important by public opinion. It is viewed as science fiction, deriving from the psychological tension between rationality and sentimentality in life, between conscious and unconscious, between the world of reason and irrationality. Our way of thinking and acting is so influenced by the new technology that our lives can be lived on different levels; that of everyday life and that of the stage, that of individual fantasies and that of collective myths and fantasies. Paradoxically, it is technology and the achievements of science that often contribute to this elimination of a surrealistic dimension to our existence, thus causing a sort of schizophrenia of the collective psyche: in this way surrealistic aspects are located and placed in reality, whereas actual needs and necessities are transferred into dreams. Science fiction novels are always placed in the future, in a time that in physical terms is different, since it is located in other worlds.

Science fiction is based on scientific methods, that is to say, it is closely connected with modern scientific culture. Without entering into a debate over various aspects of science fiction, on its pessimistic and optimistic characterisation, it can be pointed out that there is no modern technology that can symbolise better than space travel the sense of adventure and widespread faith in scientific discoveries and in a better world, but also the fear of catastrophe that would mark the end of the world. The true motivation for space travel, excluding the military one, is irrational and derives from our culture that, since the days of the Renaissance, has been based on expansion. It starts in Italy, with Dante Alighieri, who in his *Divina Commedia* does not let Ulysses go back home, to his wife and hearth – as Homer did – but sends him again to sea, towards the African coast, to Spain, to the pillars of Hercules and the frontiers of the ancient world. And the hero turns toward the Atlantic:

> behind the sun, of that deserted country
> remember your true nature:
> you were not born to live like brutes
> but to strive for virtue and knowledge.

Dante's Ulysses is no longer the man described by Homer, who suffers in silence, but rather a Titan of the new age, a Faust of the sea, a monomaniac who anticipates Columbus. With him geographical limits are no longer respected, a reckless expansion begins, which puts the new rational technology at the service of imagination. Even today we do not ask ourselves about our needs, but about what we can do. The utilisation of space travel goes unnoticed, while the projects follow the imagination. This does not mean that space technology cannot be useful for certain things, such as weather forecasting, telecommunication and astronomy. But the fall-out of space travel is less important than its symbolic value. The fact that the American flag was the first to be raised on the Moon is more important than many other social programmes which attempt to increase national pride.

Finally, some further conclusions may be drawn.

- We must apply to outer space and all the celestial bodies the *res comunis omnium*, which is that space must be internationalised and not belong to a single nation.
- It follows that space technology should become an international technology. As knowledge is power, a shared knowledge does not allow one nation to enjoy too much power.
- To apply this concept, it is necessary to enforce the international regulations already stipulated, such as the Space Treaty of 1967. Article Four of this treaty forbids the orbiting around Earth, or other celestial bodies, of nuclear weapons of mass-destruction. If this treaty were respected, space militarisation would be considerably reduced.
- Astronautics must be useful and be used to solve human and ecological problems.

In a West German newspaper in 1955 one could read that in a small village in Northern Germany the fire-brigade was called to help a man who was lopping a tree. He was on the top of the tree and could not come down, as he had cut the branches starting from the bottom. Let us hope that something of this kind does not happen in space technology, as we cannot be sure that there is an outer space fire-brigade.

9 A Technical Assessment of Potential Threats to NATO from Non-Nuclear Soviet Tactical Ballistic Missiles

Benoit Morel and Theodore A. Postol

INTRODUCTION

Since President Reagan's speech in March 1983 announcing the Strategic Defense Initiative (SDI), confusion within and without the Administration has persisted over the programme's character, intent and objectives. More recently, this confusion has spread to Western Europe in the form of a 'debate' over the merits and utility of using SDI technology to provide limited non-nuclear defences for critical NATO facilities against a class of Soviet missiles called Tactical Ballistic Missiles (TBMs). As with many other aspects of the SDI debate, the discussion of this European defence issue has largely been divorced from either assessments of technical feasibility or even the simplest estimates of the threat. Prior to the inception of the European 'debate' over defences against Soviet TBMs, these missiles were thought to have roles that might best be described as similar to that of super-heavy long-range artillery. Armed with nuclear warheads, and having ranges from several tens of kilometres to nearly 1000 kilometres, even the relatively primitive and inaccurate early Soviet TBMs have been a formidable nuclear threat to NATO since at least the mid-1960s.

Several modernised versions of Soviet TBMs – the SS-21, SS-23 and SS-22 (also known as the SS-12 mod.) – have recently begun to replace their predecessors. Although there is almost no unambiguous public data yet available about these missiles, considerable improvements in Soviet guidance and propulsion technology since the introduction of first-generation TBMs has led to speculation that they

106

could be sufficiently accurate to be threats even when armed with conventional munitions. These speculations have led to still more speculation that modernised Soviet TBMs are uniquely suitable for certain critical non-nuclear missions that would be part of a larger Soviet non-nuclear pre-emptive attack on NATO.

The goals of a Soviet pre-emptive strike would, of course, be similar to those of most large-scale military operations. Simply speaking, they would be to degrade or destroy those enemy forces and functional capabilities that pose the greatest threat to one's own forces, and hence the greatest threat of defeat. In a Soviet pre-emptive strike against NATO it is generally accepted that these goals would be the destruction of NATO airbases, air-defences, nuclear storage facilities, command and control facilities, and communications. These goals are often discussed in a context that suggests they are a peculiar by-product of Soviet doctrine, but in reality they are dictated by the extraordinary fire-power, range and capabilities of modern weapon systems. These modern capabilities, not doctrine, impel the military planner toward a 'strategic' vision of warfare.

In addition to the strategic objectives of military planners, non-military policy–makers may have larger objectives that supersede and constrain the narrower objectives of planners. For example, a pre-emptive strike with nuclear weapons might, from a sufficiently narrow military perspective, be decisive, but it would be incompatible with a larger goal of minimising the risk of a nuclear exchange with NATO. Such an attack could well result in a NATO nuclear response or, worse yet, the initiation of a chain of events leading to general nuclear war. For this reason, it is widely believed that if a pre-emptive Soviet attack on NATO eventually occurs, it will most probably be initially constrained to rely on non-nuclear systems.

Since the initial goals of a determined conventional attack would be the destruction of a large number of important facilities and functional capabilities that are located deep inside NATO's territory, only Soviet Frontal Aviation would have the range and fire-power to carry out these missions. For the surprise attack to have a reasonable chance of success, it would require a massive commitment of aircraft followed up by a rapid succession of other military actions. Extensive preparations would be required for the initial phases of the attack, and still more extensive preparations would be required for the follow-up operations. If NATO surveillance systems observed an increase in Soviet military activity prior to the attack, it cannot be ruled out that its very substantial offensive and defensive ground and airborne forces would be able to respond rapidly to the Soviet action. Soviet planners would therefore face substantial uncertainties that the element of surprise could be maintained during preparations for such a large attack. Without careful and extensive preparation, however, there might be even greater uncertainty that Soviet forces could capitalise on surprise – even if it is achieved. Thus serious and difficult dilemmas are posed by any surprise attack that must rely on either nuclear weapons or massive attacks with Frontal Aviation.

In the European debate about the utility of defences against TBMs it is now being argued that the new generation of conventional-capable Soviet TBMs threatens to change the nature of the current NATO–Warsaw Pact stalemate, and that defences against these TBMs are needed to offset this threat. A sketch of the argument is necessary. Since NATO air-defences currently have no ability to defend either themselves or critical facilities against TBM attack, and TBMs can reach targets only minutes after they are launched, TBMs will be able to achieve many of the objectives of a non-nuclear pre-emptive attack with relatively high confidence and with an economy of force. TBMs might be used to cut 'corridors' through NATO air-defences, greatly aiding Frontal Aviation efforts to penetrate into NATO's rear areas; they might be used to free many Frontal Aviation aircraft from missions aimed at destroying or disrupting air operations against NATO's airfields, and they might be used for destroying or disrupting operations at major NATO command centres and at nuclear storage sites. If NATO does not respond by developing Anti-Tactical Ballistic Missile (ATBM) defences, the Soviets will be allowed to gain a critical – and perhaps decisive – advantage at the beginning of some future NATO/Warsaw Pact conflict.

These assertions about Soviet TBMs raise many questions. How much damage could conventional TBMs do against the extremely varied types of targets that would have to be neutralised in a surprise attack? Would achievable levels of damage be adequate to meet the mission objectives? Are there threats other than TBMs that could also be used to achieve the dictated mission objectives? If so, are these threats more or less plausible than those from TBMs? Even if TBMs prove to be capable of threatening certain classes of targets, are anti-TBM defences the most appropriate response to the threat? For example, a strategy of hardening fixed facilities and improved use of mobility and/or concealment when possible might buy a great deal of protection against conventionally armed TBMs, *as well as other plausible evolving threats.*

These questions point to a broader and more important set of unresolved issues that supersede those in the ATBM debate. If the TBM threat were to prove to be both real and unique, the military implications of failure to react are obvious. However, it should also be understood that an inappropriate commitment of NATO resources in response to an overstated, fictitious, or improperly characterised threat has military implications that are just as real. It will be shown in a forthcoming publication on ATBM systems that even if some level of defensive capability ultimately proves to be feasible, defensive systems will have to operate in a constantly changing environment that is dominated by the tactical and technical counter-measures of a hostile adversary.[1] As a result, the acquisition and preservation of defensive capabilities will almost certainly require vast expenditures of capital. Any defence system concept that might hope to have significant capabilities will require expensive and elaborate sensor systems, perhaps composed of ground-based, airborne and/or space-based elements, in addition to large numbers of interceptors and an extensive chain of logistics. The procurement of such systems, should it occur, will then preclude the acquisition of other military systems that may also be important to NATO's overall defence posture. Hence, a NATO decision to develop and procure an ATBM capability cannot be viewed in isolation from other modernisation questions that also must be faced.

This chapter presents a technical analysis of the threat from conventionally armed TBMs – an analysis that, remarkably, has been absent during the more than two years that this debate has raged. What emerges from this analysis is an important set of conclusions.

● A conventional Soviet TBM can be used only once, and its delivery capability is small relative to that of a typical ground attack aircraft in today's Frontal Aviation forces. More than three TBMs would be required to deliver a payload of munitions equal to that of only one sortie-load of a single Frontal Aviation aircraft.

● TBMs will have little or no capability to attack properly-run mobile targets like air-defence radars and divisional command posts. The greatest difficulty is finding these targets, not destroying them. For example, the position of a mobile radiating target like a Patriot air-defence radar might potentially be determinable to within 50 to 100 metres from a range of 100 km – with advanced time difference radio direction finding techniques. However, it is also possible to create electronically many false targets with inexpensive low-powered decoy radio transmitters. A combination of camouflage (to deny the enemy an ability to discriminate between radars and false radars with other sensors) and periodic movement of both radar and decoy transmitters can be used to create an insurmountable surveillance and targeting problem for the adversary. Future TBMs are therefore not likely to pose a greater threat to these assets than that which already exists from aircraft, drones, remotely piloted vehicles (RPVs), air-launched short- and intermediate-range anti-radiation, infrared and TV-guided homing missiles, and long-range ground- or air-launched Cruise missiles.

• As TBMs cannot carry enough conventional explosives to damage properly-hardened underground structures, they should have little or no ability to damage or disrupt operations at fixed hardened NATO command posts.

• TBMs with sub-munitions will not be capable of doing significant damage to runways at NATO airbases unless they achieve accuracies comparable to that of terminal sensing ballistic missiles such as Pershing II. Current public estimates suggest accuracies of 200 to 300 metres for new generation Soviet TBMs, as compared to the 30 to 40 metres for the Pershing II. Even if the Soviets commit substantial technical and economic resources to achieve such accuracies, many hundreds, and perhaps several thousand, Soviet TBMs might be required if a major impact on NATO tactical air operations is to be achieved during the early phases of a surprise attack.

• Cruise missiles, not TBMs, will be the most efficient means of delivering both chemical munitions and conventional sub-munitions in a surprise attack against NATO targets. Current Cruise missile guidance systems already utilise both terrain contour matching and inertial navigation technologies. These guidance technologies are readily adaptable for use in varied types of missiles. For example, an appropriately modified version of a current Cruise missile design could utilise these technologies so that it could sense local wind conditions as it approaches a target. The missile's computer could then use local wind data to determine an optimal upwind range, path, altitude and rate of release for dispersing its load of chemical agents. Most significantly, TBMs can only dispense chemical agents at a single point. On a pound for pound basis, chemical releases from advanced TBMs can therefore be expected to be considerably less controllable and efficient than those from Cruise missiles. Further-more, even if local wind conditions are known at the time of an attack, a 500-km range TBM would have to weigh more than twice as much as a Cruise missile of comparable range to deliver the same weight of chemicals. This is yet another factor that suggests the potentially greater efficiency of Cruise missiles for delivery of chemical agents.

• Simple passive defensive measures like hardening and dispersal will be very effective in negating the destructive capabilities of even a very large force of high-technology TBMs. Such a force, if sufficiently large and technologically advanced, could, nevertheless, eventually pose a threat to NATO runways. However, Cruise missiles already pose a more credible and serious future threat to runways. Evolving technologies, like that of VSTOL aircraft, should be studied as a cost-effective counter to the potentially very serious emerging Cruise missile threat.

• ATBM systems will be discussed by the present writers in a forthcoming publication.[2] It will be shown there that anti-tactical ballistic missile defences will be complex and expensive. Contrary to common wisdom, countermeasures that might be effective against very advanced defences designed to engage intercontinental range ballistic missiles (ICBMs) would be yet easier to implement against ATBM systems. Some of the findings in our ATBM systems study that are also relevant to the debate about responses to perceived threats from Soviet TBMs follow. An ICBM typically delivers only four to five per cent of its gross weight to the vicinity of targets. Half of this weight is usually in the form of a 'bus' or post-boost vehicle. Thus, only two to three per cent of an ICBMs launch weight is available for both decoys and warheads. In striking contrast, 50 per cent of the launch gross weight of a 100-kilometre range TBM is payload; 20 per cent of the weight of a 500-kilometre range TBM is payload, and about 15 per cent of the weight of a 1000-kilometre range TBM is payload. Thus, on a pound for pound basis, TBM decoys are far less costly than ICBM decoys in terms of payload loss. Of yet greater importance is that credible light decoys should be far easier to implement for TBMs. A decoy dispensed from 1 500-kilometre range TBM re-enters the atmosphere at less than one-third the speed of one that arrives on an ICBM trajectory. The drag on such a decoy is therefore ten times less than that which would act on a comparable ICBM decoy, and the heating rate is more than 40 times less as well. This not only makes it much easier to build credible light decoys, but it results in greatly reduced infra-red and radar re-entry signatures, which the defence might otherwise hope to use as means of discriminating between false and real objects. Our analysis therefore indicates that the conventional TBM threat, if it exists, is minor compared with others that NATO will have to contend with in the future. Moreover potential improvements in conventional TBM technology do not appear to forebode new, unique or more effective pre-emptive threats against NATO relative to those that already exist, or are promised by other emerging technologies. There are many other existing military technologies that already present more convincing and troubling threats to NATO. These technologies are either available, or will be accessible, to Warsaw Pact forces in the not too distant future. These conclusions are based on a technical analysis that is described in the remaining pages of this chapter.

CHARACTERISTICS OF TBMs AND CRUISE MISSILES

For Soviet TBMs to be able to accomplish some of the missions associated with a Warsaw Pact surprise attack on NATO, they should be able quickly to neutralise or delay the use of assets and operational capabilities that would otherwise be used to oppose large-scale follow-on attacks. Since an initial attack would have to be both decisive and rapidly developing, with TBMs presumably inflicting widespread and intense damage on at least some key NATO targets, their utility is in part determined by their ability to inflict damage against targets. Accordingly we begin our evaluation of their abilities to inflict damage by first presenting an assessment of those TBM and Cruise missile characteristics that determine some of their military capabilities. The types of munitions that can be used to arm TBMs and their effects on different types of targets will then be discussed. A brief comparison of the delivery capabilities of Soviet TBMs with Frontal Aviation and Cruise missiles will then be presented – so that the emerging threat from TBMs can be framed against existing ones from Frontal Aviation and evolving Cruise missile technologies. Discussion of some general principles for determining the destructiveness of TBMs against targets of varied hardness and mobility will follow. These principles will be used to assess the destructive capabilities of TBMs armed with single high-explosive warheads. Results of specific calculations for two classes of targets of potentially great importance – soft-mobile air-defence radars and semi-hard above ground structures – will be described, assuming TBMs of varied accuracy. TBMs armed with sub-munitions will be described and evaluated in later sections. Issues and analysis that lead to conclusions about the effectiveness of TBMs against airfields, fixed very-hard command centres, and division-level command posts will then be treated. We finish with a brief concluding section.

The ability of a particular conventionally-armed TBM to damage a specific target is strongly affected by many factors Some of these factors are determined by properties of the TBM while others are influenced by those of the target. The factors that affect a particular TBM's ability to do damage is its accuracy, the weight of munitions that it can deliver, and the effectiveness of munitions that can be carried. The properties of the target that determine the likelihood of damage are the precision with which its location is known, its 'hardness' to effects from munitions and its dimensions.

Publicly available information about the characteristics of Soviet TBMs remains limited and contradictory. Table 9.1 shows the accuracies – or Circle of Equal Probability (CEP)[3] – of currently deployed American and Soviet ballistic and Cruise missiles with ranges between 30 and 5000 kilometres as reported by the International Institute for Strategic Studies and the Congressional Research Service. These sources report accuracies of 250 to 350 metres for the new generation of Soviet TBMs. However, certain other sources have reported accuracies almost ten times better – that is, about 30 metres.[4] This suggests either uncertainty or confusion (or both) about the accuracy of new-generation TBMs.

The most modern of the inertially guided NATO TBMs (SS-21, SS-23, SS-22, Lance, Honest John, Pershing I) all have accuracies between 300 and 400 metres. The Pershing II, however, uses an inertial guidance system in combination with a radar area correlator. As the Pershing II warhead approaches a target, the radar area correlator can scan the terrain around it, allowing the warhead to locate nearby terrain features. Since the locations of the target and terrain features are stored in the warhead's guidance computer, the target can be readily located and inertial measurement errors that have accumulated during launch and flight to the target can be corrected. The updated inertial guidance system in the Pershing II re-entry vehicle can then instruct the warhead to make small manoeuvres as it falls towards the target, achieving a greatly enhanced accuracy of between 30 and 40 metres. Although there are no public reports of this type of guidance on the new Soviet TBMs, a Pershing II-type guidance system cannot be rejected for future generation Soviet TBMs or, less likely, as an upgrading or modernisation of those systems currently being deployed.

Table 9.1 also shows data on the accuracy of the ground- and air-launched Cruise missile (GLCM and ALCM), which is fundamentally different from a TBM. The Cruise missile is an air-supported vehicle that is guided during most of its flight by inertial means. This accuracy is achieved by a novel hybrid scheme of inertial guidance that is occasionally updated with external position fixes. Because Cruise missiles can fly over many paths to a target at low altitudes, planners are free to choose terrain locations with features that are suitable for very accurate terrain contour measurements. As a result of this guidance system, and the high degree of missile-control that can be achieved by a missile that approaches its target at subsonic speeds, Cruise missiles can achieve accuracies of less than 15 to 20 metres.

In contrast to the circumstance of a Cruise missile, a terrain sensing ballistic missile like Pershing II – with its radar area correlator – is constrained over most of its flight path to trajectories governed by the gravitational field of the earth and Newton's laws of motion. Terrain measurements are restricted to high altitudes and to areas around intended targets. In addition, since ballistic missiles arrive at much higher speeds than Cruise missiles, this results in considerably different guidance and control conditions immediately before impact at a target. For example, the speed of a 500-km range TBM when it arrives at a target is about two km/sec. If the average misalignment of the missile is one degree in its last second of flight before impact, aerodynamic forces will result in a miss of about 35 metres. The same misalignment in the last second of Cruise missile flight will result in a miss of no more than five metres. Hence, for the immediate future, it is not appropriate to assume that TBM accuracies will reach those that seem to have been achieved by Cruise missiles. We will therefore confine our analysis of Soviet TBM capabilities to missiles with accuracies no smaller than the 30 to 40 metres which have been achieved by Pershing II technology.

Since we have now described the reasoning leading to our assumptions about the limits of TBM accuracy, we shall next consider the reasoning that leads to our estimates of the weight of munitions that TBMs might deliver. If it is assumed that the SS-21 dimensions, shape and launch weight are about the same as those of the Frog-7, a dynamical model of the flight performance of the missile is easily constructed from the known properties of propellants and aerodynamic characteristics of missiles. We estimate that a fully loaded SS-21 could deliver about 800 to 900 kilogrammes to a range of about 100

Table 9.1 Surface-to-surface missile characteristics

	First Deployed	Warheads	Yield	Range (Miles)[1]	Propellant	CEP	Launch Platform	CEPs IISS 1986-7
United States								
Pershing II[2]	1983	1	5-50 KT	1000 +	Solid	0·02	Wheeled Vehicle	40 m 0·025 miles
GLCM[2]	1983	1	200 KT	1500	Solid	0·01	Wheeled Vehicle	20 m 0·013 miles
Lance	1972	1	1-100 KT	70	Liquid	0·25	Wheeled, tracked	
Pershing I	1962	1	60-400 KT	450	Solid	0·20	Wheeled Vehicle	
Honest John	1953	1	5-25 KT	20	Solid	0·45	Wheeled Vehicle	
Soviet Union								
SS-23	1981	1	100 KT	300	Solid	0·20	Wheeled Vehicle	350 m 0·22 miles
SS-20[3]	1978	3	150 KT	3000	Solid	0·23	Mobile (No survey)	
SS-21	1977	1	100 KT	75	Solid	0·15	Wheeled Vehicle	300 m 0·19 miles
SS-22	1977	1	500 KT	550	Solid	0·20	Wheeled Vehicle	300 m 0·19 miles
SS-12 Scaleboard	1965	1	1 MT	500	Solid	0·40	Wheeled Vehicle	
FROG	1965	1	200 KT	37	Solid	0·25	Wheeled Vehicle	
SS-1c Scud B	1965	1	1 KT	185	Liquid	0·50	Wheeled Vehicle	
SS-5 Skean	1961	1	1 MT	2500	Liquid	0·60	Launch pad/silo	
SS-4 Sandal	1959	1	1 MT	1200	Liquid	1·25	Launch pad/silo	
SS-1b Scud A	1957	1	40 KT	50	Liquid	0·50	Wheeled, tracked	

[1]IRBM range is 1500-3000 nautical miles; MRBM range is 600-1500 nautical miles; SRBM range is less than 600 nautical miles.
[2]The United States will deploy 108 Pershing II and 464 ground-launched Cruise missiles in Europe by the late 1980s, if plans proceed on schedule. Pershing II missiles will replace Pershing I missiles. Soviet officials claim the Pershing II missile has a range of 1500 miles.
[3]The SS-20 is the only TNF surface-to-surface missile launcher having more than one warhead. Soviet officials claim the SS-20 missile range is only 2200 miles.

Source: Congressional Research Service, *US–Soviet Military Balance, 1980–1985* (Washington, DC, 1985) p. 194.

kilometres. The SS-23 is a 500-kilometre range TBM that is replacing its 300-kilometre range predecessor, the SS-1c SCUD B, and the SS-12 mod. is a 900-kilometre range missile replacing the SS-12 'Scaleboard'. Various public sources suggest the SS-1c SCUD B weighs around 6000 kilogrammes while the SS-12 is suggested to weigh more than 7000 kilogrammes. In the analysis that follows, both baseline SS-23 and SS-12 mod. missiles have been sized so that they can carry an interchangeable payload of munitions of about 1000 kilogrammes. Figure 9.1 shows our three baseline missiles along with our assumptions about the characteristics of their rocket motors and the performance estimates that follow from the assumed characteristics.

TYPES OF CONVENTIONAL MUNITIONS AND THEIR EFFECTS AGAINST TARGETS

An evaluation of TBM capabilities also requires an assessment of the damage-inflicting capabilities of different types of 'tailored' munitions that might be carried as TBM payloads. Among the types of munitions that might be chosen for TBM warheads may be warheads designed to maximise damage from blast, or from fragments accelerated to high speed from a blast. Other munitions, like 'shape-charge warheads' might be designed to 'direct' explosive effects to penetrate deep into a target. Other munitions again, like Fuel-Air Explosives are designed to 'spread' the effects of an explosion, so that the largest possible area is covered with enough blast to destroy intended targets. Yet other types of warheads might utilise incendiary, chemical or biological agents as means of damaging targets or disrupting military operations. TBM payloads may also be divided into clusters of smaller munitions (sometimes called sub-munitions), which under certain conditions can greatly increase the probability that a target would be damaged or destroyed. The choice of munition for a TBM is decided by its effectiveness in disrupting or destroying an intended target. This, in turn, depends on which munitions effects are most damaging to each generic class of target.

Figure 9.2 shows blast pressure versus range curves for ten, 100 and 1000 kilogramme weights of TNT/RDX-type high-explosives. The 1000 kilogramme charge causes a blast pressure of about 40 pounds per square inch (276 kPa) at a range of 15 to 16 metres, and diminishes to ten pounds per square inch (69 kPa) at 38 metres and five pounds per square inch (35 kPa) at 55 metres. A 40 pound per square inch

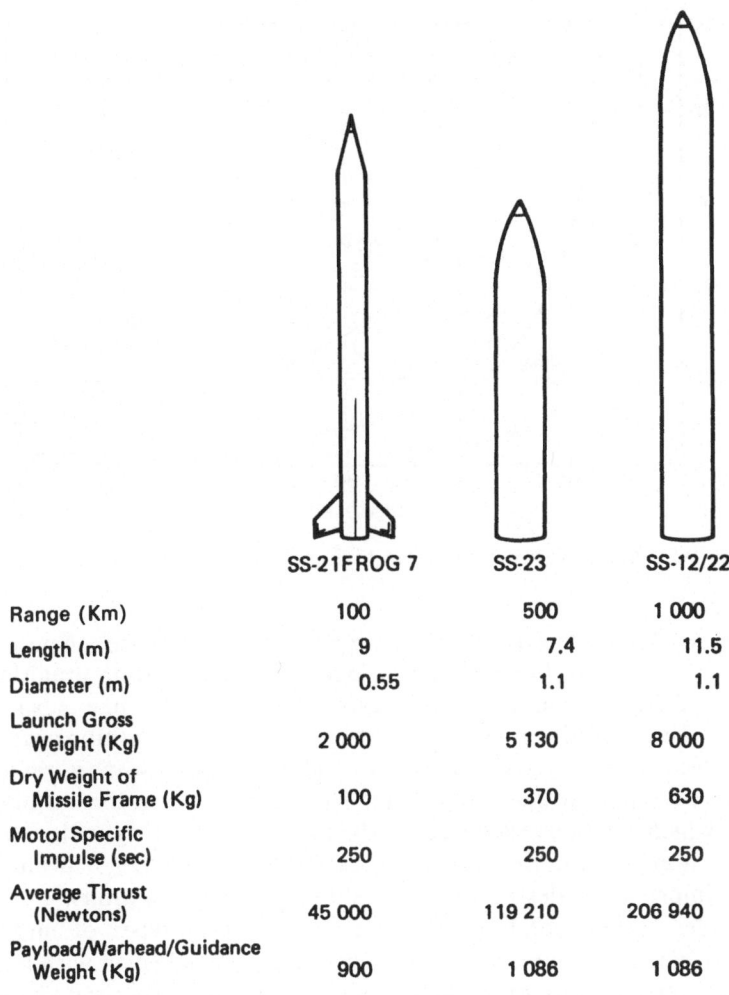

	SS-21FROG 7	SS-23	SS-12/22
Range (Km)	100	500	1 000
Length (m)	9	7.4	11.5
Diameter (m)	0.55	1.1	1.1
Launch Gross Weight (Kg)	2 000	5 130	8 000
Dry Weight of Missile Frame (Kg)	100	370	630
Motor Specific Impulse (sec)	250	250	250
Average Thrust (Newtons)	45 000	119 210	206 940
Payload/Warhead/Guidance Weight (Kg)	900	1 086	1 086

Figure 9.1

blast would completely demolish a typical unreinforced building, but it would probably not be able to destroy properly hardened above-ground bunkers. Such bunkers might be constructed by an army in the field with heavy beams and a thick overburden of soil, or they might be reinforced concrete and steel hangars in which aircraft at military airfields are sheltered. Other types of targets, like radars, trucks and missile launchers, are much 'softer'. These units would be

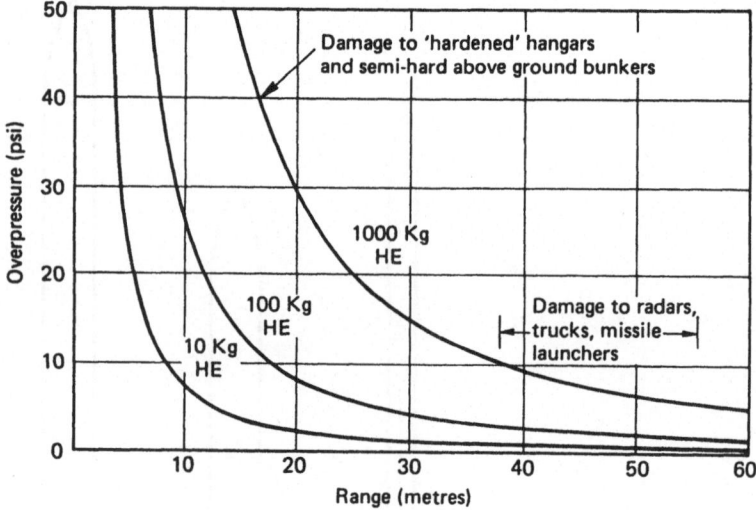

Figure 9.2   Blast over-pressure from high-explosive munitions

likely to suffer heavy damage from exposure to only five to ten pounds per square inch of blast. Common blast or fragmentation munitions derive their destructive power from high-explosives like TNT or RDX. The explosive power per unit weight of such munitions can be improved by mixing the primary explosive agent with a rapidly oxidising material like aluminium. However, because the distances over which weapons effects cause damage change slowly with explosive power, such improvements in explosive power do not dramatically improve the destructive capabilities of these munitions.

Under certain conditions, and against certain types of targets, improvements in munition damage range can be achieved with high-fragmentation types of warheads. Because hot gases generated from the detonation of a high explosive expand so rapidly, they can carry fragments to very high speeds. Common high-explosive general purpose bombs, for instance, consist of 50 per cent explosives and 50 per cent casing by weight. When such a bomb is detonated, fragments are generated by the break-up of the casing. These fragments typically weigh tens to hundreds of grammes, and achieve initial speeds of more than two kilometres per second – high enough to be able to penetrate light armour. Figure 9.3 is a plot of the number of 45-gramme fragments per square metre versus range from a detonation for different size munitions that are 80 per cent fragments by weight. Such

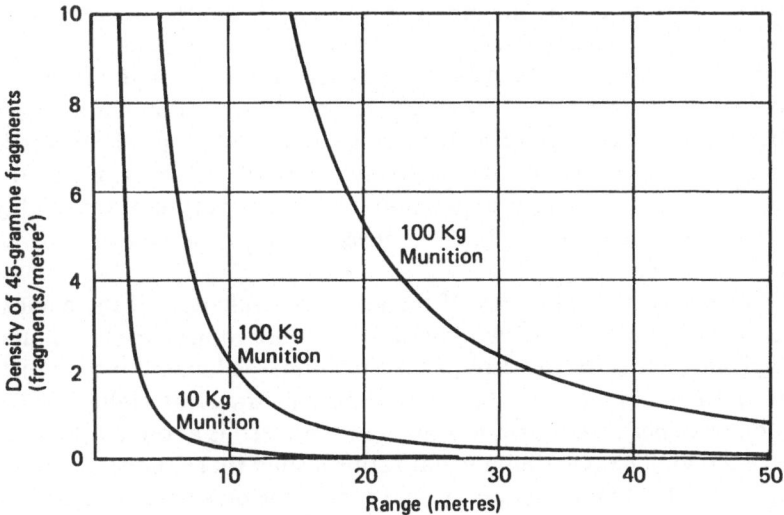

Figure 9.3   Metal loaded explosive (80% metal by weight)

munitions can generate about twice as many fragments relative to the bomb discussed earlier. The 1000-kilogramme munition only achieves a density of one fragment per square metre at a range of 50 metres. Hence a truck or radar might only be struck with ten or 15 fragments at that range. Since considerably less than half the weight of the high-fragmentation munition is in high explosives, fragments will initially have a speed of only 700 to 800 metres per second. By the time they travel 50 metres, air-drag will reduce their speed to about 500 metres per second. Since the fragments are not large, and will not strike their targets at exceptionally high speed, such high-fragmentation munitions are mostly useful against relatively soft exposed targets like troops in the field. They therefore have little utility as an improved munition payload for TBMs.

Still another form of conventional munition technology that might be used to arm a TBM are various forms of 'shaped charges'. A shaped charge is a munition that consists of a cylinder of high-explosive material that has a conical depression at one end and a detonating fuse at the other. After the fuse starts the detonation at one end of the high-explosive, a self-sustaining blast front propagates through the explosive column towards the other end. Since the detonation front travels through the explosive at about 20 times the speed of a shock in air, when the inverted conical depression is encountered, a strong shock is created in the air at the centre of the depression while the detonation front races ahead through the explosives at the cone's edge. This results in a shock front that has a higher speed at its edge relative to that at its centre. This speed difference causes the outer part of the blast wave to bend inwards towards the centreline of the conical depression, creating a very hot and highly focused jet of gas at a distance of three to four cone diameters from the end of the charge.

The destructive efficiency of the jet can be enhanced if the conical depression is covered with a liner of metal like copper or aluminium. Typically, about 20 per cent of the metal liner is focused into the jet. At the front of the jet, the metal vapours travel at eight to nine kilometres per second, while at its rear the speed is about a kilometre per second. The remaining metal forms a plug that follows the main jet at about .3 kilometres per second.[5] Since the pressure at the leading edge of the jet is well in excess of 100 000 pounds per square inch, it can penetrate deep into the interior of a target. Penetration distances of three to four cone diameters can be achieved through steel, and as much as ten to 12 diameters can be achieved through concrete.[6]

Shaped charge munitions have a number of features which could make them difficult to use with TBMs. Since their very great destructive efficiency is achieved by directing blast energy, they must essentially 'hit' their intended target if they are to achieve damage. As will be shown in discussion to follow, even a very accurate TBM will rarely make a direct hit on a target. Depending on the trajectory used by the attacker, a TBM is likely to be oriented at angles of 20 to 60 degrees from horizontal when it arrives at a target. The jet from the shaped charge will therefore have to penetrate a distance ten to 100 per cent larger than the thickness of the barrier it is intended to penetrate. If the munition is made to detonate prematurely by an overburden of extra soil, or by light structures that are above the target, the result will also be greatly reduced penetration of the target. Finally, damage to a target from a shaped charge is likely to be very severe but very localised. If, for instance, the target is a properly constructed large bunker, its interior may well be divided into many separate hardened rooms. Damage from a hit might then be localised to only those units at or near the point of penetration.

Yet another possible payload for TBMs would be sub-munitions, clusters of smaller munitions, rather than a unitary warhead. Under specialised conditions, the details of which will be discussed shortly, this type of payload can considerably increase the likelihood that a TBM will destroy or severely damage a relatively 'soft' type of target. The sub-munitions would generally be dispersed in a pattern designed to maximise destructive effects or impact with a target. The actual mode of destruction might be any of the mechanisms discussed above, perhaps even a combination of them.

A last type of explosive munition that might be considered for use on a TBM is a Fuel-Air Explosive (FAE). A fuel-air explosive is a munition that disperses an explosive aerosol that is then detonated by a delayed fuse. The delayed detonation fuse is set off several tenths to several seconds after aerosol dispersal, detonating the explosive aerosol. The use of FAEs on TBMs that arrive at very high speed would present a challenging set of applications problems to missile designers. A 500-kilometre-range TBM arrives at almost two kilometres per second. This speed is so great that each unit of TBM mass has a kinetic energy equal to half its equivalent weight of TNT/RDX. Canisters must be devised to control the dispersal of explosive aerosol and fusing mechanisms must be able to detonate the explosive mixture after a proper delay. Mechanisms for achieving these objectives may well be complex, heavy and undependable, reducing the overall reliability of a TBM as well as the fraction of payload that can be devoted to explosives.[7] In addition, since FAEs achieve destruction by spreading an explosive over a large area, an FAE explosion does not achieve very high peak overpressures. The complications imposed by the high delivery speed of TBMs and applications limitations created by the spreading of explosive effects would therefore greatly limit the utility of FAEs as an efficiency-enhancing payload for conventional armed TBMs.

It therefore appears that there are few, if any, new and exotic munitions that will dramatically alter conventional TBM capabilities. Each conventional TBM will be able to deliver the equivalent of a single 1000-kilogramme general purpose bomb, or perhaps an equivalent weight of smaller sub-munitions. In addition, such TBMs might also deliver chemical or biological agents. However TBM-delivered chemical and biological agents will be spread unpredictably by the local wind. Current Cruise missile technology could be used to design airborne vehicles that could sense local wind and disperse these agents with far more deadly precision and control. Since a higher fraction of the launch weight of these vehicles would be payload relative to that of a TBM of comparable range, it is the present writers' opinion that delivery of chemicals by such air-supported vehicles would be more efficient, deadly and controllable than by TBMs. Thus, while TBMs could certainly be used to deliver such agents, they do not introduce a new delivery technology that is more threatening or preferable to that which is already available for such purposes.

TBM, FRONTAL AVIATION, AND CRUISE MISSILE DELIVERY CAPABILITIES

It makes little sense to assess a threat to military forces in the battlefield from a single type of weapon system without also assessing the capabilities of other systems that might pose similar threats to such forces. For this reason, Soviet Frontal Aviation, as well as many rapidly evolving forms of Cruise missiles, should be examined along with TBMs as potentially potent and devastating sources of fire-power for use against NATO's (and the Warsaw Pact's) combat forces. Figure 9.4 shows a very rough comparison of the conventional

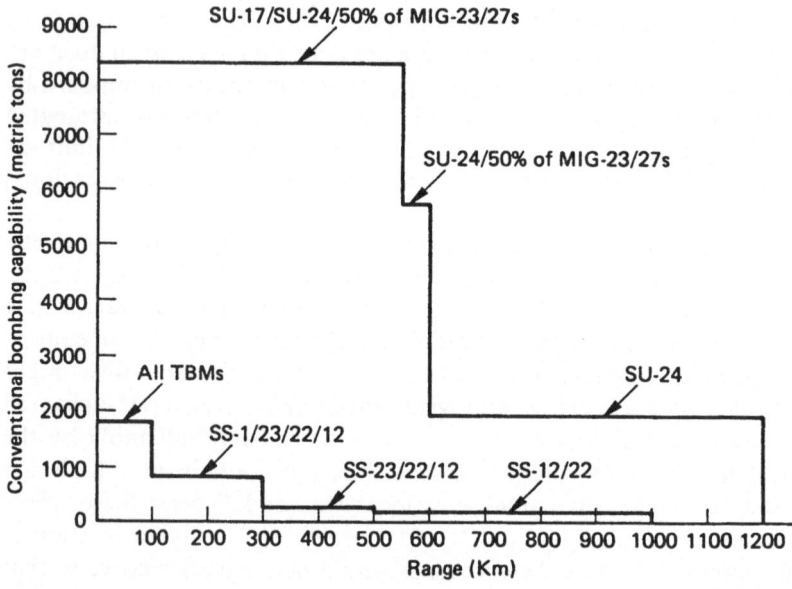

Figure 9.4   TBM bombing capabilties relative to a single sortie delivered by ground attack aircraft

bombing capabilities of Soviet Frontal Aviation relative to that from TBMs – as measured in metric tons of ordnance that can be delivered. The upper curve is constructed from the numerical estimates of the number of SU-17, SU-24 and MIG-23/27 aircraft shown in Table 9.2. The data in Table 9.2 is taken from published Congressional Research Service tables that provide estimates of the number of Frontal Aviation aircraft that might be available to the Warsaw Pact if forces are generated in preparation for war.[8] The weapons loads and combat radii assumed for each aircraft are those from *Soviet Military Power 1986*.[9] The TBM curve in Figure 9.4 is also constructed from Congressional Research Service data. It is constructed by assuming that every TBM launcher currently possessed by the Soviets is stationed in Europe and available for a surprise attack on NATO.

As can be seen from the plotted data, a single Frontal Aviation sortie could deliver upwards of eight million kilogrammes (8000 metric tons) of munitions to ranges beyond 500 kilometres from operating bases, and about two million kilogrammes (2000 metric tons) to ranges as great as 1200 kilometres. About 8000 SS-23 class TBMs would be needed to deliver a comparable load of munitions against targets at 500-kilometre range and about 2000 SS-12/22 class TBMs would be needed for similar missions at ranges between 600 and 1000 kilometres.

If all SS-1 launchers are assumed either to be replaced or modified so that they can carry modernised SS-23s, the resulting Soviet TBM force would still have no more than 700 to 800 SS-23 launchers. The weight of ordnance that could be delivered by these units to 500 kilometres would be one-tenth that deliverable by a single sortie of Frontal Aviation. If it is also assumed that all SS-12 launchers can launch SS-22s (the SS-12 mod.), and that all available launchers are armed with the more modern missile, there would be less than 200 launchers that could be used for attacks against targets 900 to 1000 kilometres distance. This too results in a threat that can deliver no more than one-tenth the weight of munitions of a single sortie of Frontal Aviation's longer range ground attack and strike aircraft.

Also worthy of note is that the delivery capabilities shown in Figure 9.4 do not include ordnance that could be delivered by TU-16, TU-22, and TU-22M intermediate-range bombers (see Table 9.2). Adding the payloads from these forces to those of Frontal Aviation suggests that 13 000 SS-23 class TBMs would be required to deliver the equivalent of a single sortie load to ranges of 500 to 600 kilometres, and 7000 SS-12/22 class TBMs would be needed for the longer ranges.[10]

Table 9.2

Ground attack aircraft	Maximum weapon load (kg)	Number of aircraft	Total load capability per sortie (kg)
MIG-23/27	3 000	2 510	7 530 000
SU-24	3 000	630	1 890 000
SU-25	2 000	70	140 000
	Subtotal	9 560	000
Intermediate range bombers[1]			
TU-16 Badger	9 000	287	2 583 000
TU-22 Blinder	5 500	136	748 000
TU-22M Backfire-B	12 000	130	1 560 000
	Subtotal		4 891 000
	Total[2]		14 451 000

[1]All these aircraft can also carry stand-off air-to-surface missiles (ASMs).
[2]Load equivalent of about 14 000 Tactical Ballistic Missiles.

Yet another source of fire-power that might be compared with that of TBMs is from Cruise missiles. Consider Figure 9.5, which shows a sub-munition dispensing Cruise missile and its F-16 launch platform.[11] As this missile weighs only 1500 kilogrammes (about one-fourth the weight of an SS-23 TBM), the F-16 could carry four such missiles. If the F-16 flies a 'lo-lo-lo' attack profile (penetrate enemy air-defences at low-altitude, release the Cruise missiles at low-altitude, and return to base at low-altitude), it could carry these four missiles to a range of 300 kilometres if its air speed is Mach ·85 or to 450 kilometres if its speed is Mach ·7. Since the particular configuration of Cruise missile shown in the diagram can carry its 550-

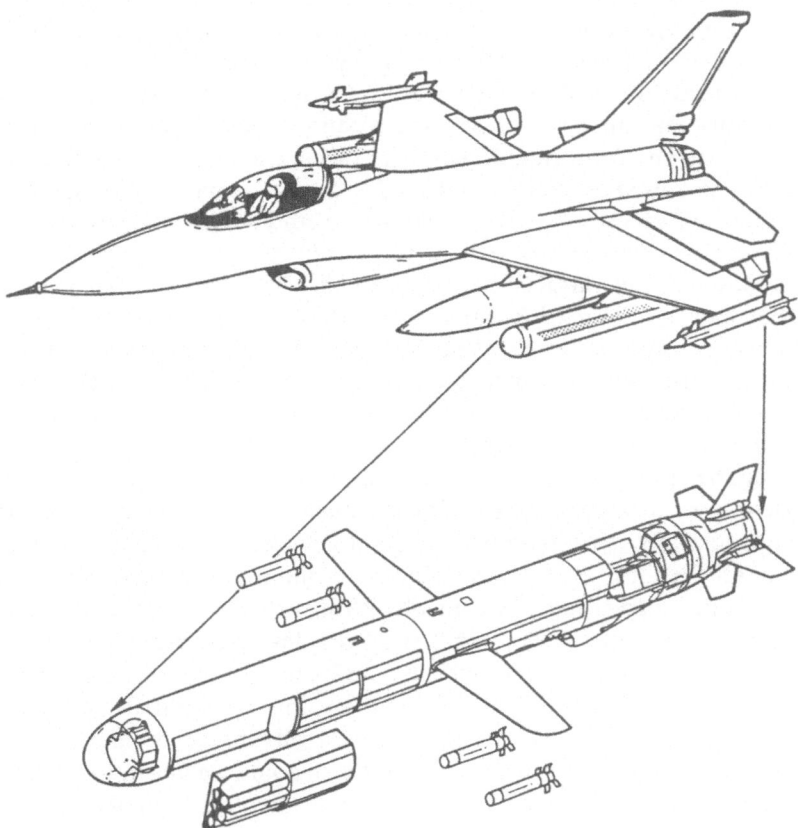

Figure 9.5

kilogramme payload to a distance of 450 kilometres, the overall range of delivery is between 750 and 900 kilometres. This allows ample extra ground track for the choice of optimal penetration routes to targets at 500' to 600-kilometre range.

The Cruise missile shown in Figure 9.5 is only one example of an unmanned air-supported vehicle that could be built for varied military applications with currently available technology. Modern guidance technology also makes it possible to fit Cruise missiles with terminal sensing systems in addition to already very accurate Tercom guidance systems. Missile effectiveness in some applications might be improved still further with the addition of sophisticated decision-making computers. The particular missile shown in Figure 9.5 is designed for use against airfield runways. It is fitted with a sensor that allows it to 'look-down' and recognise a runway, adjust its flight path, and dispense the 40 or so sub-munitions it carries along the length of the runway. There are other examples of missions that might be performed by yet different types of Cruise missiles. Radio sensors mounted on missiles can be used to allow them to home on radio transmissions from radars, jammers or communications equipment. Such missiles are sometimes referred to as anti-radiation homing. Air-supported missiles can carry television and infra-red sensors in a reconnaissance role, as well as for localisation and attack of non-radiating mobile targets. Missiles can also be fitted with jammers so that they can be used in support of aircraft or ground-force operations. Yet another application for which Cruise missiles are ideally suited is the remote dispensing of chemicals and other agents. Since Cruise missile guidance systems can utilise both inertial measurement and external terrain recognition technologies, their guidance systems could be designed to be able to determine local wind conditions in the vicinity of a target. Such a Cruise missile, under the control of an appropriately programmed internal decision-making computer, would then be able to choose an optimal cross-wind path for controlled dispensing of chemicals. With such great adaptability and flexibility, it appears likely that such missiles could provide a significantly more controllable, efficient and reliable means of chemical delivery than that of TBMs.[12] Since these missiles could also choose a path of approach that takes maximum advantage of masking by terrain features, warning of their approach could well be non-existent. Thus it appears that many types of Cruise missiles, along with Soviet Frontal Aviation, could well pose threats to NATO that are perhaps more troubling than those from advanced TBMs.

CAPABILITIES OF SINGLE WARHEAD TBMs AGAINST TARGETS OF VARIED HARDNESS AND MOBILITY

The probability that a TBM will damage its target – that is, the 'damage expectancy' – is itself the product of several 'conditional' probabilities. These include the probability that the TBM survives to be launched; the probability that all missile systems function reliably in flight; the probability that the missile penetrates opposing defences; and the probability of its warhead killing the target. Since the scenario being considered here is a Warsaw Pact surprise attack on NATO, and NATO has no defences capable of engaging TBMs, the pre-launch survivability of Soviet TBMs is one, and the probability of penetrating defences is also one. The probability that all missile systems function reliably during flight can be estimated from engineering experience with missile technology. Experience suggests that relatively high system reliabilities of between 0·7 and 0·9 can be expected of current and future Soviet TBMs. However, unlike the other probabilities that are needed to construct damage expectancies, TBM probabilities of kill vary drastically with the properties of targets and accuracies of delivery. Hence, for the case of a Warsaw Pact surprise attack against a NATO that does not possess anti-TBM defences, the primary source of variations in damage expectancy will be due to variations in the probability of kill. Since it is clearly both an important and complex quantity, the probability of kill merits detailed analysis and discussion.

The probability of kill (P_k) depends on the hardness of the target to weapons effects, the weight and explosive power of the ordnance delivered by a TBM, and the accuracy of TBM delivery. The effects of target hardness and explosive power can be mathematically described in terms of a quantity called the 'lethal radius', which is the range at which a detonating weapon creates effects sufficiently intense to destroy a target. Since the lethal radius is a property of the 'hardness' of a target to weapons effects, it will vary greatly with the nature of the target. For example, a blast of five to ten pounds per square inch (35 to 70 kPa) is likely to be sufficient to destroy or damage a radar's antenna. As shown in the curves plotted in Figure 9.2, such blast pressures occur at ranges of 38 to 55 metres from the detonation of a 1000 kilogramme high-explosive TBM warhead. If the target is instead a lightly fortified above-ground bunker, its equipment and occupants might not be seriously disrupted or destroyed unless it is subjected to blast effects in excess of 40 pounds per square inch

(276 kPa). Since such a blast occurs at about 16 to 17 metres from the detonation point, the TBM must fall much closer to this target to destroy it. In this case, the target might be said to be 40 pounds per square inch hard, and the lethal range of a 1000-kilogramme munition against it would be about 17 metres.

Still another quantity used in calculating the probability of kill is the accuracy of the TBM. As has already been noted, the accuracy of a missile is conveniently described in terms of the CEP, which is the distance from a target within which half of the missiles aimed at it will impact. Like the lethal radius, the CEP can be thought of as being determined by properties of both TBMs and targets. To see this, consider a target that has a location which is not precisely known. This can come about because it is mobile, or because of the imprecision of methods used to locate it (such location uncertainties can be due to measurement errors in radio direction finding, or navigation and measurement errors on reconnaissance vehicles, and the like). If a target's location is not precisely known, even a perfectly accurate missile will, on the average, miss the target by a distance equal to the uncertainty in target location. For missiles with either finite or perfect accuracy, this effect can be accounted for by assuming the missile has an effective CEP. The effective CEP (CEP_{eff}) is equal to the square root of the sum of the squares of the CEP and the average uncertainty in location.[13] Figure 9.6 shows probability of kill curves versus missile 'effective' accuracy for targets of different hardness and for missiles armed with 1000–kilogramme high-explosive warheads. Since the effective accuracy includes not only target miss due to missile guidance and control errors, but also miss due to uncertainties in the location of the target, even a perfectly accurate missile that is shot at a target whose location is uncertain will have a non-zero effective accuracy (CEP_{eff}).[14]

A missile that has been fired at a target whose location is known to only 30 metres therefore has the same effective accuracy as one that is used against a target whose location is perfectly known, but has guidance and control errors resulting in an accuracy of 30 metres. As can be seen from Figure 9.6, even if a TBM has accuracy of 30 metres, comparable to that of the Pershing II missile – with its radar area correlator – the probability of kill against a fixed above-ground semi-hardened bunker is no better than 0·2. If missile reliability of 0·8 to 0·9 is factored in, the damage expectancy per missile drops to below 0·2, and if the target's location is not known to better than 20 or 30 metres, the damage expectancy drops to well below 0·1!

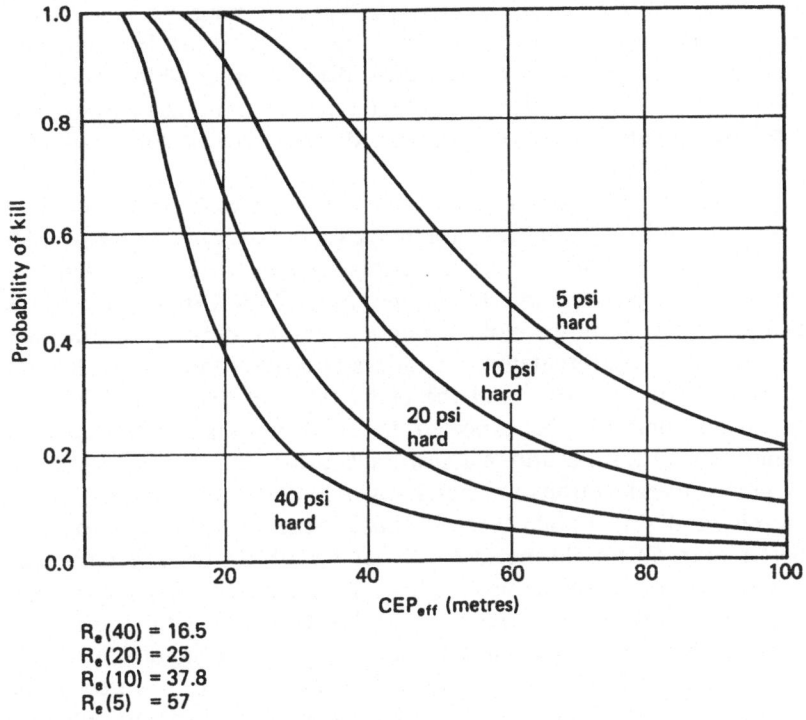

R_e(40) = 16.5
R_e(20) = 25
R_e(10) = 37.8
R_e(5) = 57

Figure 9.6 Probability of kill of 1000-kilogramme HE warhead against targets of different hardness

A radar, unlike a semi-hard bunker, is only likely to be hardened to five or ten pounds per square inch (35 to 70 kPa). For this reason, radars, like many other military assets, are intentionally designed to be mobile – to make it hard for an enemy to locate and target them. Since radio direction finding is possible against radars, they must be operated in a way that will confound enemy attempts to localise them. This can be done by building numerous decoy transmitters that radiate signals which cannot be distinguished from those of the radar. In addition, both the radar and the decoy transmitters must be moved often enough to overwhelm the enemy's ability to cope with such circumstances. The incorporation of decoys and mobility into modern air-defence design has resulted in highly mobile systems like NATO's Patriot system and the Soviet SA-12. Because accurate information on the location of properly operated air-defence radars can be difficult to obtain, an attacker might have to resort to tactics that provoke the air-defence to turn on its radars. This can be done with combined aircraft attacks, or with drones in combination with aircraft and anti-radiation and other types of homing missiles. The 1982 Israeli suppression of Syrian air-defences is an example of such an operation.[15]

As can be seen from Figure 9.6, even a perfectly accurate TBM would be unlikely to achieve a damage expectancy much in excess of 0·1 to 0·2 unless the locations of critical *mobile* air-defence elements are known to better than a 100 metres. Since TBMs are not adaptable to tactics that could provoke a radar to reveal its location, they are considerably less threatening to radars and other mobile air-defence elements than aircraft – and the many types of drones and homing munitions that can be expected to be commonplace in the next generation of NATO and Warsaw Pact forces. Moderately hardened targets whose locations are not known to better than a few tens of metres, and soft-mobile targets whose locations are not known to better than 50 to 100 metres, will not be subject to especially high likelihood of damage from attacking TBMs armed with single high-yield warheads. Thus it appears that TBMs armed with single high-explosive warheads would not be especially effective against moveable semi-hardened bunkers of the type that might be constructed in the field. Moreover it appears that they would be even less effective against softer mobile-targets – like the critical elements of mobile air-defences.

CAPABILITIES OF TBMs ARMED WITH SUB-MUNITIONS

Since it is so difficult to damage targets with a single high-yield warhead, it is relevant to ask how much the likelihood of success might be altered if the unitary TBM payload is exchanged for one of sub-munitions. As can be seen from Figure 9.3, the detonation of ten kilograms of high-explosives results in a five pound per square inch blast at 12 metres from the explosion and a 40 pound per square inch blast at 3·5 metres. This suggests that if a 1000-kilogramme TBM payload is divided into 100 ten-kilogramme sub-munitions, and the sub-munitions can be uniformly dispersed, a roughly circular area of radius 35 metres could be covered with blast pressures of 40 pounds per square inch or more. If a less dense pattern of dispersal is chosen instead, a circular area of about 120 metres radius could be subjected to five psi or more. Thus a TBM that can carry and disperse 100 such sub-munitions can achieve more than twice the lethal range of one that carries a 1000-kilogramme unitary warhead. This suggests that, when certain types of targets are to be attacked, a significant improvement in TBM effectiveness might result from the use of sub-munitions.

Figure 9.7 illustrates the trade-offs that govern a sub-munitions-dispersing TBM's ability to inflict damage. Since the location of the impact point of an arriving missile will on the average be different from that of a target, the larger the area over which sub-munitions can be spread, and the more likely the target will be encompassed within the area where they fall (Figure 9.7a). However, the larger the sub-munition impact area, the lower will be the density of munitions, and the less likely it is that the target will be hit by one or more of them (Figure 9.7b). For missiles of specific accuracy, and for targets of specific hardness, there is a well defined value of sub-munition dispersal radius which results in a maximum probability of hit.

A simple probabilistic theory provides good estimates of target hit probabilities for sub-munition dispensing TBMs.[16] Figure 9.8a shows a typical result of calculations using this theory. The plotted results of calculations assume a sub-munition-armed TBM with a CEP of 100 metres and a target that can only withstand blast pressures of five pounds per square inch or less. The TBM is assumed to evenly disperse 100 sub-munitions within a circular impact region of radius R_d. Figure 9.8a shows the probability of kill ($P_k = P_D X P_H$) as a function of the dispersal radius R_d. As can be seen from an inspection of the graph, the probability of kill has a maximum value when the dispersal radius is between 70 and 90 metres.

The reason for this maximum can be understood from the behaviour of two quantities from which the probability of kill is constructed: the probability that the distance between the TBM impact point and a target is closer than the dispersal radius (P_H); and the probability that a target within the region of dispersed sub-munitions is subjected to lethal weapons effects by at least one of the 100 dispensed sub-munitions (P_D). Since a larger value of dispersal radius results in a lower density of sub-munitions, the probability that a sub-munition will destroy (or 'hit') a target (P_D) within the region where sub-munitions land decreases as the dispersal radius (R_d) increases. However, as the dispersal radius (R_d) increases in magnitude, it is more probable that a target will be within the dispersal radius (R_d). Hence the probability of a TBM impacting at a distance

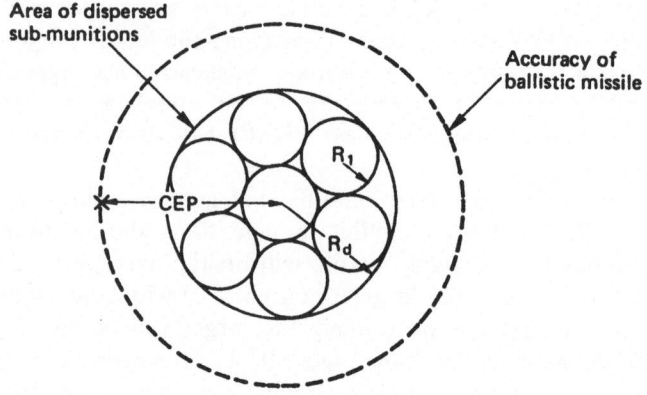

(a) Tight dispersal of sub-munitions

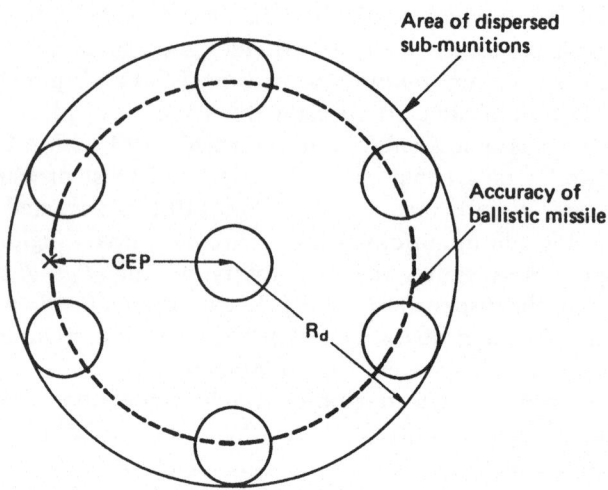

(b) Loose dispersal of sub-munitions

Figure 9.7   (a) Tight dispersal of sub-munitions; (b) Loose dispersal of sub-munitions

closer than the dispersal radius (P_H) increases with the value of the dispersal radius (R_d). Since the probability of kill is equal to the product of these two probabilities $(P_k = P_H X P_D)$, the monotonically increasing value of P_D with dispersal radius R_d and the monotonically

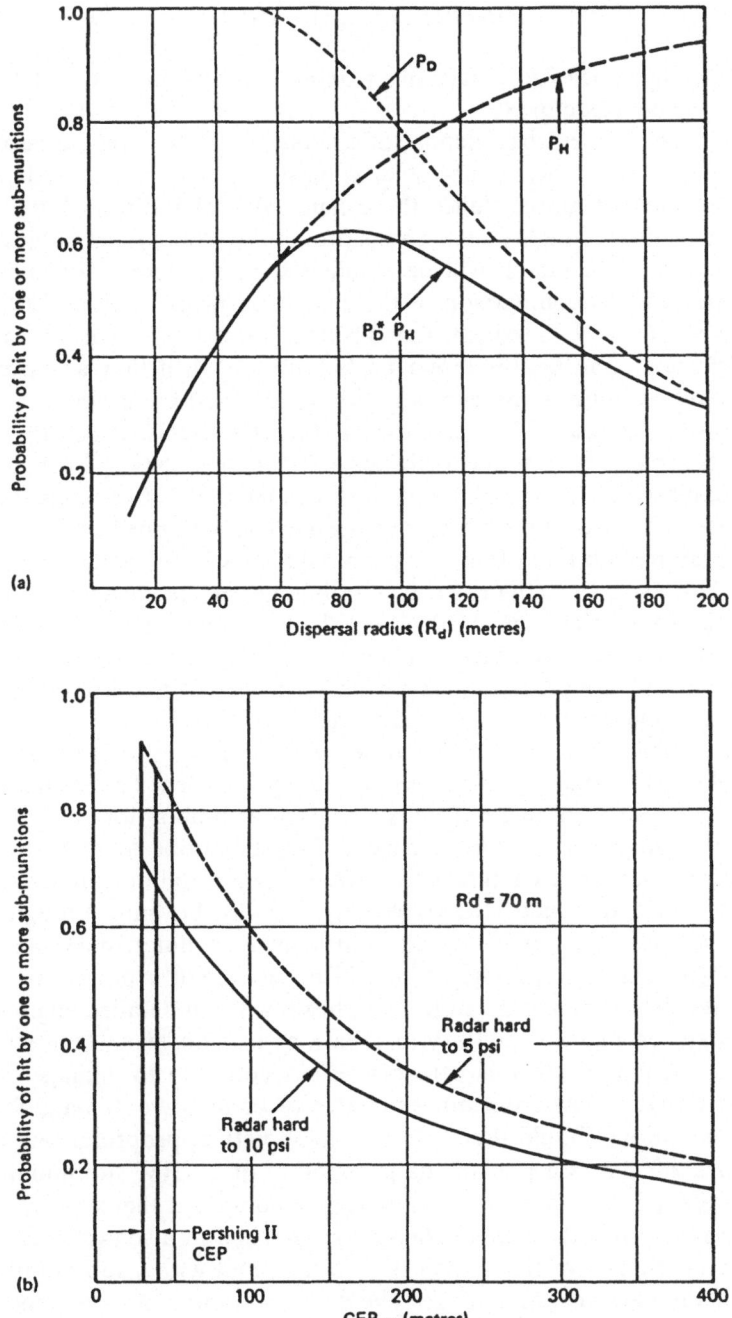

Figure 9.8 Probability TBM sub-munition destroys a radar

decreasing value of P_H results in a maximum probability of kill (P_k) at a particular value of R_d.

As noted earlier, the antennae of a radar unit are unlikely to be able to function after being subjected to blast effects of five pounds per square inch or more. Hence the calculated probabilities plotted in Figure 9.8b are roughly equal to the probability that a radar subjected to such a TBM attack will be eliminated as an element of an air-defence unit. A comparison of the calculated probabilities of kill in Figures 9.6 and 9.8b indicate that TBMs of similar 'effective accuracy' will have a considerably larger probability of kill if they are armed with sub-munitions rather than with single 1000-kilogramme high-explosive warheads. For example, if the effective CEP (CEP_{eff}) of TBMs armed with unitary warheads is 100 metres, they will achieve a probability of kill (P_k) of 0·2 against a radar that is five pounds per square inch hard. If the TBM is armed instead with one hundred ten-kilogramme sub-munitions, the probability of kill can be increased to 0·6. However, since the radar is a mobile target, the meaning of this increase in efficiency is ambiguous. The probability of kill could be greatly reduced, back to 0·2, if technical and tactical measures are devised and adopted to obscure its position so that it cannot be located to better than about 400 metres.

It is also of interest to ask how effective a sub-munition-armed TBM might be against an above-ground semi-hardened structure that can be destroyed incrementally by each sub-munition hit. Such facilities might be buildings needed for operations at airbases or modestly hardened command centres of perhaps minor importance. However, many such structures may actually be hard enough to ensure that single hits will result in little or no cumulative damage of this type. As each sub-munition will have a small explosive power relative to that of a large unitary warhead, without detailed engineering data on relevant military installations it is unclear which structures might fall into this category of susceptibility to damage. The results that follow from our estimates of cumulative damage from sub-munitions should therefore be viewed with appropriate caution. Figure 9.9 shows a plot of the probability of a TBM sub-munition hitting the roof of a semi-hard above-ground bunker that is presumed to have a roof area of about 300 square metres. It is also assumed that the construction of this shelter is such that a hit with a sub-munition results in incremental damage. The dispersal radius (R_d = 60 metres) is chosen to maximise the probability that at least one sub-munition will hit the shelter. Since a single sub-munition hitting the shelter

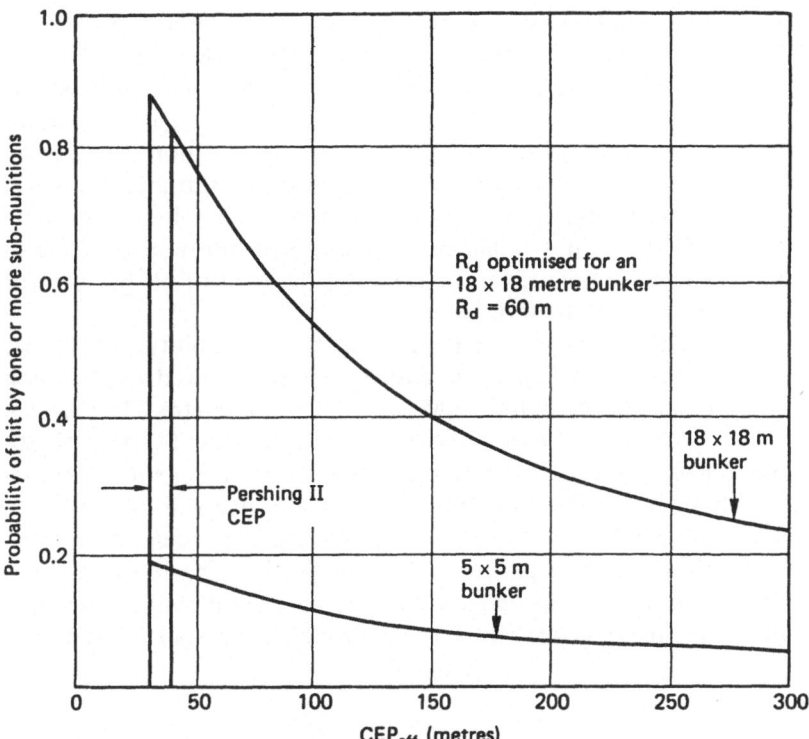

Figure 9.9   Probability of TBM sub-munitions hitting a bunker

would not necessarily destroy it, it is necessary to assess the levels of damage that multiple hits with sub-munitions might inflict.

In order to do this, we can consider the bunker to be made of perhaps 12 different 'sub-bunkers' which are about five by five metres on a side. We assume that each sub-bunker will be completely destroyed if it is hit by a ten-kilogramme sub-munition. The probability of at least one sub-munition hitting a sub-bunker is also plotted in Figure 9.9. If a Soviet TBM were to achieve Pershing II accuracy, and there is no uncertainty in the attacker's knowledge of the target's location (CEP_{eff} = 30 metres), then the probability that the bunker will be hit with at least one sub-munition is about 0·88 and the probability that each sub-bunker will be hit is about 0·19. Since there are 12 such sub-bunkers, the average number of sub-munitions expected to hit the bunker would be $12 \times 0·19 = 2·28$ sub-munitions.

Two TBMs would raise the average number of munitions hits to
2·28 × 2 = 4·56, and three would result in 2·28 × 3 = 6·84 hits.
Although the structure might be hit by 6·84 sub-munitions on the
average, some of the time it might be hit by more sub-munitions,
while at other times it might be hit by less. In addition, for a given
number of hits on a bunker, some fraction of the sub-munitions will
hit sub-bunkers that have already been destroyed. These additional
hits will therefore not inflict more damage on the bunker. It is of
interest to know how much of the time the goal of 50 per cent or
higher damage is achieved.[17]

Figure 9.10 shows our predictions of the probability that 25 per
cent, 50 per cent and 75 per cent damage will be achieved as a function
of the number of attacking TBMs with 30- or 60-metre CEP. As can
be seen from Figure 9.10, although three attacking TBMs with 30-
metre CEPs will result in more than 50 per cent damage to the target
more than half of the time, between four and five will be needed to
achieve at least that level of damage more than 75 per cent of the time.
If the CEP of attacking TBMs is 60 metres, then nearly five missiles
will be required to achieve more than 50 per cent damage more than
half of the time, and six will be needed to achieve 50 per cent damage

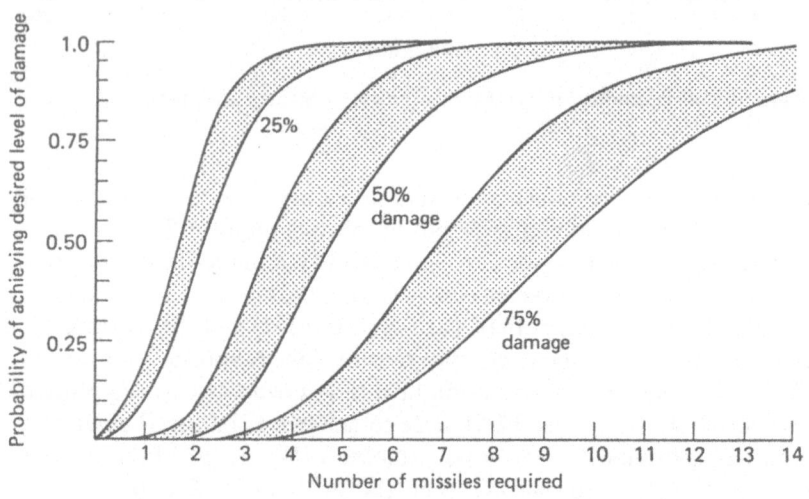

Target consists of 12 sub-units
P_x per sub-unit = 0.15–0.20 for one missile

Figure 9.10

more than 75 per cent of the time. Accordingly, if redundancy built into the target bunker requires that more than 50 per cent of it be destroyed before it ceases to be able to function, a large number of TBMs will be required to attack such a bunker before it is no longer able to continue functioning as a militarily useful unit.

TBM EFFECTIVENESS AGAINST AIRBASE RUNWAYS

Since even very large numbers of accurate non-nuclear TBMs could not destroy a large fraction of aircraft housed in hardened and dispersed airbase shelters, a TBM surprise attack might instead be aimed at damaging runways. The objective of a runway attack would be to deny, or sufficiently slow, the launch of interceptors that would be used to defend the bases from massive follow-on Frontal Aviation attacks. In order to implement these objectives, TBMs must be able to do enough damage to runways to drastically slow the rate of interceptor take-off. This task could be quite demanding. Two characteristics of modern fighters that make them so suitable for air combat is their low wing loading and high thrust to weight ratio. These characteristics – which result in a high manoeuvre and acceleration capability in air combat – also result in an aircraft that can take off using only very small sections of runway. An F-15 interceptor, for instance, needs only 275 metres (0·275 kilometres) of take-off run before its wheels leave the runway.

Another planning factor – which could lead to a potentially disastrous failure to achieve surprise – stems from the need to prepare Frontal Aviation for rapid and massive follow-on base attacks. In order to deliver a decisively successful and lasting blow to NATO aviation, Frontal Aviation would have to be able to control the air over NATO bases while also destroying facilities on the ground. If large-scale preparations for such attacks are not made, unprepared Frontal Aviation forces might not be able to gain a decisive advantage over those NATO defending interceptors that make it into the air. Preparations for adequately rapid and massive follow-on base attacks might, however, be difficult to hide from NATO's prying long-range surveillance sensors. Frontal Aviation would therefore have to plan for the possibility that observed preparations would cause NATO to put aircraft into the air or on quick reaction alert, disperse them to civilian bases – or possibly even to strike first.

Since many NATO airbases are sized to handle large heavy-lift aircraft in addition to interceptors and fighters, this type of TBM surprise attack would have to decisively deny NATO a very large fraction of its available runway space. A detailed evaluation of how many TBMs might ultimately be needed to achieve this objective requires consideration of the following factors:

- The length of runway required by interceptors for take-off.

- Total length of runways available at NATO airbases.

- The average width of these runways.

- The accessibility of the runway sections to aircraft (that is, the existence or non-existence of runway shoulders for the ferrying of aircraft to open runway sections during emergencies).

- The number of TBMs required to deny a section of runway to NATO interceptors.

A determination of this quantity requires a knowledge of:

- the accuracy of TBMs that might be used to try and cut the runways;

the reliability of the TBMs;

the type of munitions used by the TBMs.

We begin our analysis by assuming that each tactical ballistic missile is either an SS-22 or SS-23 which has an as yet undetermined accuracy (CEP) and reliability (P_{RE}). Each TBM can carry a payload of about 1000 kilogrammes to a range of 1000 or 500 kilometres respectively. Furthermore the payload of either a single 1000-kilo-gramme warhead or 100 sub-munitions weighing ten kilogrammes each can be delivered to any of the nearly 50 NATO airfields shown in Figure 9.11.

A 1000 kg warhead delivered by such tactical ballistic missiles could create a crater of 12 to 17 metres diameter (see Figure 9.12) – assuming that the warhead can survive impact and penetrate one to two metres into a runway's pavement before detonating in the subgrade (in spite of its arrival speed of two to three kilometres per second). Since upheaval around the crater periphery can potentially extend its effective 'blocking' diameter an additional 20 to 30 per cent (Figure 9.12), an impacting TBM armed with a unitary warhead could

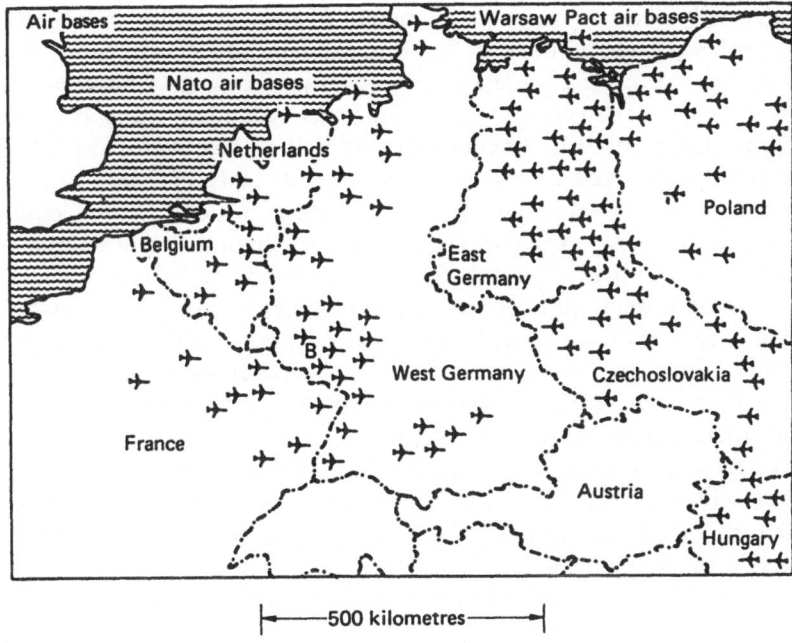

Figure 9.11

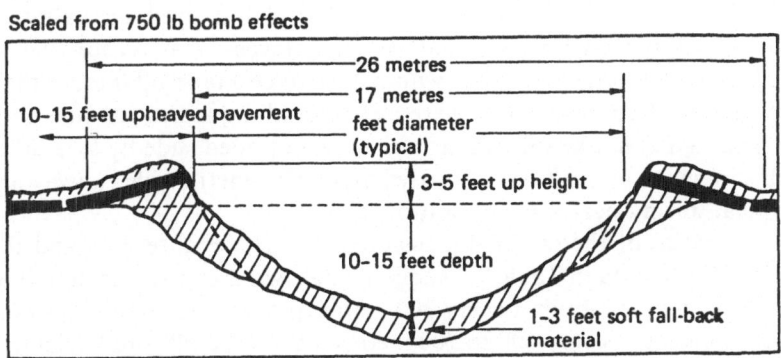

Figure 9.12 A typical runway crater made by the explosion of a 2200 lb bomb

Source: *The Military Engineer*, October 1984.

potentially deny as much as 26 metres of runway width to aircraft. Runways at large military air facilities are typically 50 metres or more wide and 1·5 to 2·5 kilometres long. A large airbase may have two or more large main runways and numerous secondary runways that might also be used in emergencies. Since the objective of the TBM phase of the attack is to deny rapid take-off to NATO interceptors and fighters, or to slow take-off rates to the extent that fighters could not get into the air to defend their bases, enough of the runway area of each airbase must be destroyed for this objective to be met.

During normal peacetime operations, fighters can be launched from a typical 50-metre wide runway two abreast, and with standard safety lapse times of about ten seconds between pairs of aircraft.[18] Hence, if only a single section of runway is used for fighter launch, and there are no bottlenecks to prevent aircraft from assembling to take their turn taking off, 12 aircraft per minute launch rates can be routinely achieved. In wartime operations, the standard safety lapse time might be reduced from ten seconds to perhaps six or eight, raising the launch rate to 15 or 20 aircraft per second.

Table 9.3 shows wheelbase, take-off run, and landing run data for typical NATO interceptor/fighter aircraft currently in service. The Northrup F-5E/F requires a take-off run of 600 to 700 metres, the McDonald Douglas F-15 needs 275 metres, and the General Dynamics F-16 requires about 300 metres. Since the wheelbase of these aircraft is between four and six metres (Table 9.3), the main impediment to take-off and landing on a runway that is not totally blocked will be control of the aircraft in cross-winds. In wartime, faced with the choice of being caught on the ground and destroyed, or taking off to engage approaching Frontal Aviation forces, it is reasonable to speculate that pilots would be willing to chance a take-off if more than 15 metres of runway width were available along a straight line path over a fighter's take-off run (as already mentioned, side by side take-offs, in which aircraft occupy only half of a 50-metre wide runway are typical during peacetime operations).

Appraisal of the crater dimensions shown in Figure 9.12 and the take-off run data in Table 9.3 suggests that two craters on each half of the runway are required every 300 metres if F-15 and F-16 take-off is to be denied, or every 600 to 700 metres if F-5 take-off is to be denied. However each TBM fired at a target on a runway will not necessarily impact at a location that results in optimal runway blockage. In order to determine the number of TBMs that might be required to cut the runway at chosen points, it is necessary to derive a mathematical

Table 9.3   TBM Attacks on NATO Command and Control Elements

Aircraft	Take-off run (metres)	Landing run (metres)	Wheelbase (metres)
F4	900–1300	1000 (with parabrake)	—
F5	600 (2600 kg fuel)	750–800 (with parabrake)	4·3
F5	1800 (max. wt.)	—	4·3
F15	300	800 (without parabrake)	5·5
F16	320³	—	4·0
F111	990–1200	900	
A10	1000 (max. wt.)	610	5·4
A10	450 (forward airstrip wt.)	400	5·4
AV8B	350 (max. wt.)	10–20	0·5
F5E¹ (at 7053 kg)	610 m	762 m at 5230 kg wt & with brake-chute	
F5F (at 70371 kg)	701 m	792 m at 5554 kg wt & with brake chute	
F15 (Interceptor) (about 18 829 kg)	274 m	762 m (without parachute)	
F16A² (Air-to-Air) 10 800 kg (no external tanks)	~ 300 m	?	
F16B (Air-to-Air) (10 568 kg – no external tanks)	~ 300 m	?	

¹F5E   Empty weight 4 410 kg
Max TD Wt 11 214 kg
F5F   Empty weight 4 797 kg
Max TD Wt 11 409 kg
²F16   Gross area of wings 300 ft²
Thrust/Wt ratio (clean) = 1·1
³Estimated from data on wing loadings, take-off weight and engine thrust.

F15   Gross Area of Wings 608 ft²
Two Pratt & Whitney F100-PW-100 engines (turbo??)
23 930 lb thrust with afterburning for take-off

June 1983–84
pp. 447–448

relationship for the probability of TBM impact as a function of distance from the runway centreline.

If it is assumed that the distribution of TBM impact points around intended targets can be represented as a normal Gaussian probability function, then the probability that a target of area A (in this case, a runway) will be hit is given by

$$P_{runway} = \frac{1}{2\pi\sigma^2 A} \int dA \, e^{-\left(\frac{r^2}{2\sigma^2}\right)}$$

where r is the miss distance between an impact point and the intended target and σ is the standard deviation of miss distances. σ is simply related to the CEP by the constant

$$\sigma = \frac{CEP}{\{2\ln(2)\}^{1/2}} = (0 \cdot 849322)(CEP)$$

A missile will be very unlikely to hit a runway unless the magnitude of its CEP is roughly equal to the runway's width. Since runway lengths are very much larger than their widths, a runway target can be treated as if its length is infinite relative to the magnitude of the TBM's CEP. If the x-axis of integration is parallel to the runway length, the y-axis is parallel to its width, and the runway's width is w, then the probability of a runway hit (P_{runway}) can be written in the form of the following integral:

$$P_{runway} = \frac{1}{2\pi\sigma^2} \int\limits_{-\infty}^{\infty} \int\limits_{-w/2}^{w/2} dx \, dy \, e^{-\left(\frac{x^2+y^2}{2\sigma^2}\right)}$$

Upon integration this integral reduces to the Gaussian error function

$$P_{runway} = ERF\left(\frac{w/2}{\sqrt{2}\sigma}\right)$$

where

$$ERF(x) = \frac{2}{\sqrt{\pi}} \int\limits_{0}^{x} e^{-t^2} dt$$

Hence the probability of a TBM impact at a distance less than b from the runway centreline is

$$P_{runway} = ERF \left(\frac{b}{\sqrt{2}\sigma} \right)$$

and the probability of an impact at a distance *greater* than b is

$$1 - P_{runway}$$

Figure 9.13 is a plot of the probability that a missile will impact at a range *greater* than a distance R from the centreline of a runway. The distance is plotted in units of missile CEP. If it is assumed that a TBM achieves a Pershing II CEP of 30 metres, and a runway is 50 metres wide, then the curve in Figure 9.13 indicates that more than 30 per cent of missiles fired at the runway will have impact points that are *not* on the runway (CEP = 30 metres, R = 25 metres or 0·83CEP results in a 0·33 probability of impact at a range in excess of 0·83CEP). For a TBM armed with a 1000-kilogramme warhead to block half the runway, it might have to impact at a distance between perhaps eight and 16 metres from either side of a runway centreline. The probability

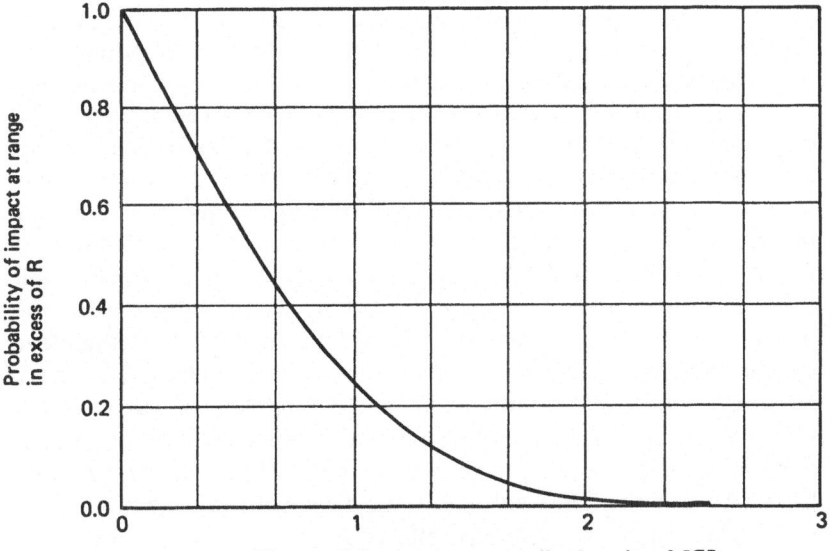

Figure 9.13 Probability of missile impact at a range in excess of R from a runway centreline

of landing in this band can be estimated using the curve plotted in Figure 9.13. It is simply the probability that the missile will impact at a range in excess of eight metres from the centreline (about 0·74) subtracted from the probability that it will impact at a range in excess of 16 metres from the centreline (about 0·49), that is 0·74 − 0·49 = 0·25. The probability that a second TBM will impact at the same distance from the centreline, but on the other side of it, is half the previously calculated probability. Hence the probability that two TBMs aimed at the same point on the runway will cut it so that aircraft will not be able to use it is (0·25)(0·125) = 0·03! This indicates that, if TBMs are to be able to deny sections of runway to taking off aircraft, they will either have to be considerably more accurate than Pershing II missiles or they will have to be armed with sub-munitions. Each ten-kilogramme sub-munition that hits a runway might block a width of two to three metres to aircraft. Hence, if 20 to 30 sub-munitions could be distributed over the width of each 300-metre section of runway, even F-15s and F-16s might not be able to use those sections for take-off.

Figure 9.14 illustrates the trade-offs that would have to be made if a sub-munition-armed TBM is used in the attack instead. Again, it is instructive to assume that the attacking TBMs have CEPs of 30 metres. If the sub-munitions are dispersed over a circular area of radius 55 metres (1·8 CEPs) around the TBM impact point, then about 25 per cent of the time a missile will land too far from the centreline for the entire width of the runway to be covered. Even if the dispersal radius is chosen to be 85 metres (2·8 CEPs), one to two per cent of impacting missiles will fail to disperse their sub-munitions over the entire width of the runway. If a runway must be cut in four places, and each attempt to cut the runway has a 0·25 probability of failure, then the probability that at least one part of the runway will remain open is 0·68. If it must be cut in six or eight places, the probability of an open stretch will be 0·82 to 0·90! It is therefore evident that since fighter/interceptor aircraft have such extraordinary abilities to take off rapidly from small stretches of runway, the objective of the attack cannot be met with confidence unless there is a relatively high probability of runway cut per TBM fired. This can only be achieved if the TBM sub-munition dispersal radius is more than two TBM CEPs – to offset miss effects due to the innaccuracy of the TBMs – and if each runway section is attacked by at least two TBMs – to offset the effects of finite missile reliability (P_{RE}).

Figure 9.15 shows a plot of the expected number of TBM sub-

145

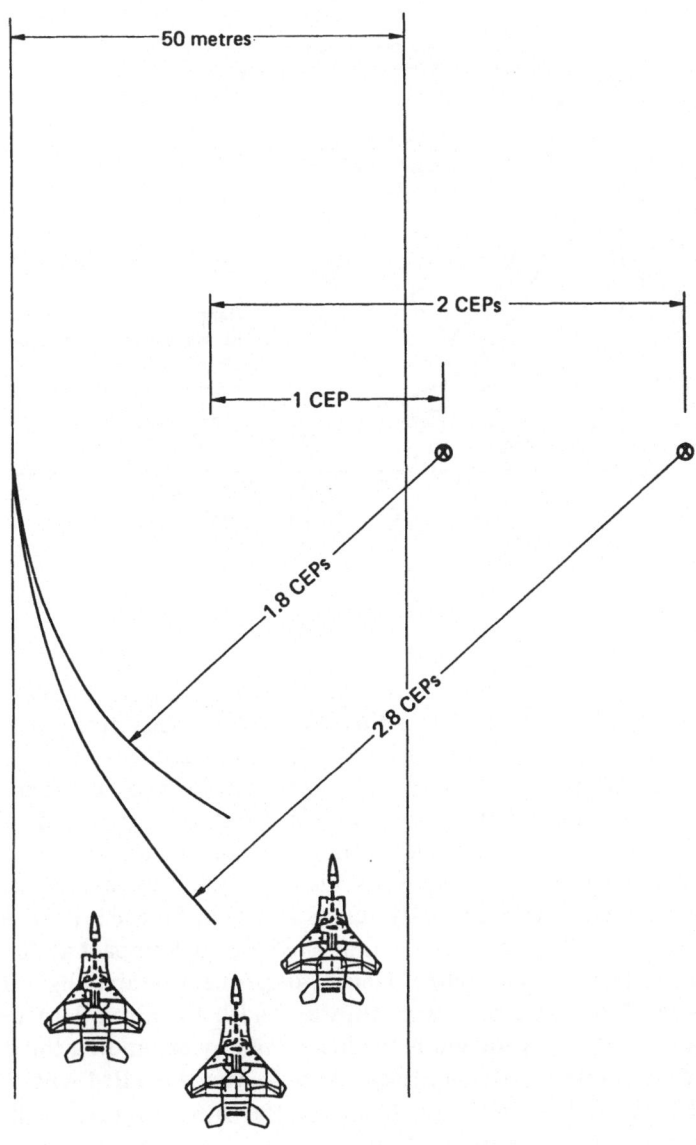

CEP = 30 metres

Figure 9.14

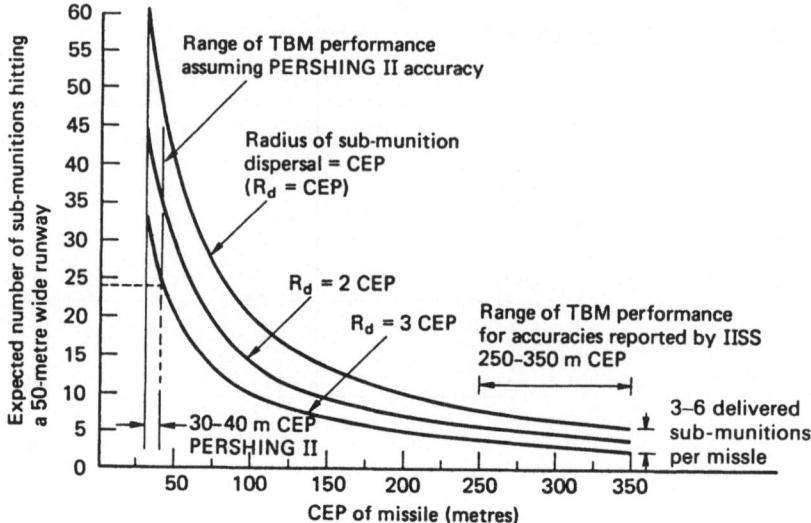

Figure 9.15

munitions that will hit a runway as a function of both missile accuracy and choice of dispersal radius. As can be seen from the plotted calculations, if a TBM has both a dispersal radius R_d and CEP of 30 metres ($R_d = $ CEP $= 30$ metres), then the average number of sub-munitions that will hit the runway is about 60. Because the dispersal radius is so small, whenever a missile strikes the runway, a very large fraction of its sub-munitions also hit the runway. However an inspection of the probability curves shown in Figure 9.13 reveals that more than 40 per cent of the time the TBM will impact at 20 or more metres from the centreline. Hence, 40 per cent of arriving TBMs will leave 15 or more metres of runway width for aircraft to use when taking off! This problem is readily solved by expanding the dispersal radius. However, the net result is that, even if a TBM with 30-metre CEP is used, only 30 to 40 of its 100 sub-munitions will hit the runway. An inspection of the Cruise missile shown in Figure 9.5 suggests that this much smaller and less expensive missile could deliver the same number of sub-munitions to the runway. Unlike the TBM, the Cruise missile might also be able to choose a dispersal path along the length of the runway, dropping sub-munitions as it flies. Sub-munitions delivered in this manner would not have to be heavily armoured to survive the tremendous two to three kilometre per second impact with the runway, as they would if they are to be dispensed from a TBM.

It therefore appears that Cruise missiles may well be considerably more efficient than TBMs as means of delivering sub-munitions against runways. Furthermore Cruise missiles would be a far more appropriate platform for delivering a wide variety of advanced 'smart' sub-munitions against runways. These microprocessor-controlled sub-munitions could be designed to behave unpredictably, sometimes detonating when they are moved, and perhaps even detonating when they sense the presence of nearby aircraft. The result of such an attack would therefore be a 'mined' runway that would be very hazardous for aircraft to use, and time-consuming for base personnel to clear. Since TBM-delivered sub-munitions would strike runways at speeds between Mach 6 and Mach 9, microprocessor-controlled sub-munitions, if they could be delivered at all, would have to be very heavily armoured to survive impact with the runway. This last consideration suggests a still greater reduction of TBM delivery capacity relative to that of Cruise missiles.

Even if each TBM that is launched against a runway section could achieve a probability of runway blockage of one, missile in-flight reliabilities would probably result in one to two out of every ten missiles failing during boost – or during other phases of flight, re-entry, and sub-munition deployment. Since a high probability of runway blockage must be achieved – or it will be very likely that sections of runway will be left for interceptor launches – two missiles would have to be assigned to each section of runway to be attacked if a high probability of success is to be possible. If each base has one to three runways of average length 1·5 kilometres, five to ten accurate TBMs (CEP = 30 metres) would be needed to guarantee denial of each runway to F-15/16 interceptor launch. Fifteen to 30 TBMs would therefore be required per base, and 750 to 1500 would be needed for the 50 bases shown in Figure 9.11. If 30 per cent of the TBM force is assumed to be withheld for nuclear strikes (or counter-strikes) against NATO, or if still another 30 per cent are withheld for non-nuclear contingency missions that might require TBMs, a force of 1000 to 2000 Pershing II-equivalent TBMs might be needed!

TBM EFFECTIVENESS AGAINST SEMI-HARD ABOVE GROUND AND SUPER-HARD UNDERGROUND COMMUNICATIONS AND COMMAND FACILITIES

Since a very large number of non-nuclear TBMs are required before the Warsaw Pact could have any hope of denying NATO interceptors the ability to defend their bases, a better strategy might be to use TBMs to attack NATO communications and command facilities. If TBM attacks could cause enough damage to these facilities to disrupt their critical communications and command operations, many NATO units might not conceivably get orders during the early phases of a Warsaw Pact attack. If this resulted in an improper allocation of NATO forces during this period, heavy damage might be inflicted on NATO forces before communications and command functions could be re-established.

Examination of Figure 9.12 indicates, however, that even a direct hit with a TBM that carries a single 1000-kilogramme unitary warhead will only result in a crater that is three to five metres deep. Hence, if a command centre has an earth overburden of more than five metres, and its roof is reinforced with enough concrete and steel to prevent its collapse, a direct hit on such underground facilities should not result in its destruction. A command centre that has six or seven metres of earth overburden on top of a two-metre thick roof of concrete and steel might therefore be expected to survive a direct hit from a TBM. Another way that damage against such shelters could be achieved is with a gigantic shape-charge warhead mounted on the top of the missile. A one-metre diameter shape-charge warhead that is detonated at an optimal stand-off distance above a bunker could create a hot penetrating jet of explosion reaction products and metal that might penetrate from ten to 12 metres of overburden, concrete and steel. However, the jet from a shape-charge warhead will travel in a straight line parallel to the axis of the shape-charge lens, and an arriving TBM is likely to be oriented at about 45 degrees relative to the vertical when it strikes the target. As a result, the jet will only penetrate to a depth of seven to nine metres – barely deep enough to penetrate into a shallow buried command centre and not enough to penetrate into one buried at even moderate depths (a command centre with perhaps ten metres of overburden and a two-metre concrete and steel roof could survive a hit from such a warhead).

Still another factor that could degrade the effectiveness of non-nuclear TBMs against very hard bunkers is their accuracy. Figure

9.16 shows a set of curves for the probability that TBMs with different CEPs will hit the roof of a square bunker with dimensions between 0 and 100 metres per side. For example, if a square bunker covers as much area as half a soccer field, a TBM with CEP of 30 metres will have a probability of hit of about 0·47, and if the CEP is 60 metres, the probability of hit will be about 0·15. The probability of a hit on a very large command centre of 20 metres per side, would only be 0·1 if the TBM had a CEP of 30 metres and would be of the order of 0·02 if the CEP were 60 metres. It is therefore clear that modest measures such as deep burial of command posts, increases in the depth of overburden on shelters, as well as hardening and dispersal of shelter sub-unit elements, could readily negate the most severe non-nuclear TBM threats, even if Soviet TBMs manage to achieve Pershing II accuracy and are armed with exotic (and perhaps unfeasible) shape-charge

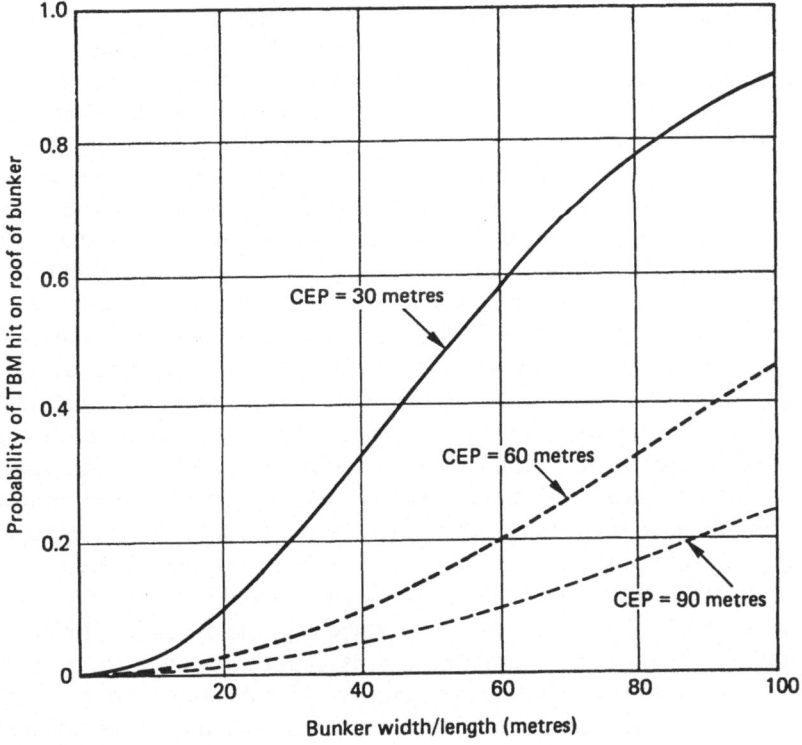

Figure 9.16

warheads. Such exotic and advanced TBMs could not therefore hope to achieve significant kill probabilities against facilities that utilise even a modest degree of low-technology hardening.

TBMs AGAINST ARMY DIVISION COMMAND POSTS

If non-nuclear TBMs could be used to destroy or disrupt many of the division-level command posts that belongs to NATO's 29 or so ground-force divisions, this could well result in a disastrous collapse of NATO.[19] As Warsaw Pact forces strike behind the onslaught of a TBM barrage, whole divisions critical for defence of NATO's front lines – each manned by about 20 000 fighting personnel – might be denied their division-level leadership. However it has long been recognised that in large-scale co-ordinated military operations, chaos and confusion that can follow the loss of key command elements can be used to advantage, and that such advantage can make the difference between victory or defeat. The enormous range, fire-power, and surveillance capabilities latent in modern military forces has accordingly resulted in equally extensive measures to reduce the likelihood and consequences of successful long-range strikes against key command elements.

Figure 9.17 shows the approximate arrangement of the elements of an American Army Deployed Main Divisional Command Post (DMAIN) deployed in the field. Since this is the most well staffed and well equipped of three command posts associated with the division, the division commander – a two-star general – will usually be in residence when the command post is not in motion. The different elements of the command post are housed or dispersed around armoured vehicles or trucks that transport personnel and equipment associated with each functional sub-unit of the post. The units are typically spread over an area that is about 300 metres per side. Each of the sub-units may make extensive use of natural cover and/or camouflage to make them difficult to observe from the air. In Europe, communications between units in the field can be affected over commercial telephone lines, to reduce the likelihood that the command post's location will be revealed by enemy radio direction finding. When radio communications cannot be avoided, remote radio transmitters that are connected to the command post by temporary land-lines are used. These transmitters may use directional low-sidelobe microwave antennae, to reduce the likelihood of enemy

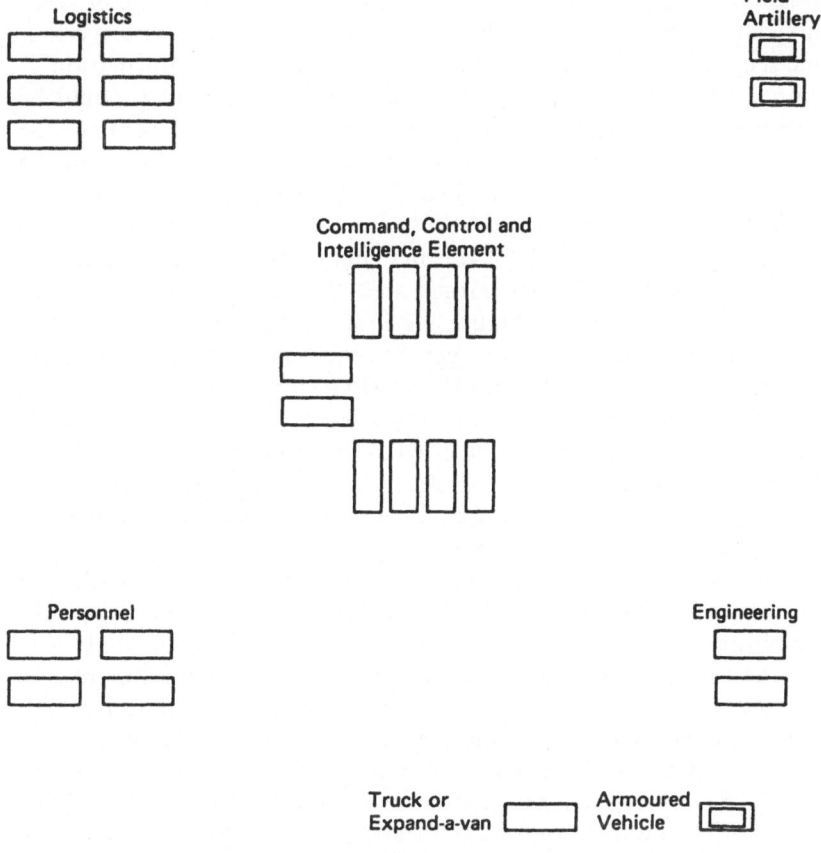

Figure 9.17 Approximate physical arrangement of Deployed Main Divisional Command Post (DMAIN)

cross-fixing on signals, and they may be displaced from seven to ten kilometres from the command post.

The DMAIN is backed up by two other command posts, the Tactical Operations Command Post (DTOC) and Jump Command Post (JUMP). Each of these posts is commanded by a one-star general who, if circumstances dictate, can take over command of the division. During periods of combat, or perhaps crisis, one of the three posts will be performing command functions for the division, one will be passively monitoring the status of the active post in case it is cut off and successfully attacked, and the final unit will be in motion. It is therefore clear that divisional command functions could be very difficult to attack with TBMs, as they reside in numerous dispersed, hidden and semi-mobile elements. Nevertheless, if information is gained about the location of a command post element, it could be subjected to devastating levels of damage, as all the elements of the post are relatively soft. It is of some interest to know how well the command post must be located for it to be endangered by TBMs.

Since the command post is both soft and dispersed, sub-munitions are the obvious choice of TBM armament, as they spread potentially lethal effects over a much larger area than that of a single large warhead. The curves in Figure 9.2 indicate that a ten-kilogramme sub-munition will destroy a target that can only withstand five to ten pounds per square inch blast (35 to 69 kPa) if it lands within eight to 12 metres of it. Since we are only interested in an estimate of a TBM's capability, we can consider each truck or armoured vehicle to be a target of area equal to $A \approx \pi R_L^2$, where R_L is the lethal radius associated with an overpressure of five or ten pounds per square inch.

Figure 9.9 is a plot of the probability that a TBM sub-munition will hit an 18 by 18 metre bunker. Since the area associated with that bunker is nearly the same as the 'effective' area swept by a sub-munition that lands near a soft command vehicle, the calculated curve for the 18 by 18-metre bunker is suitable for our rough estimates. As can be seen from the curve in Figure 9.9, if the location of the command, control and intelligence element is perfectly known, and a sub-munition-armed TBM with CEP of 30 metres is available, 90 per cent of the vehicles collected in that unit could be destroyed in a TBM attack. If, however, only the location of the 300 by 300-metre deployment area is perfectly known, then the command sub-unit location might be uncertain to 150 metres. In this case, the effective CEP for a missile with 30-metre CEP is slightly more than 150 metres, and a single TBM attack might destroy 40 per cent of the vehicles in the command unit. Hence it is clear that the unit must keep its location uncertain to many hundreds of metres if it is to be relatively safe from TBM attack.

It is perhaps worth reiterating that, if the location of these important soft command units is revealed to the degree of accuracy assumed above, many threats other than TBMs would pose problems for the unit and its personnel. For this reason, the use of multiple command posts with great mobility is required if continuity of command at the divisional level is to be preserved. While TBMs could clearly pose a threat to these elements if their locations are revealed, it should be clear that if measures already taken to protect such units are successful, these same measures will also be more than adequate for countering TBMs armed with sub-munitions. However, it has long been recognised that the chaos that can follow the destruction of an enemy's key command elements can greatly enhance the likelihood of victory. Since the range, fire-power and surveillance capabilities in modern military forces are so extensive, equally extensive measures have evolved to reduce the likelihood and consequences of loss of command elements.

CONCLUSIONS

In the analysis that has just been presented a wide variety of military/ technical issues related to NATO's potential vulnerabilities to accurate conventionally armed Soviet TBMs have been analysed. This analysis clearly demonstrates that many NATO assets can maintain a high degree of invulnerability to a Warsaw Pact pre-emptive attack if mobility can be maintained in a crisis. The analysis also indicates that careful and determined efforts must be made to operate air-defence radars in such a way as to make their location prior to an enemy pre-emptive attack so difficult and time-consuming as to render it impossible. Dispersal of airfields, unfortunately, is more difficult.

Our analysis of attacks on airfields assumes that there are no critical nodes that can shut an airfield down if a small attack succeeds in destroying them. In our judgement, this assumption is reasonable. But military experience is replete with examples of planning errors, neglect, or simply failures to recognise which problems create the greatest vulnerabilities and, therefore, those which require the most immediate attention. Hence, it should be recognised that, while our estimates of Soviet TBM force size and capabilities indicate what *could* be required to shut down airbases, it does not necessarily indicate what *is* required to shut down airbases. Of still greater importance are questions raised by our analysis about the future viability of air-power that depends on runways in the European theatre. While TBMs are unlikely to emerge as a major threat in the near term, it appears that a wide variety of stand-off and Cruise missiles, in combination with 'smart' sub-munitions, could well play havoc with air operations in a future large-scale NATO/Warsaw Pact conflict. If NATO's air-power is to offset Warsaw Pact numerical advantages, as some would argue, the question of a continuing dependence on airfield runways must be addressed. Even if it can be demonstrated that future military technology will reduce NATO's dependence on air-power, it seems that the issues raised by the dependence of air-power on airfield runways must still be resolved.

We stated our major conclusions regarding the conventional TBM threat at the outset of this chapter, as a way of introducing our analysis. We reiterate them here. First, though there is no publicly available evidence that the Soviets have as yet achieved spectacular improvements in TBM guidance, even if we assume that by the 1990s they will achieve 30 metre CEPs, TBMs will still have little or no capability to attack mobile targets such as air-defence radars and divisional command posts, if mobility, deception and camouflage are properly exploited for protection. The greatest difficulty lies in finding these targets, not destroying them. Secondly, given their small pay-loads, TBMs cannot carry enough conventional explosives to damage properly hardened underground structures, hence, they should have little or no ability to damage or disrupt operations at fixed, hardened NATO command posts. Thirdly, TBMs with sub-munitions will not be capable of doing significant damage to runways at NATO airbases unless they achieve accuracies of the order of 30 to 40 metres. Even with such accuracies, many hundreds, and perhaps several thousand TBMs might be required for the Soviets to achieve a major impact on NATO tactical air operations during the early phases of a surprise attack. In our judgement, both Cruise missiles and aircraft, not TBMs, will continue to be the most efficient means of delivering conventional munitions or chemical/biological weapons in a surprise attack against NATO targets. Perhaps at the risk of stating the obvious, we do not want to end this discussion without making one final point. It should be understood that diplomatic agreements limiting the number of TBMs can clearly have as real an impact on NATO's future military security as they can on its political security.

Notes

1.   Benoit Morel and Theodore A. Postol, *Anti-Tactical Ballistic Missile Technologies and NATO*, Manuscript in progress, to be published by the Stanford University Center for International Security and Arms Control.
2.   Ibid.

3. The CEP or 'Circle of Equal Probability' is a fictitious circle within which half the weapons fired at a target are expected to fall.

4. Manfred Worner, 'A Missile Defense for NATO Europe', *Strategic Review*, Winter 1986, pp. 13–20.

5. For a qualitative discussion of some aspects of shape charge technology, see K. J. W. Goad and D. H. J. Halsey, *Ammunition (Including Grenades and Mines)* (London, 1982) pp. 110–21; C. L. Farrar and D. W. Leeming, *Military Ballistics* (London, 1983) pp. 146–9; R. G. Lee, *Introduction to Battlefield Weapon Systems and Technology* (London, 1985).

6. C. I. Hudson and P. H. Haas, 'New Technologies: The Prospects', in Johan J. Holst and Uwe Nerlich (eds), *Beyond Nuclear Deterrence* (London, 1977) pp. 127–8.

7. These remarks also apply to TBM dispensing mechanisms associated with the use of sub-munitions, since sub-munition deployment must be accomplished at very high TBM approach speeds of two to three kilometres per second. However, in our analysis of sub-munitions-armed TBMs, we will ignore these significant engineering, reliability and technical problems posed by sub-munitions dispersal.

8. Congressional Research Service, *US–Soviet Military Balance, 1980–1985* (Washington, DC, 1985). For numerical trends on the SS-23 and SS-22 see Table 12 (Medium-Range Theater Nuclear Forces) p. 190. For trends on the SS-21 see Table 13 (Short-Range Nuclear Forces) p. 192. Estimates of the SS-21, SS-23, and SS-22 CEPs can be found at the end of Table 14 (Theater Nuclear Force Recapitulation) p. 194.

9. Congressional Research Service, *Soviet Military Power 1986*, (Washington, DC, 1986) pp. 72–9; and *Soviet Military Power 1987* (Washington, DC, 1987) pp. 76–80.

10. These rough comparisons are, of course, only indicative of gross trends. The curves in Figure 9.4 overstate the number of Frontal Aviation aircraft that might be available for attacks on NATO, as logistics and maintenance problems are not considered. (For a discussion of this point, see Joshua Epstein, *Measuring Military Power* (Princeton, New Jersey, 1984).) However, the assumption that all TBM launchers possessed by the Soviets would be available for use in an attack against NATO also overstates the TBM count.

11. Data on Cruise missile ranges and weights, and F-16 range capabilities for different speeds and penetration profiles taken from General Dynamics Briefing Material, 'Medium-Range Air-to-Surface Missile: A New Dimension in Conventional Airpower' (1983).

12. In order to get an estimate of how accurately a Cruise missile might measure local wind conditions using currently available guidance technology, consider a hypothetical Cruise missile that can fly upwind for one minute, instantaneously turn around, and fly back along its original path for one minute. If the missile is flying in a wind of 10 knots (about 5 metres per second), the centre of its path will drift 300 metres (60 seconds times 5 metres per second) during its upwind run and another 300 metres during its downwind run. Thus when it arrives at the same place in the moving frame of the air-mass, its location over

the ground below it will be displaced 600 metres from the point at which it began its run. The TERCOM update should be able to achieve tens of metres or better resolution. The missile is also equipped with an inertial guidance system and, if need be, it can also be equipped with sensors to measure its air speed. Such a system *would* therefore use 'current Cruise missile technology' and *would* be able to measure the local wind field to an accuracy of a few per cent.

13.   If the area of the target is small relative to the area of circles with radii equal to the CEP or the lethal radius, the probability of kill is well approximated by the following expression:

$$P_k = 1 - 0.5(R_L/CEP)^2$$

(CEP = average distance within which half the missiles fall and detonate; R_L = range at which effects of the detonation result in the loss of the target as a militarily useful entity.)

14. $$CEP_{eff} = (CEP^2 + R_u^2)^{1/2}$$

(R_u = the radius of uncertainty. That is, the radius of a circle within which the target resides with a 0.5 probability.)

15.   For a fairly detailed discussion of the June 1982 Israeli air-defence suppression operations in the Bekaa Valley see M. de Arcangelis, *Electronic Warfare* (London, 1985) pp. 265–72.

16.   For a point target and a TBM impact closer to the target than the dispersal radius R_d, the probability that a single sub-munition will destroy the target is equal to the ratio between the area destroyed by the sub-munition and the area in which the sub-munition might impact, that is

$$\pi R_L^2/\pi R_D^2$$

and the probability of a miss is

$$1 - R_L^2/R_D^2$$

If N sub-munitions fall uniformly within the dispersal radius, then the probability that none of the sub-munitions will hit the target is

$$(1 - R_L^2/R_D^2)^N$$

and the probability of at least one hit is

$$P_D = [1 - (1 - R_L^2/R_D^2)^N]$$

A successful hit also requires that the target is within the sub-munition impact area. The probability P_H that a TBM impacts within a distance R_d from the target is

$$P_H = [1 - 0.5(R_D/CEP)^2]$$

and the probability that at least one sub-munition will hit the target is the product of the probabilities P_D and P_H, destruction

$$P_k = P_D P_H = [1 - (1 - R_t^2/R_D^2)^N][1 - 0 \cdot 5(R_D/CEP)^2]$$

17. The probability that $k > 6$ of the $n = 12$ sub-bunkers are hit is easily calculated. If p is the probability that a sub-bunker is hit then the probability it will not be hit is $1 - p$, and the probability that k out of n sub-bunkers are hit is:

$$\binom{n}{k} = \frac{n!}{k! \, (n-k)!}$$

where

$$P(k,n) = \binom{n}{k} p^k (1 - p)^{n-k}$$

The probability that at least N out of n sub-bunkers will be hit is:

$$P_{cum}(N,n) = \sum_{i=N}^{n} P(i,n)$$

Each missile attack is a probabilistically independent event. Hence, the probability that any single sub-bunker will be destroyed after M missile attacks is:

$$p_M = 1 - (1 - p_1)^M$$

where $p_1 = p$ is the probability that a sub-bunker will be destroyed per missile attack.

18. The authors would like to thank Commander John Stevens (USN-retired) for his extensive explanations of aircraft ground operations during normal and emergency conditions.

19. The authors would like to thank Sgt Richard McNeal (USA-retired) for his extensive explanation of army division, brigade, and battalion level command operations.

Part III
Civilian Nuclear
Technologies and Nuclear
Weapons Proliferation

Part II

Relevant Dual-Use
Technologies and Nuclear
Weapons Proliferation

10 Civilian Nuclear Technologies and Nuclear Weapons Proliferation

John P. Holdren

INTRODUCTION

Links between civilian nuclear technologies and nuclear weapons proliferation should be viewed in the larger context of the interactions of three sets of issues: nuclear-energy options and choices; the potential proliferation of nuclear weapons capabilities to additional nations (and perhaps sub-national groups); and the status of major-power nuclear armaments and arms control. Each of these sets of issues influences the others and is influenced by them.

Among these mutual influences it is possible to distinguish stronger and weaker ones. Proliferation prospects, for example, are strongly influenced by the armaments/arms-control regime (which is a major determinant of proliferation *incentives*) and by nuclear-energy options and choices (which, it will be argued here, are important contributors to proliferation *opportunities*). Nuclear-energy options could be strongly influenced by an arms-control regime radical enough to include a cut-off in production of nuclear-explosive materials. Weaker, but not necessarily negligible, influences have been or could be exerted by proliferation prospects on nuclear energy choices (for example, deferral of plutonium recycle to reduce proliferation opportunities); by proliferation prospects on the arms-control regime (for example, insofar as the spread and growth of nuclear weapons capabilities in other countries may make the major nuclear-weapons states reluctant to shrink their own arsenals); and by nuclear-energy options on arms-control prospects (for example, supporting confidence in deep cuts by providing the means, through power reactors, to 'burn up' the nuclear-explosive material from dismantled weapons).

Although emphasis in this chapter is on the influence of nuclear-energy options and choices on proliferation opportunities, the other two 'strong' links also receive particular attention: the influence of the armaments/arms-control regime on proliferation prospects begins the analysis, and the possible influence of a radical arms-control regime on nuclear-energy options concludes it.

THE ROLE OF THE NUCLEAR ARMS RACE IN PROLIFERATION INCENTIVES

Introduction

It is a truism that if there were no incentives to acquire nuclear weapons, there would be no need for concern about the existence and character of technical means for acquiring them. It is equally true but perhaps less obvious that the nature of the incentives (and disincentives) can influence the choice of technical means. For both reasons, it is appropriate to begin a discussion centred on means with at least a brief inquiry into the issue of motivations.[1]

Factors Favouring Weapons Acquisition

Among the factors that could motivate a country's interest in acquiring nuclear weapons (or a capacity to produce them quickly upon perception of need) are the following:

● *Deterrence* of nuclear-armed potential adversaries from the use of their nuclear weapons either in combat or as tools for coercion; or 'extended deterrence' (after the manner of North Atlantic Treaty Organisation (NATO) policy) in the form of a threat to respond with nuclear weapons against conventional attack by potential adversaries whose conventional military power it is uneconomical or inconvenient to match.

● *Perceived military utility* of nuclear weapons, that is the perception that, beyond the deterrent value of such weapons, they may bring actual military advantage if used in combat.

● *Political leverage*, in the form of international bargaining power deriving from explicit or implicit threats to use the nuclear weapons one possesses.

● *Autonomy* in the form of reduced reliance on security guarantees provided by other countries (a particularly important factor, perhaps, for politically-isolated or so-called 'pariah' states that do not enjoy such guarantees at all).

● *Status* either in the domestic political context, or internationally, or both, associated with the perception that possession of nuclear weapons signifies membership in an exclusive and prestigious club or constitutes a symbol of modernity and technological competence.

● *Insurance* against a possible future need for any of the above 'benefits', even if no compelling need for them is evident at the moment.

The attractions associated with these incentives have, of course, been greatly magnified by the behaviour of the existing nuclear-weapons states (most importantly, the behaviour of the United States and the Soviet Union) over the past few decades. Given the tremendous resources continuing to be devoted by the two most powerful countries in the world to the refinement of their already vast nuclear arsenals – and given the emphasis and importance attached to nuclear-weapons issues in the foreign policies of most of the countries that possess them – the non-weapons states could hardly be blamed for concluding that these weapons represent an exceptionally valuable currency in the market-place of international power.

Two aspects of major-power nuclear-weapons policies must be viewed with special alarm in the context of proliferation incentives. The first is the failure to agree on a comprehensive ban on testing nuclear explosives (Comprehensive Test Ban or CTB), despite repeated affirmations of this goal by the United States and the Soviet Union over most of the past 25 years, and despite the circumstance that such a measure is regarded by most of the 125 non-weapons-state parties to the Non-Proliferation Treaty of 1968 as an essential first step in the adherence of the weapons-state parties to their side of the bargain.[2] Continuation of nuclear testing, of course, reinforces the (mis)perception that nuclear weapons have useful military functions towards which their designs need continuing refinement, and thus adds to the allure of the 'perceived military utility' incentive for other countries to acquire nuclear weapons.

Even worse, in a way, is the continued adherence of NATO to its long-standing policy of threatening 'first use' of nuclear weapons if such use seems necessary to stop a conventional attack. By insisting that the threat to use nuclear weapons is appropriate and credible as a way to deter conventional attack by a powerfully armed adversary, NATO is setting a dismal example that cannot fail to impress other countries facing heavily-armed neighbours.[3] If nuclear weapons are a suitable way for NATO to deter the Warsaw Pact, why not for Taiwan to deter China, Pakistan to deter India, Iraq to deter (or defeat) Iran, and so on?

It is clear, then, what kinds of measures offer the greatest potential for reducing incentives to proliferation: conclude a Comprehensive Test Ban; reduce the reliance and emphasis on nuclear weapons in the military strategies of the nuclear-armed powers (including, on the NATO side, moving away from its first-use posture); and begin a process of considerable reductions in nuclear arsenals to replace the process of continuing build-up of capabilities that has characterised the major-power nuclear-arms competition since its inception. It must be admitted, at the same time, that even reductions to 'minimum deterrent' levels of 1000 warheads or less on each side would not eliminate incentives for proliferation entirely, since as long as the United States, the Soviet Union, Great Britain, France, and China profess to need such deterrents, other countries may be expected to want similar 'insurance policies'. Not until we somehow reach a world in which no country feels the need to possess nuclear weapons at all will incentives for proliferation be reduced to zero.

Factors Discouraging Weapons Acquisition

To talk only of the incentives for weapons proliferation, of course, is to present only part of the picture. Important disincentives are surely also operating, since otherwise the process of proliferation would certainly have proceeded much further and faster than it actually has done so far.[4] Important disincentives to proliferation include:

● *Becoming a nuclear target*, in the sense that a country possessing nuclear weapons must reckon with the possibility of pre-emptive nuclear attack on those weapons in time of crisis (all the more so given the vulnerability of the sorts of nuclear forces likely to be possessed by many potential proliferators).

● *Disruption of security guarantees* offered by major-power protectors, who may decide to withdraw or weaken those guarantees in reaction to the acquisition of nuclear weapons by the protectee.

● *Other international reactions* to the act of proliferation, which may include opprobrium, political and economic sanctions of various kinds, and military countermeasures by potential adversaries (who may even acquire nuclear weapons themselves in response, or form alliances with others who possess them).

● *Domestic political costs* of a decision to acquire nuclear weapons (insofar as important opposition groups exist and can give voice to their objections), including, in some cases, the possibility that a government choosing to acquire nuclear weapons could be removed from power.

● *Diversion of resources* of money and technical talent from other perceived needs, both military and civilian.

● *Unattainability of 'sufficiency' in nuclear weapons* in the sense that the limitations on resources are seen to prevent the acquisition of sufficient nuclear-weapons capability (measured in number and yield of bombs and/or in delivery capability) to accomplish whatever goal is seen as the main motivation for such an effort.

One aspect of the last disincentive could be troublesome under a major-power arms-control regime involving very deep cuts in nuclear arsenals: such cuts might give the impression to potential proliferators that 'sufficiency' in some sense was now within easier reach. This concern should not be given too much weight, however, both because cuts deep enough to produce such a syndrome are not yet in sight (meaning there is time to think about how to counter it), and because the many benefits of such cuts – including their beneficial effects on other incentives and disincentives to proliferation – greatly outweigh this one negative element.

Reinforcing an important subset of these disincentives is the Non-Proliferation Treaty (NPT), which formalises the international 'norm' against acquisition of nuclear weapons, links international assistance in civilian nuclear technologies for non-weapons states with their refraining from acquiring such weapons, and is associated with a set of safeguards operated by the International Atomic Energy Agency (IAEA) intended to detect diversion of nuclear technology from civilian into military applications.[5] Strengthening this Treaty – for example, by persuading such important countries as France, China, India, Pakistan, Brazil, and Argentina to join, and by strengthening IAEA safeguards and increasing the resources available for implementing them – is one obvious way to bolster the disincentives against nuclear-weapons acquisition. Strengthening the NPT will surely be difficult, however, unless and until the nuclear-weapons powers show signs of taking seriously their obligations under Article VI of the Treaty 'to pursue negotiations in good faith on effective measures relating to cessation of the nuclear arms race at an early date and to nuclear disarmament'.

Is Proliferation Becoming More Likely?

The circumstance that proliferation has proceeded much more slowly than many analysts had feared has by now engendered a rather widespread opinion that the array of disincentives discussed here tends to dominate the array of incentives and probably will continue to do so. The prevalence of this opinion has reduced the general sense of urgency about proliferation dangers and, accordingly, it has contributed to a widespread impression that attaining a Comprehensive Test Ban is not very urgent and that the non-proliferation benefits of other possible nuclear-arms agreements do not deserve much weight in considering whether to reach such agreements.

This complacency about proliferation, in the view of the present writer, is unjustified and hence the consequent undervaluing of the anti-proliferation benefits of arms agreements is unfortunate. The predominance of disincentives over incentives, which has slowed proliferation so far, is not necessarily either universal or permanent. In the first place, the incentives for particular countries to acquire nuclear weapons will inevitably fluctuate with local conditions (regional conflicts and crises, changes in government), so that over time more and more occasions in which incentives exceed disincentives at a particular time and place can be expected to arise. In addition, there is reason to think that the predominance of the disincentives is being eroded systematically and generally in at least two ways: first, as already indicated, some of the most powerful incentives are being enhanced by the continuing reluctance of the major weapons states to relinquish their own nuclear arsenals; second, the steady increase in general technological and military capabilities of countries around the world is spreading and simplifying the possibilities for acquiring nuclear weapons and credible means to deliver them (that is reducing the 'resource diversion' disincentive as well as, in some instances, the 'unattainability of sufficiency' disincentive); and, third, the spread of civilian nuclear technologies has been spreading opportunities to acquire or approach a nuclear-weapons capability while preserving sufficient ambiguity about this condition to avoid the paying of the political costs that constitute some of the most important disincentives.

It is this last question – the links between civilian nuclear technologies and opportunities for acquiring a nuclear-weapons capability – that will now be addressed in more detail.

THE ROLE OF CIVILIAN NUCLEAR TECHNOLOGIES IN PROLIFERATION OPPORTUNITIES

Introduction

As a starting-point for this discussion, Table 10.1 provides a taxonomy of technical issues relating to the nature of a nuclear-weapons capability and how it may be acquired. With respect to the two main types of ingredients of a weapons capability, knowledge and materials, proliferation concerns have tended to focus mainly on the latter. This has been so because the general kinds of competences relating to fission weaponry (such as basic knowledge of nuclear

Table 10.1   Nuclear-weapons capabilities and their acquisition: a taxonomy of the technical issues

WHAT are the ingredients of a nuclear-weapons capability?
— Knowledge
    general competence in technical disciplines
    insights and capabilities relevant to particular weapons types
— Materials: nuclear-explosive materials or their precursors

related to fission weapons ('atomic bombs') or thermonuclear weapons ('hydrogen bombs')
WHO might acquire these ingredients?
— Nations
— Sub-national groups

acting as insiders (with authorised access to the needed facilities) or outsiders (gaining unauthorised access to the facilities)
WHENCE might the ingredients come?
— Programmes and facilities dedicated to weapons acquisition
— Pre-commercial nuclear-energy programmes and facilities
— Commercial nuclear-energy programmes and facilities

where the relevant energy programmes and facilities may relate to fission, fusion, or fusion–fission hybrid technologies
HOW might the ingredients be acquired?
— From dedicated facilities: through normal operation, theft or forcible take-over
— From nuclear-energy facilities: through diversion (by insiders) or theft by outsiders) of ingredients normally present, or by conversion (open or clandestine) to produce ingredients not normally present

physics, chemical engineering, conventional explosives, and electronics) are of such widespread usefulness and availability that it would be futile and counterproductive to try to contain them; and because the key *specific* insights about the design and construction of fission weapons, given the nuclear-explosive materials, are by now readily obtainable by most nations and even some sub-national groups. Thus, lack of access to suitable fissile materials – which for a long time only major industrial nations had the wherewithal to obtain – has been considered the main technical barrier to the acquisition of fission weapons. The case of thermonuclear weapons is somewhat different, in that the specific insights and techniques required are much more difficult and much less widely available, making knowledge more important in relation to materials than is the case with fission weapons; the implications of this difference for proliferation concerns are taken up below.

With respect to the 'who' of proliferation, the main focus of concern has been on weapons acquisition by additional nations, and that will be the main focus here. Nonetheless it cannot be ignored that possession of nuclear weapons by certain sub-national groups could be seen by them as being convertible into large amounts of money and influence, as well as capacity for terrorism or revenge against other segments of society. The problem of proliferation to sub-national groups is surely less important today than that of proliferation among nations, due both to the limited technological capabilities of most such groups and to their preoccupation with other means of achieving their aims. It is possible, however, that an increase over the next few decades in the wisdom and restraint shown by nations with respect to nuclear weapons, sharply reducing the prospects and dangers of proliferation among states, could be accompanied by an increase in the nuclear-weapons interests and capabilities of sub-national groups, propelling the sub-national proliferation problem into prominence over the international one. Accordingly, some attention is given in what follows to the way technological developments may affect future proliferation possibilities for groups other than states.

With respect to the last two categories of issues in Table 10.1 – the potential sources of the ingredients of a weapons capability and the means by which these sources might be exploited – the most contentious (and important) questions are the relevance of nuclear-energy programmes and facilities as opposed to those openly dedicated to nuclear weaponry; and the relative vulnerability of different types of nuclear-energy facilities compared to one another. For purposes of addressing these questions, Table 10.2 summarises the main routes to acquisition of nuclear-explosive materials for fission weapons from both weapons-dedicated and nuclear-energy facilities, while Table 10.3 provides some quantitative information about fissile materials that will be useful in comparing these possibilities.

Comparisons of the pathways indicated in Table 10.2 often begin with the observation that use of dedicated facilities to acquire a weapons capability is faster, cheaper and technically less demanding than using nuclear-energy facilities for this purpose. This observation is correct in a narrow sense, but does not tell the whole story. Let us consider first the respects in which it is true.

Characteristics of Dedicated Weapons Facilities

The simplest of the 'dedicated' routes is surely the first one listed in Table 10.2, a production reactor plus a reprocessing plant. The difficulty and expense of uranium-enrichment technology is the main reason for preferring the production-reactor/reprocessing route over enrichment of uranium to weapons grade, so it makes sense to choose a reactor type that avoids the need for uranium enrichment altogether, that is one that can be fuelled with natural uranium. Of the

Table 10.2   Main routes to nuclear-explosive materials for fission weapons

---

(A)   Using 'dedicated' facilities
    (1) Generating plutonium in a production reactor and extracting it from the irradiated fuel in a simple reprocessing plant. The most convenient reactor types for this purpose are:
       (a) graphite-moderated, natural-uranium reactor
       (b) heavy-water-moderated, natural-uranium reactor
    (2) Enrichment of uranium to weapons grade using
       (a) centrifuges
       (b) nozzle technology
       (c) gaseous diffusion
       (d) laser-enrichment processes
(B)   Using 'civilian' (commercial or pre-commercial) nuclear-energy facilities
    (1) 'Front end' of the fuel cycle: enriching uranium to weapons grade
    (2) 'Back end' of the fuel cycle: reprocessing spent reactor fuel to obtain plutonium or uranium-233
(C)   Purchase or theft of material originating in (A) or (B).

---

two options that meet this condition – the graphite-moderated and the heavy-water-moderated reactors – the first is probably preferable because the needed high-purity graphite is more easily and cheaply obtained than heavy water. (Producing heavy water involves another form of isotopic enrichment technology, albeit an easier one than that for uranium.)

It was estimated in the mid-1970s by the Office of Technology Assessment (OTA) of the US Congress that a graphite-moderated reactor of about 30 megawatts thermal output – a size sufficient to produce about 9 kilogrammes of weapons-grade plutonium per year – could be built in about three years by a modest crew of engineers and technicians for 15 to 30 million dollars. The same report placed the cost of a reprocessing plant adequate to extract this plutonium from the reactor fuel at under $25 million. Detailed plans for both reactor and reprocessing plant exist in the open literature. Scaled-up versions of these plants, capable of producing 100 kg of weapons-grade plutonium per year, could be built in five to seven years by a crew of 50–75 engineers and 150–200 technicians at a cost in the range of $175 to $300 million, according to the OTA study.[6]

The alternative to a production-reactor/reprocessing-plant combination for dedicated production of nuclear explosive materials is a facility that can enrich natural uranium (0·7 per cent U-235) to weapons-grade (above 90 per cent U-235 is preferred, but somewhat less will suffice, as indicated in Table 10.3). Of the variety of uranium-enrichment technologies that exist or are under development, centrifuges would appear to be the most suitable for most potential proliferators: gaseous diffusion plants are large, expensive, technologically sophisticated, and highly consumptive of energy; the Becker nozzle process requires a large number of stages and is even more energy-consumptive than gaseous diffusion plants; and the laser processes will remain technically out of reach for any but the most advanced nations for some time. The mid-1970s OTA study estimated that a centrifuge enrichment plant capable of producing 30 kg of weapons-grade uranium per year would contain in the range of 1000 centrifuges and would cost $30 to 50 million if built from scratch (in contrast to adding stages to an enrichment plant constructed for producing reactor fuel at 2 to 3 per cent U-235, as discussed below). A centrifuge plant capable of producing 300 kg of weapons-grade uranium per year would cost $120 million to $240 million, the OTA estimated.[7]

Table 10.3   Some quantitative aspects of nuclear-explosive materials

Description	Fissile isotopes	Dilutants	Fissile fraction in mixture (%)	Critical mass for mixture[1]
weapons-grade uranium	U-235	U-238	94	15 kg
high-enriched uranium	U-235	U-238	60	37 kg
minimum-enrichment U for weapons	U-235	U-238	20	250 kg
weapons-grade plutonium	Pu-239, Pu-241	Pu-240, Pu-242	94	5 kg
reactor-grade plutonium (LMFBR)	,,	,,	80	5.5 kg
reactor-grade plutonium (LWR, once-through)	,,	,,	70	6.7 kg
reactor-grade plutonium (LWR, multiple Pu recycle)	,,	,,	50	9.6 kg
weapons-grade uranium-233	U-233	U-238, Th-232	94	5 kg
minimum enrichment U-233 for weapons	U-233	U-238, Th-232	12	100+ kg

[1]Critical masses vary with weapon design. Figures shown are for spherical masses at normal density with natural-uranium reflectors.
Sources: American Physical Society, 'Report to the APS by the Study Group on Nuclear Fuel Cycles and Waste Management', *Reviews of Modern Physics*, vol. 50, no. 1, part II, January 1978; Office of Technology Assessment, Congress of the United States, *Nuclear Proliferation and Safeguards* (Washington, DC, 1977); and T. B. Cochran, W. M. Arkin, and M. M. Hoenig, *Nuclear Weapons Data Book, Vol. I: U.S. Nuclear Forces and Capabilities* (Cambridge, Massachusetts, 1984).

Shortcomings of Energy Facilities for Weapons Acquisition

By comparison, constructing nuclear-energy facilities for electricity generation at commercial scale is much costlier and more demanding of time and technological skills. The smallest reactors that have been considered economical to build for electricity generation in the last decade are in the range of 400 electrical megawatts (MWe) capacity, corresponding to 1200 or more thermal megawatts. Even at mid-1970s prices (for comparison with the OTA estimates cited above) such reactors cost $500 million and more; and a modern, 1000-MWe power reactor of the light-water moderated type most commonly marketed today costs typically $1.5 to $2.0 billion in mid-1980s money. The time required from inception of a power-reactor project to first production of electricity has typically been ten years, although in some circumstances the task has been accomplished in six.

A reactor alone is not enough to provide access to nuclear-explosive materials, moreover. For a commercial nuclear-electricity-generation operation to provide such access, it must include either a uranium-enrichment plant or a fuel-reprocessing plant. The smallest and cheapest uranium enrichment plant that might be contemplated would be a centrifuge plant capable of servicing one 400-MWe commercial reactor; the OTA study cites a mid-1970s Japanese estimate of $165 million for such a plant. Much larger plants (say, capable of serving ten 1000 MWe reactors) could be a few times cheaper per unit of output, but the total investment would be larger – probably $600 million or more.[8] Costs of commercial-scale fuel-reprocessing plants are highly uncertain, but would appear to be not less than $50 million per reactor served in the case of a large plant (handling the output of 50 1000-MWe reactors) and much more per reactor served by a small plant.[9]

All of these technologies – reactors, enrichment plants, reprocessing plants – must be built and operated to a higher technological standard if they are to be used for economical electricity production than would be necessary for facilities intended solely for producing weapons materials. In addition, the nuclear-explosive materials that might be derived from electricity-production facilities would be, in at least some cases, more difficult to use in weapons than materials derived from dedicated facilities. In particular, the plutonium ordinarily produced in the most common types of power reactors – Pressurised water Reactors (PWRs) and Boiling-water Reactors (BWRs) – contains an inconveniently high proportion of the even-numbered isotopes, Pu-240 and Pu-242. These isotopes are not dilutants in quite the same sense as U-238 dilutes U-235, but their presence does make the design and manufacture of nuclear explosives using plutonium more difficult.[10] Most importantly, excessive Pu-240 and Pu-242 in a plutonium weapon may cause premature initiation of the chain reaction when the weapon is detonated (because of the high spontaneous production of neutrons in these isotopes), reducing the explosive yield. As a result, weapons made from 'reactor-grade' plutonium (see Table 10.3) tend to have both a lower expected yield and a larger uncertainty about yield than do weapons made with plutonium having a higher Pu-239 content. Although these shortcomings can be largely overcome with sufficient sophistication in weapons design, use of 'weapons-grade' plutonium provides more flexibility to the advanced designer as well as a higher and more reliable yield to the less advanced one.[11] 'Reactor-grade' plutonium is also radiologically more dangerous and generates more heat than the 'weapons-grade' material does, and these characteristics further complicate the task of those who would divert reactor plutonium to weapons use.

Attractions of Nuclear-energy Facilities for Weapons Acquisition

Notwithstanding the problems just described, there is a number of reasons for potential proliferators choosing to exploit nuclear-energy facilities for weapons purposes rather than relying strictly on weapons-dedicated facilities. In the first place, the technical and economic barriers to the use of civilian facilities are not quite as formidable as superficial analysis suggests. As already noted, reactor-grade plutonium is not ideal for weapons, but it is usable; and for some potential proliferators, reductions in expected yield, reliability of yield, and efficiency compared to the standards of the major nuclear-weapons powers may be quite acceptable. For proliferators who insist on higher-quality plutonium, moreover, it is available from power reactors under certain circumstances: most importantly, power reactors that can be refuelled without being shut down, such as the Canadian heavy-water-moderated CANDU, can be operated to produce weapons-grade plutonium without significant loss of electrical output; and even PWRs and BWRs can produce high-quality plutonium for weapons if the owners of the reactor are willing to sacrifice some electricity.[12] If, instead of reprocessing spent reactor fuel, the chosen route to weapons is use of a 'civilian' enrichment plant to enrich uranium to weapons grade, there is no handicap at all in the quality of the resulting weapons material.

On the economic side, direct comparison of the cost of dedicated weapon-production facilities with the cost of a civilian nuclear-energy system is not completely appropriate. For one thing, the nuclear-energy system can repay much of its cost through the sale of electricity. In addition, a commercial-scale nuclear-energy system can produce far larger quantities of bomb-usable material than the modest-sized dedicated production facilities whose costs were mentioned above: a single 1000-MWe PWR produces in a year between 200 and 250 kilogrammes of reactor-grade plutonium, enough for at least 30 to 40 bombs.[13] This is a torrent of material, compared to the dribble of one to two bombs' worth per year from a small production reactor.

The most informative economic figure for the cost of bomb material from civilian nuclear-energy facilities is the marginal cost of producing this material, not the total cost of building and operating facilities that make electricity as well. In the case of using a civilian uranium-enrichment plant to produce weapons-grade uranium-235, the marginal cost is approximately the value of the 'separative work units' diverted from the production of reactor fuel: at current and near-future enrichment costs, this figure would be in the range of $200 000 to $400 000 per bomb.[14] If plutonium is to be diverted from a reprocessing plant, the marginal cost for a bomb's worth of reactor-grade plutonium (7 to 10 kg, from Table 10.3) is approximately the value of this material as reactor fuel, which again turns out to be a few hundred thousand dollars.[15] If weapons-grade plutonium is desired and a commercial light-water reactor is operated in the manner required to minimise Pu-240 production, one can obtain perhaps 30 bombs' worth in a year with a loss of about a third of the reactor's normal electrical output; this translates under current circumstances to about $2 million per bomb.[16]

To compare these costs with those for dedicated facilities, one can use the OTA figures cited above, with an inflation factor of 1·9 to convert mid-1970s dollars to mid-1980s dollars.[17] Then applying a fixed-charge rate of 10 per cent per year (appropriate to government activities) to convert capital investment costs into annualised costs, and dividing by the annual numbers of bombs producible with the output of these facilities, gives a range of $1.1 million to $5.2 million per bomb. These calculations are very approximate, of course, and they neglect some components of the total costs (such as labour to operate the facilities and, in some cases, costs of the uranium ore to feed the process). Rectification of these omissions would not, however, change the overall conclusion, which is that use of civilian facilities to acquire nuclear-explosive materials is *not* inherently more expensive than using weapons-dedicated facilities for this purpose – indeed, in many cases it can be considerably cheaper – given only that the costs on the civilian side are calculated, as they should be, as the *incremental* costs of generating the bomb materials from facilities that also are producing commercial electricity.

On the technical and economic side, the possibility of 'mixed' (that is, partly civilian and partly weapons-dedicated) operations also needs to be mentioned. For example, a country with a power reactor as a source of plutonium may build a small, relatively crude reprocessing plant in order to extract some of the plutonium from the reactor fuel for use in a few bombs. Such a reprocessing plant can be much simpler and cheaper than one designed for large-scale commercial plutonium recycling for reactor fuels; and it could be concealed more easily than the dedicated production-reactor/reprocessing-plant combination that would be needed to achieve the same result in the absence of a power reactor to serve as the plutonium source. Alternatively, a country that possesses a commercial uranium-enrichment plant to fuel its power reactors could add a smaller, clandestine supplementary enrichment facility to perform the extra enrichment steps between reactor fuel and bomb material. The larger, commercial facility could be open to international inspectors, who would see that it was being operated to produce only the 3-per cent enriched fuel used in typical PWRs (and not usable for bombs). The dedicated facility, operating on a diverted fraction of the commercial plant's output, would gain benefits in size, cost, and concealability by virtue of starting with 3-per cent enriched material rather than with natural uranium. (Contrary to intuition, obtaining 90-per cent U-235 starting from 3-per cent U-235 requires 80 per cent less input mass and almost 65 per cent less separative work than producing 90-per cent U-235 from natural uranium.)[18]

Also relevant to an overall technical/economic assessment of the proliferation potential of civilian nuclear power, of course, are the scale and distribution of nuclear-power operations world-wide. Table 10.4 shows the number of nuclear power reactors operating and pending (that is under construction or on order) as of 1 January 1987, as well as their geographic distribution and the electric generating capacities involved (in electrical gigawatts, GWe). As can be seen, both the operating and pending plants are concentrated heavily in the most industrialised regions. The 268 GWe of operating nuclear capacity represents, using a rough average figure of 250 kg of 'reactor grade' plutonium per GWe per year and a conservative estimate of 7 kg of this material per bomb, an annual plutonium production of nearly 10 000 bomb quantities per year. (Under present circumstances, most of this plutonium remains intimately mixed with fission products and uranium in unreprocessed spent fuel. In this state, it represents a *potential* for conversion to weapons uses, becoming an

Table 10.4   Nuclear energy worldwide, 1 January 1987

	Number of reactors		Capacity, GWe	
	In operation	Pending	In operation	Pending
Europe	161	78	104·7	68·1
United States and Canada	117	33	96·4	38·2
Soviet Union	46	27	29·2	27·1
Japan	33	16	23·6	16·1
Other Asia	19	11	10·7	6·4
Latin America	3	7	1·6	5·4
Africa	2	—	1·8	—
Oceania	—	—	—	—
Totals	381	172	268·0	161·3

Source: *Nuclear News*, World List of Nuclear Power Plants, February 1987.

immediate threat only where and when reprocessing is undertaken.) Table 10.5 shows the share of total electricity generation accounted for by nuclear power in various regions and world-wide. Table 10.6 summarises the status of uranium-enrichment and fuel-reprocessing facilities related to nuclear power world-wide.

Table 10.5   Nuclear share of electricity generation, 1986

	Nuclear TWh[1]	Total TWh	Nuclear share (%)
Europe	610	2500	24
United States	410	2500	16
Soviet Union	170	1550	11
Japan	150	600	25
Other Asia	70	1050	7
Canada	70	500	14
Latin America	10	450	2
Africa	10	200	5
Oceania	—	150	—
Totals and average	1500	9500	16

[1] 1 TWh (terawatt-hour) = 10^{12} watt-hour = 10^9 kWh
Source: Author estimates derived from *Nuclear News*, World List of Nuclear Power Plants, February 1987, and US Department of Energy, *Commercial Nuclear Power: Prospects for the United States and the World* (Washington, DC, 1986).

Table 10.6   Energy-linked enrichment and reprocessing plants world-wide, excluding plants presumed to be solely weapons-dedicated

Uranium-enrichment plants	Capacity, 10^6 SWU/yr
United States: diffusion plants at Oak Ridge, TN, Paducah, KY, Portsmouth, OH	27.30
Soviet Union: diffusion plant(s) in Siberia	ca. 10.00
Great Britain: centrifuge plant at Capenhurst	0.60
Argentina: diffusion plant at Pilcaniyeu	0.02
Brazil: jet-nozzle plant at Resende	0.03
China: diffusion plant at Lanchou	0.08
France: diffusion plants at Tricastin, Pierrelatte	11.20
Japan: centrifuge plants at Ningyo-toge	0.15
The Netherlands: centrifuge plant at Almelo	1.00
Pakistan: centrifuge plant at Kahuta	0.005
South Africa: centrifuge/nozzle plant at Valindaba	0.30

Fuel-reprocessing plants	Capacity, tonnes/yr
United States: Barnwell, SC (mothballed)	1500
Great Britain: Windscale/Sellafield	1500
Belgium: Mol (mothballed)	100
Brazil: Resende, São Paulo	2+
France: La Hague, Marcoule	1000
Germany: Karlsruhe	40
India: Trombay, Tarapur, Kalpakkam	255
Japan: Tokai Mura	210
Pakistan: Rawalpindi	ca. 10
South Africa: Pelindaba	?

Note: 1 million separative work units (SWU) per year is enough to provide low-enriched uranium (2–3 per cent U-235) to 7–8 large light-water reactors. 100 tonnes/yr of reprocessing capacity would handle the spent fuel from 3–4 such reactors.
Sources: *Nuclear Engineering International*, December 1987, pp. 47–8; Leonard S. Spector, *Going Nuclear* (Cambridge, Massachusetts, 1987).

Probably more important than any of the details of technical and economic potentials of civilian versus weapons-dedicated facilities as sources of weapons materials, however, is a key political dimension of the issue that favours the civilian route: the electricity-generation rationale provides a legitimating 'cover' for the acquisition and operation of technologies that otherwise would be seen as unambiguously weapons-oriented. This legitimacy – or at worst (from the potential proliferator's standpoint) ambiguity – is a rather effective shield against both the domestic political costs and the international sanctions that could result from an unambiguous commitment to acquiring nuclear weapons. In fact, this phenomenon of legitimacy/ ambiguity has an even stronger implication: not only may some countries choose from the outset to pursue a nuclear-weapons capability along the politically safer path offered by civilian nuclear-energy programmes, but countries initially ambivalent about acquiring weapons – and even countries with no initial intention to do so at all – may find the temptation irresistible once acquisition of nuclear-energy facilities has both lowered the technical barriers and provided at least a partial political cover.[19]

Of course, the legitimating 'shield' of a civilian nuclear-energy programme will be effective only as long as the weapons dimension is not revealed by an observable nuclear test, by indisputable detection (for example, through international safeguards) of diversion of nuclear-explosive materials from the civilian facilities, or by a voluntary announcement that the weapons capability exists. This limitation is not much consolation, however. A shield against political repercussions is probably most important *before* the weapons capability is in full flower, because it is in that early time-frame that the potential proliferator is likely to feel least confident and most vulnerable, and because it is then that dissent or sanctions might actually be effective. Furthermore it may suit the purposes of some proliferators never to test or to announce a weapons capability at all; in such cases, and failing detection by other means, there will be no confirmation of the existence of the weapons unless and until they are used in hostilities. (It is now quite clear that a sophisticated proliferator can construct an arsenal of reasonably reliable fission weapons without nuclear-explosive testing.)[20]

What History Does and Does Not Tell Us

Proponents of the view that the connection between civilian nuclear energy and weapons proliferation is tenuous often assert that history makes their case: after all, none of the six countries known to have tested a nuclear weapon has used commercial nuclear-energy facilities as the source of the nuclear-explosive material.[21] This contention is misleading in two important respects. First, it oversimplifies to the point of distortion the nature of the past linkages between commercial energy programmes and weapons capabilities. Secondly, it ignores or undervalues the trends tending to increase, over time, the relative importance of the energy-linked pathways to nuclear weapons.

It is true that the first three nuclear-weapons powers – the United States, the Soviet Union, and Great Britain – developed their nuclear weapons through programmes wholly focused on that military goal, while their interest in the electricity-supply application of nuclear energy only came later. Already in the case of France, the fourth country to acquire nuclear weapons, the situation was less clear-cut, however. In the early part of the French nuclear programme, sufficient ambiguity was maintained about the goal of the work for many of the participants themselves not to know whether they were working towards electricity or bombs.[22] It seems fair, then, to consider the French case the first instance where the *prospects* of nuclear power provided a partial cover for an embryonic weapons programme. The fifth nuclear-weapons power, China, may also have benefited from the ambiguity made possible by prospective civilian applications of nuclear energy: differing accounts exist about whether the assistance in nuclear technology provided to China by the Soviet Union in the 1950s was aimed at development of civilian nuclear-energy systems, was consciously dual-purpose, or was all along understood to be a weapons programme until the Soviets phased out this assistance in 1959–60.[23]

The sixth confirmed addition to the nuclear-weapons 'club' came in 1974 with India's announcement that it had exploded a 'peaceful' nuclear device. (There is no practical difference between such a device and a bomb.) The plutonium for this nuclear explosive came from a research reactor provided by Canada, which used heavy water supplied in part by the United States. That the source of the plutonium was not an actual *power* reactor should give scant consolation to those who would like to insist on the irrelevance of nuclear energy to nuclear weapons, for it is hard to believe that Canada and the United States would have provided a research reactor, heavy water, and other nuclear assistance to India for any other reason than to facilitate the development of nuclear energy as a source of electricity.

Two additional countries widely believed to possess nuclear weapons or the capacity to assemble them very quickly (although no nuclear-explosive test by either has been confirmed) are Israel and South Africa. The capabilities of South Africa in nuclear technology were built up with substantial assistance from the United States in terms of training programmes and technology transfer (a research reactor complete with fuel, high-enriched uranium fuel and heavy water for a second such reactor) extending from the late 1950s to the mid-1970s, from France in the form of the sale of two large power reactors in the mid-1970s, and from West Germany in the form of assistance from a German firm in the construction of the South African enrichment plant (not subject to IAEA safeguards) at Valindaba. These forms of assistance presumably were motivated not only by commercial interest but by the belief that South Africa's nuclear capabilities were intended for civilian rather than military applications. The case of Israel, whose nuclear programme benefited very substantially in its early years from assistance from the United States and from France, is much less clear-cut in terms of whether the assistance was tied to expectations about applications of the technology for energy: Israel has never announced any specific plans for acquiring a nuclear electricity-generating capacity; and the French provision of the key research/plutonium-production reactor at Dimona took place under a secret agreement that did not require IAEA safeguards, and the agreement also transferred plutonium-reprocessing technology whose weapons-oriented purpose must have been clear to the French.[24]

When one turns, finally, to the further cases of latent or impending proliferation that seem most relevant in the late 1980s – namely, Pakistan, Brazil, and Argentina – the legitimating and/or masking role of nuclear-energy applications is, in contrast to the Israeli case, much more pronounced. All three countries have conducted vigorous programmes of nuclear-technology development with the announced aim of civilian electricity generation; all operate commercial nuclear power plants supplied, variously, by West Germany, Canada and the United States; all have pursued uranium-enrichment and fuel-reprocessing capabilities with assistance, in various forms, from France, West Germany, Italy and the United States; and all have constructed – or are constructing – enrichment and/or reprocessing facilities that employ insights or components acquired from industrially more advanced nations without, however, being subject to the IAEA safeguards that apply when such technologies are supplied under formal agreements.[25] In all three cases, despite substantial reasons to believe that weapons capabilities as well as electricity-generating capacity are at issue, the electricity-generation aspect has provided sufficient ambiguity and legitimacy to shield the countries involved from the full weight of criticisms and sanctions that open and unadorned pursuit of nuclear weapons presumably would have engendered.

What the history actually reveals, then, is a more or less steady trend towards increased relevance, in the proliferation process, of the intertwining and blurring of civilian and military goals of national nuclear-energy programmes – starting with the unambiguously military nuclear-weapons programmes of the United States, the Soviet Union and Great Britain in the late 1940s and early 1950s, continuing with the initially ambiguous but later clearly weapons-oriented programmes of China, France and Israel in the late 1950s and 1960s, and ending up with the continuing and deliberately ambiguous programmes of India, Pakistan, Argentina and Brazil in the 1970s and 1980s. There is every reason to expect the recent salience of the energy-weapons ambiguity in the proliferation context to persist: the technological, economic and political disincentives to proliferation are all lowered in one way or another, in today's world, by possession of a nuclear-energy capability; and the most dramatic of these barrier-lowering phenomena relate precisely to reducing the certainty and severity of the *political* costs of a weapons decision, which costs are to an ever-increasing extent the most important of the three types of disincentives.

Comparative Vulnerability of Various Fission Energy Systems

If, as just argued, the potential use of nuclear-energy facilities to contribute to acquisition of nuclear-weapons capabilities is and will remain a real possibility, then it becomes important, in the context of energy-technology assessment as well as of anti-proliferation policy, to understand the relative susceptibility of different nuclear-energy technologies to being misused in this way. This relative susceptibility depends on four types of factors, as follows:

● The *quality* of the nuclear-explosive isotopes present, meaning the relative ease of design, construction, maintenance, and use of nuclear weapons utilising these materials.
● The *quantity* of nuclear-explosive materials produced or circulated for a given quantity of electricity generation (measured as, say, number of nominal nuclear weapons producible from the material associated with a gigawatt-year of electricity generation).
● The *barriers* – most importantly chemical and radiological – that must be overcome in transferring the material from the forms and places in which it occurs in the nuclear fuel cycle into a condition suitable for use in a nuclear weapon (an additional 'isotopic' barrier is considered under 'quality'; 'physical' barriers such as fuel-cladding,

buildings, or burial would have some relevance against sub-national diversion threats but do not have much relevance against national ones).

• The likely *detectability* (by, for example, international safeguards measures) of attempts to divert nuclear-explosive materials from the nuclear fuel cycle into weapons.

It is convenient to first judge these factors in relation to the way nuclear fuel cycles would be operated in the absence of proliferation concerns – that is, without the imposition of additional physical or institutional barriers motivated by the desire to reduce proliferation risks. The idea, in other words, is first to evaluate the *intrinsic* vulnerability of the different nuclear technologies. The potential of additional, imposed barriers to modify these intrinsic vulnerabilities is taken up subsequently.

Table 10.7 summarises the main points of vulnerability of a variety of fission fuel cycles and provides a tentative and necessarily some-what judgemental ranking of these fuel cycles relative to one another on a scale of 1 to 5 (from least severe to most severe) with respect to the different factors they contribute to susceptibility to national diversion of fissile material for weapons purposes. The following sub-paragraphs briefly describe the considerations that were used in arriving at the rankings.

• *Quality*. The two categories under this heading relate to before and after possible isotopic enrichment beyond the state in which the material occurs ordinarily in the fuel cycle. The rankings are: $5 =$ uranium with U-235 > 90 per cent; $4 =$ uranium with 60 per cent < U-235 < 90 per cent or U-233 > 40 per cent, or plutonium with over 75 per cent fissile isotopes; $3 =$ plutonium with less than 75 per cent fissile isotopes; 2 would be reserved for uranium with 20 per cent < U-235 < 60 per cent or 12 per cent < U-233 < 40 per cent; and 1 would relate to material, such as tritium or uranium with lower fissile concentrations than those already listed, which can play useful supporting roles in nuclear weapons but cannot by itself initiate a nuclear explosion.

• *Quantity*. Here the number of critical masses per 1-GWe reactor per year is the key to the rankings: $5 = > 100$ critical masses (that is less than 1 per cent/yr diversion yields a 'bomb quantity' of material); $4 = 30$ to 100 critical masses; $3 = 10$ to 30; $2 = 3$ to 10; $1 = < 3$.

• *Chemical barriers*. $5 =$ fissile material in metallic form and not mixed with effective dilutant; 4 = fissile material in oxide form and not mixed with effective dilutant; 3 = plutonium mixed with significant non-fissile uranium; 2 = plutonium mixed with fission products and non-fissile uranium; 1 = plutonium or uranium-233 mixed with fission products and thorium.

● *Radiological barriers.* 5 = radiation levels associated with high-enriched U-235, or lower; 4 = those associated with various plutonium mixtures; 3 = those associated with uranium-233 and associated iso-topes; 2 = those associated with low-burn-up reactor fuel; 1 = those associated with high-burn-up reactor fuel.

● *Detectability.* While a more refined indexing scheme could cer-tainly be developed, only two factors have been considered here: first (from easiest to hardest to detect), whether diversion requires qualita-tively new operations (for example, reprocessing from an otherwise once-through fuel cycle), significant modification of existing opera-tions (for example, use of an enrichment facility to attain a much higher U-235 percentage than for reactor fuel), or simply the redirec-tion of existing process streams (as in plutonium diversion from a fuel cycle that is already recycling it); and, second, whether the radiologi-cal signature of the material is unusually helpful for monitoring.

Table 10.7 is intended to be illustrative rather than definitive; a more comprehensive approach would look at additional points of vulnerability for each fuel cycle (in Table 10.7, only the points of greatest vulnerability are listed) and at additional fuel cycle options. How to combine and weight such factors to reach an overall judge-ment about relative proliferation susceptibility, moreover, is likely to be even more controversial than the individual rankings, and no particular scheme is offered here. Still, it is clear enough from a glance at the table that, as one would expect from other reviews of this topic,[26] the once-through (PWR and CANDU) and denatured High-temperature Gas-cooled Reactor (HTGR) fuel cycles are the least susceptible, while those fuel cycles that use highly-enriched uranium (conventional HTGR) or that recycle plutonium are the most suscept-ible.

Engineered and Institutionalised Safeguards

A tremendous amount of attention has been devoted over the years to the attempt to devise ways to reduce the susceptibility of fission fuel

Table 10.7 Relative proliferation-susceptibility of fission fuel cycles

Fuel cycle and point of vulnerability	Quantity of fissile material & main diluant(s) at this point (per 1-GWe reactor/yr)	Further processing required from this point for use in nuclear explosives	Indices of relative susceptibility (5 = worst, 1 = best)					
			Quality As is	*Enrch*	*Quantity*	*Barriers* Chmcl	*Radgl*	*Detection*
PWR/ONCE-THROUGH								
enriched uranium	855 kg U235 in 28500 kg U238 (3% enrch)	extensive further isotopic enrichment	1	5	4	4	5	3
spent-fuel storage	250 kg Pu (69% fiss) in 26 000+ kg U, FP	chemical separation from U & FP	3	4	4	2	1–2	1
PWR/PU-RECYCLE								
reprocessed Pu	440 kg Pu (61% fiss), possibly mixed with U	chemical separation from U (if present)	3	4	4	3–4	4	4
CANDU/ONCE-THROUGH								
on-line removal of fuel rods	345 kg fissile Pu in 128 000 kg U, FP	chemical separation from U & FP	4	4	4	2	2	2

CANDU/PU-RECYCLE

reprocessed Pu	188 kg fissile Pu, possibly mixed with U	chemical separation from U (if present)	4	4	4	3–4	2	4

HTGR/U233 RECYCLE

enriched uranium	325 kg U235 with 25 kg U 238 (93.5% enrch)	minor chemical processing at most	5	5	3	4	5	5
reprocessed U233	190 kg U233 + 50 kg U235	minor chemical processing at most	4	4	3	4	3	4

HTGR/DENATURED

fabricated fuel	650 kg U233, 235 in 5500 kg U + 10 000 kg Th	chemical separation from Th, further enrichment	1	4	4	3	3	1

LMFBR/NATURAL U FEED

reprocessed Pu	2350 kg Pu (80% fiss), possibly mixed with U	chemical separation from U (if present)	4	4	5	3–4	4	4

Abbreviations: chmcl = chemical, enrch = enriched, fiss = fissile, FP = fission products, GWe = electrical gigawatt, radgl = radiological

Sources: Report of the American Physical Society Study Group on Nuclear Fuel Cycles, *Reviews of Modern Physics*, vol. 50, no. 1, part II, January, 1978; Office of Technology Assessment, *Nuclear Proliferation and Safeguards* (Washington, DC, 1977)

cycles to misuse for weapons purposes.[27] The most enticing idea – which, however, has never borne fruit – is to find a 'denaturant' for plutonium-239 that would accompany it throughout the fuel cycle and would make it unusable for nuclear weapons. The trouble is that no isotopes of plutonium are really effective in this respect, and additives consisting of other elements can be rather readily removed by potential proliferators using chemical means. Uranium-235 and -233 can be 'denatured' by mixing them with high proportions of uranium-233, which is removable only by relatively difficult isotopic enrichment technologies, but the presence of U-238 subject to fission neutrons assured the production of some plutonium; this is the shortcoming of the 'denatured' versions of the HTGR fuel cycle, although here the problem at least can be minimised by not recycling the plutonium but leaving it mixed with fission products.

A related idea is 'spiking' the fissile materials in nuclear fuel cycles with powerful emitters of penetrating gamma rays (as happens automatically in the form of the fission products that accompany plutonium in fuel that is not reprocessed), so that the material can be handled by potential bomb-makers only with great technical difficulty and/or at great risk to their lives. The problem here is that the extra barrier to misuse of the material, while fairly great from the viewpoint of sub-national groups of bomb-makers, is not much of an obstacle to a nation with access to facilities for removing the gamma emitters, and in any case comes at a high cost in money and occupational radiation exposures in the commercial nuclear fuel cycle.

A third idea is the colocation of 'sensitive' parts of the fuel cycle – those involving high-enriched uranium or plutonium unaccompanied by fission products – in well-monitored and well-guarded international fuel cycle centres, while only low-enriched uranium (including U-233) and highly radioactive spent fuel travels to and from 'national' satellite reactors outside these centres. This idea has some merit but also substantial shortcomings: the degree of 'internationalisation' involved (ratio of size and cost of international facilities compared to the national satellites) tends to be high, and the political willingness of nations to place such important components of their electricity-supply system under international control seems to be low.[28]

With no panacea in sight, the world continues to settle for the useful but none the less limited set of safeguards institutionalised under the rubric of the International Atomic Energy Agency in connection with the Non-Proliferation Treaty. These involve monitoring of materials/flow records and the use of automatic cameras and other instruments at safeguarded facilities, tamper-proof seals on stored fissile materials, and periodic site visits by IAEA inspectors. These measures are much better than nothing, but their weaknesses are widely recognised: there are too few inspectors, and their powers are too limited; at some kinds of facilities, continuous inspector presence during construction and operation would be required to detect diversion reliably, but such presence does not happen today; not all countries are parties to the Non-Proliferation Treaty, and not all nuclear-energy facilities in countries that are members are subject to the safeguards; countries can withdraw from the Treaty (and hence from the safeguards) if their governments decide the national interest requires this step; and the safeguards at best *deter* diversion through the threat of detection, but they do not and cannot physically prevent it.[29]

Vulnerability of Fusion and Hybrid Energy Systems

Two further classes of nuclear energy systems – fusion reactors and fusion–fission hybrid energy systems – are on the drawing-boards but do not yet exist in concrete form.[30] It is quite appropriate to consider their possible proliferation characteristics now, however, for these may have some bearing on the incentives for investing the money to bring these systems to fruition.

Fusion–fission hybrid breeder reactors would use high-energy neutrons from deuterium-tritium reactions in a fusion-plasma 'core' to generate fertile-to-fissile conversion reactions in a surrounding breeding blanket. The fissile material produced would be used to fuel fission converter reactors, eliminating the dependence of these on high-grade uranium deposits and uranium-enrichment facilities. This role is analogous to that envisioned for fission breeder reactors, with the difference that each fusion–fission hybrid breeder plant could support far more converter reactors than could a fission breeder of comparable thermal power rating. The hybrid breeders could produce plutonium-239 from uranium-238 fertile material, or uranium-233 from thorium-232 fertile material. Most recent studies have favoured the latter option, inasmuch as it permits a higher 'support ratio' (number of converter reactors per hybrid breeder of equivalent thermal power). It has also been argued by some analysts that the U-233 option offers significant anti-proliferation advantages.

A hybrid breeder with a thermal power rating of 3000 megawatts could produce about 3000 kilogrammes of U-233 or a roughly comparable amount of plutonium per year, and, unlike the Liquid Metal Fast Breeder Reactor (LMFBR) characterised in Table 10.7, would not require the use of any of this material to sustain its own continued operation. Accordingly, it could provide the fissile input requirements of some 10 to 15 'client' reactors (depending in detail on their characteristics). The flow of bomb-usable material *per electrical gigawatt of system capacity* would therefore be in the same range as the figures given for the non-breeder reactors characterised in Table 10.7.

The fissile content of plutonium produced in fusion–fission hybrid breeders would be in the vicinity of 90 per cent – even higher than that from LMFBRs – hence particularly well suited for use in nuclear explosives. On the other hand, hybrid plutonium would contain more Pu-236 than that from LWRs or LMFBRs, leading to penetrating gamma emissions from the U-232 decay chain and tending to decrease the attractiveness of the material for bomb-makers.

If the U-233/Th-232 fuel cycle had significant anti-diversion advantages over the Pu-239/U-238 alternative, this would add an important fringe benefit to the potential of hybrid breeders to make the U-233/Th-232 cycle economically attractive. As suggested above and in many other reviews, however, the anti-diversion advantages sometimes claimed for the U-233/Th-232 fuel cycle – based mainly on radiological hazards to bomb-makers and on the possibility of 'denaturing' the U-233 to be used in client reactors with U-238 – are not overwhelming, tend to be offset by corresponding disadvantages, and come at substantial cost in money and convenience.[31]

Fusion energy systems other than fusion–fission hybrid breeders would not ordinarily produce or contain fissile materials, so the weapons-linkage concern that is most acute for fission energy systems – that the technology would provide access to this 'limiting ingredient' for producing fission weapons – would be far less acute for such fusion systems. The only remnant of this particular concern in 'pure' fusion reactors would be that fertile material not ordinarily present could be introduced into the reactor and exposed there to fusion neutrons in order to breed fissile isotopes – either openly or clandestinely. That is, any fusion reactor could, with greater or lesser difficulty, be converted into a 'hybrid breeder' on some scale if the persons wishing to do so had prolonged access to the reactor and possessed the necessary technical skills and resources.

If a government or major industrial concern in possession of a fusion reactor were to decide to undertake such a step openly, they certainly would be able to produce significant quantities of fissile material in a rather short time. In the Brookhaven study of proliferation and safeguards issues in future technologies, it was estimated that insertion of uranium-carbide 'breeder' modules in a hypothetical fusion reactor design of recent vintage would yield about 14 kg of fissile plutonium per square metre of module per year, and that dissolving uranium in the reactor coolant could yield 160 kg of fissile plutonium per year.[32]

On the other hand, it would be extremely difficult if not impossible for a government or industrial concern to achieve significant fissile material production *clandestinely* in an ostensibly pure-fusion reactor subject to international monitoring of the sort that can be readily envisioned based on fission practice: the modifications to the fusion reactor and its operating procedures needed to make and extract fissile material would be easy to detect, especially given that one is looking for fertile or fissile material in an environment where none is expected (in contrast to looking for small discrepancies in large inventories, as is the case in monitoring fission or fusion–fission fuel cycles). Similarly, the necessary modifications would be too extensive and too obvious to escape detection by the operators of the plant if an attempt at clandestine fissile-materials production were being made by a subgroup of insiders; and because of the need for prolonged access as well as easily observed modifications of systems and procedures, outsiders would have no chance at all.

It follows, then, that there is a potential problem of open 'diversion of neutrons' from a fusion reactor by a government or major industrial concern to produce fissile material, but there is little prospect of this being done clandestinely by governments, industrial concerns, or smaller groups. Since, moreover, a government or industrial concern that is prepared to acquire fissile materials openly and that has access to fusion reactors would also have access to more direct and less cumbersome ways to produce fissile materials than modifying commercial fusion power plants (for example, uranium enrichment technology, fission reactors optimised to produce plutonium, fusion–fission hybrid breeders dedicated to production of weapons materials), it is hard to give much weight to open modification of commercial, pure-fusion power plants as a dangerous weapons link.

The next question that arises is whether the diversion of tritium from fusion reactors for use in fusion weapons poses problems analogous to those of diversion of fissile material from fission reactors for use in fission weapons. It is true that tritium is used in a number of modern thermonuclear warhead designs; that the inventories and throughputs of tritium in large D-T fusion reactors (typically kilogrammes in inventory and hundreds of grammes per day in throughput) are more than large enough to be significant in the weapons context (the total tritium inventory in the US weapons stockpile has been estimated to be about 70 kg);[33] and that fusion reactors could be rather readily modified to maximise net tritium output for weapons purposes. For these reasons it is to be expected that commercial fusion reactors will be subject to materials accounting procedures and other safeguards designed to minimise the chance that reactor tritium will be misused for weapons.

There are, none the less, some strong arguments for regarding the tritium safeguards issue as fundamentally less problematical for fusion than the fissile-material issue is for fission. Most importantly, tritium acquisition is not the *limiting* ingredient on fusion-weapon construction in the way that fissile materials are the limiting ingredient on fission-weapon construction: both the needed fission-bomb 'trigger' and the technical insights required for design and fabrication of fusion weapons are more important barriers than access to fusion fuels, and while access to tritium is convenient in fusion-weapon construction it is not necessary. Also, the fusion fuel cycle as it seems most likely to be practised is largely self-contained with respect to tritium: production, 'reprocessing'. and recycling are all integral to the reactor plant, eliminating the shipment of tritium except for the start-up of new plants. This circumstance greatly reduces the possibilities for unauthorised access to the material.

The spread of highly weapons-specific types of knowledge – as distinct from the spread of particular facilities and materials usable for weaponry – has not been given much weight in recent years as a liability of fission energy systems, in large part because the knowledge needed to build fission weapons (and most certainly those parts of this knowledge that are derivable from nuclear power technology) is already so widespread. In the case of fusion weaponry, by contrast, the much greater scientific and technological sophistication required is not so widespread, and the question logically arises whether the spread of fusion energy technology would spread important insights relevant to fusion weapons.

The fact that the major powers declassified their research programmes on magnetic-confinement approaches to fusion energy in 1958, and that they have maintained this field as an area of declassified and indeed highly internationalised research ever since, strongly suggests that the important insights about fusion weaponry have nothing in common with magnetic fusion energy. As is well known, however, there *are* connections between research in inertial-confinement fusion and insights relevant to the design and effects of fusion weapons,[34] and these connections have been the basis for classification of many aspects of research on inertial confinement since the inception of such work. In the present writer's view, these connections represent a serious shortcoming of inertial confinement fusion in comparison to the magnetic confinement approach.

INTERACTION OF A RADICAL ARMS CONTROL REGIME WITH NUCLEAR ENERGY OPTIONS

Recent interest in and improved prospects for more drastic approaches to nuclear arms control than any accomplished heretofore make it natural to ask how such drastic arms control regimes might interact with nuclear energy options. Two specific questions suggest themselves immediately. First, is an arms control regime incorporating a cut-off of production of fissile materials for weapons compatible with continued operation of commercial nuclear power facilities? That is, can the kinds of safeguards developed to hinder proliferation of nuclear weapons capabilities to additional nations be extended and applied to verify, within present-day weapons states, that fissile materials from commercial nuclear energy facilities will not be diverted to subvert a ban on fissile-material production for weapons? Secondly, in the event that drastic reductions in nuclear weapons stockpiles are agreed, is it feasible and sensible to use commercial nuclear reactors to 'burn up' the plutonium and highly enriched uranium from the dismantled weapons?

Concerning the first question, different analysts have given very different answers – albeit in somewhat different contexts. L. Hunter Lovins has argued often and eloquently that ridding the world of nuclear weapons will require the abandonment of nuclear power as well.[35] The essence of his argument is that the clandestine production of even a few nuclear weapons would represent a drastic and intolerable perturbation in a world from which such weapons had ostensibly been abolished altogether, and that the huge quantities of fissile materials in circulation under any sizeable use of nuclear energy could not be accounted for with sufficient accuracy to rule out such production; if, on the other hand, no nuclear energy activities at all were permitted, then detection of any such would constitute presumptive evidence of an intent to produce weapons, greatly simplifying the verification problem in comparison to the case where small diversions from a large 'legitimate' production would have to be detected.

Frank von Hippel and others have addressed the question in the less demanding context of a fissile-material production cut-off designed to stop the growth of the existing nuclear arsenals and to establish the basis for beginning a process of reductions.[36] For this purpose, they conclude, the kinds of commercial-facility safeguards already largely in place for non-proliferation purposes would suffice to verify that the weapons-material production cut-off was not being significantly violated. The essence of the argument here is the supposition that a *significant* violation would have to entail annual flows of material of the order of one per cent or more of the existing stockpiles; these amount to perhaps 500 tons of highly enriched uranium and 100 tons of plutonium on each side, and one per cent of these figures annually means quantities of material in the range of 5 to 15 per cent of what flows through the present-day nuclear energy systems of the two superpowers. Since IAEA safeguards for commercial nuclear facilities are designed to detect diversions in the range of one per cent of annual flows per year, they would appear to be easily capable – so goes the argument – of providing the needed assurances that flows 5 to 15 times larger were not being diverted into weapons. Since, however, the ability of IAEA-type safeguards to perform their *present* functions adequately is the subject of some controversy, the size of the margin of safety they could provide in verifying a cut-off of fissile material for weapons may need a closer look. All else being equal, moreover, the problem of monitoring civilian flows of fissile material with reassuring accuracy will become *more* difficult if and when the weapons stockpiles get much smaller and/or the civilian nuclear energy business gets much larger.

Concerning the possible use of commercial nuclear reactors for 'burning up' fissile material from dismantled nuclear weapons, the matter seems fairly straightforward. The *only* ways to render plutonium permanently unusable for weapons are irretrievable geologic storage for periods in excess of 100 000 years; firing the plutonium into deep space; and consuming it in fission reactors of one kind or another. The arguments for using civilian power reactors are that, from a technical standpoint, using reactors is the best understood, least uncertain of these options; that only using reactors offers some economic benefit from the investment originally made in producing the plutonium; and that the plutonium-consuming capacity of the existing inventory of power reactors is much higher than that of the world's research reactors and naval propulsion reactors. In the case of high-enriched uranium, there exists the further 'disposal' option of diluting it with uranium-238 to concentrations unusable for weapons. This would not be difficult technically, and might well be done as the first step in a scheme to use the material as reactor fuel; but *not* to use the material in the end in reactors would, as in the case of the plutonium, throw away a considerable economic resource.

A rough idea of the rate at which fissile material from weapons could be dumped into commercial nuclear reactors emerges from the consideration that a 1000-MWe light-water reactor on a once-through fuel cycle takes in about 900 kg of U-235 per year (about 30 tons of uranium fuel enriched to 3 per cent U235; see also Table 10.7). Therefore, if mining and enrichment of fresh uranium as a source of this material were stopped entirely and the fuel for the 85 000 MWe of existing US reactors were supplied instead by adding weapons U-235 and plutonium to depleted uranium from past enrichment operations, the weapons material would disappear into these reactors at a rate of 85×900 kg/yr; with correction for the 0·20 to 0·25 per cent U-235 content of the depleted uranium, let us say 70 tons per year. Thus, the US weapons stockpile of about 500 tons of U-235 and 100 tons of plutonium could be fed into US power reactors over the course of less than a decade.[37] The Soviet stockpile, similar in size, would take perhaps 2·5 times longer to be burned up in existing Soviet power reactors, since their total capacity is smaller than that of the US nuclear-energy system by this factor. Use of power reactors outside the United States and the Soviet Union, on the other hand, would speed up the operation (see Table 10.4).

Today's power reactors do produce some additional plutonium in operation – about a third as much, net, as the fissile material 'burned' – but, as noted above, under the present once-through fuel cycle this plutonium remains mixed with highly radioactive fission products in the unreprocessed spent fuel and so is rather well protected from re-use in weapons. Still, this spent fuel would have to be safeguarded as part of the verification provisions of a fissile-stockpile reduction. The alternative would be to develop a new type of 'burner' reactor using fuel that contains no uranium-238 or other fertile material, and thus produces no further fissile material in operation. This is undoubtedly feasible, but developing the new reactor and then building a sufficient supply of them to consume the weapons material at a high rate would be time-consuming and costly, and hardly seems worth the trouble.

A programme for consuming weapons plutonium and high-enriched uranium in power reactors would require, in any case, solution of a variety of problems in processing, monitoring and verification, as well as worker health and safety; and there would be some economic dislocation associated with replacing the mining and enrichment of uranium by 'mining' of warheads. Still, it is difficult to argue that any of these problems is likely to be insurmountable, especially when compared with the costs and risks of other ways to manage the weapons material, or with the risks of leaving the weapons stockpiles intact.

CONCLUSION

The main conclusions to be drawn from this survey can be rather briefly summarised as follows:

● All nuclear-energy technologies unavoidably provide to their developers and users *some* capabilities related to nuclear weaponry. No technical solutions for rendering such energy technologies irrelevant to weapons capabilities appear to exist. The nature and magnitude of the associated weapons capabilities, however, vary with the type of nuclear energy technology (such as fission or fusion, breeder or non-breeder), with the details of the fuel cycle, and with the presence or absence of specific anti-proliferation countermeasures; and the balance of incentives and disincentives relating to whether a country chooses to *use* the weapons-related capabilities provided by nuclear-energy technologies depends also on the political environment and on the costs and availability of other pathways towards weapons goals.

● The most useful available measures to reduce political incentives to acquire nuclear weapons would be to conclude a Comprehensive Test Ban Treaty; to reduce the reliance and emphasis on nuclear weapons in the military strategies of the nuclear-armed powers; and to begin a process of deep reductions in the existing nuclear arsenals. At the same time, political barriers to proliferation could and should be strengthened by persuading key non-signatories to join the Non-

Proliferation Treaty regime and by bolstering the procedures and resources available for the associated International Atomic Energy Agency safeguards.

● A rough ranking of nuclear-energy systems in terms of intrinsic technical susceptibility to use for weapons purposes, ordered from most susceptible to least susceptible, is as follows: fission breeder reactors and reactors using high-enriched uranium fuel > fusion–fission hybrid breeders and converter reactors with plutonium or U-233 recycle > converter reactors using 'once-through' fuel cycles > pure-fusion reactors.

● If IAEA safeguards are or can be made close to adequate for their original non-proliferation purpose, then they would certainly also be adequate for detecting significant violations of a cut-off of fissile material production for nuclear weapons as a means of limiting major-power nuclear arsenals to their existing sizes. If the arsenals were reduced by large factors (tenfold or more), this task would become significantly harder.

● In the context of an agreement to dismantle large fractions of the major-power nuclear arsenals, existing nuclear-power reactors would have the capability to 'burn up' the plutonium and highly enriched uranium from the dismantled warheads rather quickly using 'once through' fuel cycles. But the spent fuel from this process would have to be safeguarded because of its residual plutonium content.

The much higher proliferative susceptibility of fuel cycles that recycle plutonium or uranium-233 compared to most of those that do not warrants the following additional observation: although it is true that recycling in some form would be required to extend large-scale use of fission energy much beyond the year 2050, it is *not* needed sooner and it does not seem likely to be economical sooner.[38] Continuing to defer a transition to recycling systems thus is not only good anti-proliferation policy; it is also sound energy policy.[39]

Notes

1.   See, for example, George Quester, *The Politics of Proliferation* (Baltimore, Maryland, 1973); Ted Greenwood, Harold A. Feiveson, and Theodore B. Taylor, *Nuclear Proliferation: Motivations, Capabilities, and Strategies for Control* (New York, 1977); Willam C. Potter,

Nuclear Power and Nonproliferation: An Interdisciplinary Perspective (Boston, Massachusetts, 1982); and Stephen M. Meyer, *The Dynamics of Nuclear Proliferation* (Chicago, 1984).

2. William Epstein, *The Last Chance: Nuclear Proliferation and Arms Control* (New York, 1976).

3. John P. Holdren, 'Extended Deterrence, No-First-Use, and European Security', *Scientia*, vol. cxx (1985) 191–201.

4. See works cited in note 1.

5. Lawrence Scheinman, *The International Atomic Energy Agency and World Nuclear Order* (Baltimore, Maryland, 1987).

6. Office of Technology Assessment (OTA), US Congress, *Nuclear Proliferation and Safeguards* (Washington, DC, 1977).

7. Ibid.

8. Office of Nuclear Energy, US Department of Energy, *Nuclear Energy Cost Data Base* (Washington, DC, 1985).

9. American Physical Society, 'Report to the APS by the Study Group on Nuclear Fuel Cycles and Waste Management', *Reviews of Modern Physics*, vol. l, no. 1, part II (January 1978); see also OTA, *Nuclear Proliferation*; and Office of Nuclear Energy, *Nuclear Energy Cost*.

10. Like U-238, Pu-240 and Pu-242 are not 'fissile' in the usual meaning of that word (that is, able to sustain a chain reaction based on thermal neutrons); but, in contrast to the situation with U-238 (which above a certain concentration makes the formation of an explosive critical mass impossible), the fission cross-sections of Pu-240 and Pu-242 for fast neutrons are high enough for an explosive critical mass to be formed from plutonium no matter what the Pu-240/Pu-242 concentration. See, for example, OTA, *Nuclear Proliferation*; and American Physical Society, 'Nuclear Fuel Cycles and Waste Management'.

11. Ibid.

12. OTA, *Nuclear Proliferation*. Plutonium quality for weapons purposes is maximised by removing fuel from the reactor sooner than would be the case if power generation were the only consideration, thereby minimising the time-dependent build-up of Pu-240 and higher-numbered plutonium isotopes. But removing fuel from light-water reactors (and most other reactor types) requires shutting down the reactor for a prolonged period. If this is done several times per year, as is needed to achieve high plutonium quality for weapons – as opposed to once per year, as is typical in a cycle optimised for power generation – electricity output necessarily falls.

13. See OTA, *Nuclear Proliferation*; and American Physical Society, 'Nuclear Fuel Cycles and Waste Management'.

14. Estimates of enrichment costs in commercial facilities fall mostly in the range of $50 to $100 per separative-work unit (ibid.). Producing the roughly 15 kg of weapons-grade uranium needed to make a bomb, starting with natural uranium, would require about 3500 separative-work units, which at the indicated unit costs would be valued at $175 000 to $350 000.

15. The value of plutonium as reactor fuel is usually placed in the range of $30 to $50 per gramme (ibid). Using the higher figure, which is the most

recent, gives $350 000 to $500 000 for the 7 to 10 kg of reactor-grade plutonium nominally required for a bomb.

16. The figure of 30 bombs' worth of high-quality plutonium at the cost of a third of the electricity output of 1000-MWe light-water reactor is from OTA, *Nuclear Proliferation*. Assuming a 'normal' capacity factor of 68·5 per cent (somewhat higher than the national average in the United States), the loss of a third of the normal annual output would amount to about 2000 gigawatt-hours, or 2×10^9 kWh. Assuming a generation cost for such a reactor of about 3·3 cents/kWh in mid-1980s money (Office of Nuclear Energy, *Nuclear Energy Cost*), one derives a cost of $66 million for 30 bombs' worth of plutonium, or $2.2 million per bomb.

17. The implicit price deflator for the US Gross National Product between 1975 and 1985, as calculated by the Department of Commerce, is 1·88. US Department of Commerce, *Statistical Abstract of the United States* (Washington, DC, 1987).

18. According to the OTA, producing 30 kg of 90 per cent U-235 requires 8000 kg of natural UF_6 feed and 6900 separative-work units, while starting with 3 per cent U-235 reduces the feed requirement to 1500 kg and the separative-work requirement to 2500 units (OTA, *Nuclear Proliferation*).

19. The situation in which acquisition of nuclear-energy capabilities inevitably brings a weapons capability closer. whether that result is initially desired by the possible proliferator or not, was termed 'latent proliferation' in an early, astute analysis; see H. A. Feiveson, 'Latent Proliferation: The International Security Implications of Civilian Nuclear Power' (PhD Thesis, Woodrow Wilson School of International Affairs, Princeton University, 1972). See also works cited in note 1; and J. P. Holdren, 'Nuclear Power and Nuclear Weapons: The Connection is Dangerous', *Bulletin of the Atomic Scientists* (January 1983).

20. See, for example, Leonard S. Spector, *Going Nuclear* (Cambridge, Massachusetts, 1987).

21. See, for example, Bernard Spinrad, 'Nuclear Power and Nuclear Weapons: The Connection is Tenuous', *Bulletin of the Atomic Scientists* (February 1983); and Alexander de Volpi, 'Technological Misinformation: Fission and Fusion Weapons', in David Carlton and Carlo Schaerf (eds), *The Arms Race in the 1980s* (London, 1982).

22. Lawrence Scheinman, *Atomic Energy in the Fourth Republic* (Princeton, New Jersey, 1965).

23. See, for example, Meyer, *The Dynamics of Nuclear Proliferation*, Appendix A, and sources therein; and OTA, *Nuclear Proliferation*, ch. 4.

24. See Spector, *Going Nuclear*; and Leonard S. Spector, *Nuclear Proliferation Today* (New York, 1984).

25. Ibid.

26. See, for example, OTA, *Nuclear Proliferation*; and American Physical Society, 'Nuclear Fuel Cycles and Waste Management'.

27. See, especially, International Nuclear Fuel Cycle Evaluation (INFCE), *Summary Volume* (International Atomic Energy Agency, Vienna,

1980); and Harold A. Feiveson, Frank von Hippel, and Robert H. Williams, 'An Evolutionary Strategy for Nuclear Power', in David Carlton and Carlo Schaerf (eds), *The Hazards of the International Energy Crisis* (London, 1982).

28. For detailed discussions of internationalisation from many perspectives, see Stockholm International Peace Research Institute, *Internationalization to Prevent the Spread of Nuclear Weapons* (London, 1980).

29. Scheinman, *The International Atomic Energy Agency*.

30. An up-to-date review of possible characteristics of fusion and fusion–fission hybrid energy systems is given in J. P. Holdren, D. H. Berwald, R. J. Budnitz, J. G. Crocker, J. G. Delene, R. D. Endicott, M. S. Kazimi, R. A. Krakowski, B. G. Logan, and K. R. Schultz, 'The Competitive Potential of Magnetic Fusion Energy: Summary of the Report of the Senior Committee on Environmental, Safety, and Economic Aspects of Magnetic Fusion Energy', *Fusion Technology* (in press).

31. In addition to references cited earlier, see J. P. Holdren, 'Fusion–Fission Hybrids: Environmental Aspects and Their Role in Hybrid Rationale', *Journal of Fusion Energy*, vol. 1 (1981) p. 197; and Committee on Fusion Hybrid Reactors, *Outlook for the Fusion Hybrid and Tritium-Breeding Fusion Reactors* (Washington, DC, 1987).

32. B. Keisch, C. Auerbach, A. Fainberg, S. Fiarmen, L. G. Fishbone, W. A. Higinbotham, J. R. Lemley, and J. O'Brien, *Long-term Proliferation and Safeguards Issues in Future Technologies* (Brookhaven National Laboratory, 1986).

33. T. B. Cochran, W. M. Arkin, R. S. Norris, and M. M. Hoenig, *Nuclear Weapons Databook, Vol. II: U.S. Nuclear Warhead Production* (Cambridge, Massachusetts, 1987).

34. See Keisch, *et al.*, *Long-Term Proliferation*; and J. P. Holdren, 'Fusion Power and Nuclear Weapons: A Significant Link?', *Bulletin of the Atomic Scientists* (March 1978).

35. See especially B. Amory and L. Hunter Lovins, *Energy/War: Breaking the Nuclear Link* (New York, 1982).

36. Frank von Hippel, David H. Albright, and Barbara G. Levi, 'Stopping the Production of Fissile Materials for Weapons', *Scientific American*, vol. 253, no. 3 (1985) 41.

37. The stockpile figures are from Cochran *et al.*, *Nuclear Weapons Databook, Vol. II*.

38. See Feiveson, von Hippel, and Williams, 'An Evolutionary Strategy for Nuclear Power'.

39. The views in this chapter are the author's and not those of his institution.

11 Nuclear Disarmament and Peaceful Nuclear Technology – Can We Have Both?

Theodore B. Taylor

INTRODUCTION

Can nuclear disarmament reduce the risks of nuclear war sufficiently to be acceptable to most people in a world in which use of nuclear technology for peaceful purposes continue to flourish? My short answer is: perhaps. The nuclear materials needed for making nuclear weapons are automatically produced, or used as fuel, or both, in all types of fission reactors, whether for military or for peaceful purposes. Plants for producing the low-enrichment uranium used as fuel in most of the world's nuclear power plants can quickly be modified to make highly-enriched uranium for nuclear weapons. Successful development of economical thermonuclear power will provide the technological means for producing nuclear weapon materials at low cost. The reason is that the relevant thermonuclear reactions yield copious quantities of neutrons that could be captured in ordinary uranium to produce plutonium. Furthermore the non-nuclear materials and equipment needed to make nuclear weapons are accessible world-wide.

If the principal goal of nuclear disarmament is world-wide elimination or sharp curtailment of existing stockpiles of nuclear weapons and the facilities for their production, effective safeguards to prevent the diversion of peaceful nuclear technology to use for making nuclear weapons will therefore also be needed world-wide. Otherwise, it is quite possible for major nuclear disarmament actions in some nations to be under way at the same time that other nations are in the process of acquiring nuclear weapons for the first time, or increasing the numbers and diversity of their nuclear weapons. In short, major nuclear disarmament actions by the nuclear superpowers will not, by

199

themselves, ensure that further proliferation of nuclear weapons will cease.

HISTORICAL CHANGES IN THE CHARACTER OF NUCLEAR WEAPON PROLIFERATION

Nuclear weapon proliferation, which will henceforth be abbreviated to 'proliferation', can be characterised in two different ways. The first distinguishes among 'horizontal', 'vertical', and 'non-national' proliferation. Horizontal proliferation refers to the acquisition of nuclear weapons by more countries. Vertical proliferation refers to increases in the numbers and diversity of nuclear weapons of a nation after it has initially acquired them. Non-national proliferation refers to the acquisition of nuclear weapons by terrorists or other criminal groups not formally affiliated with a national government. The present writer knows of no evidence that proliferation in this last category has yet taken place, but international concerns about it are increasing. The second categorisation of proliferation is according to whether it is publicly disclosed, secret, ambiguous, or 'latent'. Harold Feiveson of Princeton University has adopted the term 'latent proliferation' to mean substantial progress towards acquisition of nuclear weapons, whether or not a decision has been made to produce them. Examples include the possession of significant quantities of nuclear-weapon materials, in chemically separated form or not, produced or used in the course of using nuclear energy for peaceful purposes. The closeness of approaches of particular states of latent proliferation to actual possession of nuclear weapons form a continuous spectrum, ranging from possession of no nuclear facilities (but the knowledge and resources needed to acquire them), through large stockpiles of reactor plutonium that has been chemically separated from spent fuel, to production of all the components needed for nuclear weapon manufacture, without having actually assembled and tested any weapons.

All historical proliferation, both horizontal and vertical, has started secretly. Horizontal proliferation was publicly announced, after a first nuclear test, by the United States, the Soviet Union, Great Britain, France, and China. All these countries, with the possible exception of China, have, at least to some extent, used 'dual-purpose' reactors for producing both electric power and plutonium for weapons, although the primary function of the reactors has been plutonium production. It is not clear from official disclosures to what extent uranium enrichment plants have been used for producing both highly-enriched uranium for weapons and high- or lower-enrichment uranium for peaceful purposes. Important details concerning vertical proliferation in these five countries have remained secret, although the numbers and many of the characteristics of their newer weapons are fairly well known publicly.

Although India's publicly disclosed nuclear test has been consistently characterised by the Indian government as for peaceful purposes, it clearly demonstrated, at the very least, a fully completed state of latent proliferation. Many would classify the present state of India's nuclear weapon capabilities as ambiguous *and* secret. There has been a growing belief, starting in the late 1960s, that Israel has nuclear weapons, but the Israeli government has consistently said, for more than two decades, that 'Israel will not be the first to introduce nuclear weapons in the Middle East'. This is literally true, since the United States placed nuclear weapons in Turkey nearly three decades ago. Public disclosures in October 1986 by Mordechai Vanunu, a former employee at Israel's Dimona nuclear facility, are consistent with the existence of a much more elaborate nuclear weapon production facility at Dimona than was generally believed to exist previously.[1] Nuclear materials for as many as 200 nuclear weapons may have been produced there since the late 1960s. The credibility of Vanunu's revelations is enhanced by the fact that he was recently brought to trial in Israel for treason. In sum, proliferation in Israel has very probably been horizontal, then vertical, and both secret and ambiguous. South Africa and Pakistan are generally believed to have begun to stockpile nuclear weapons or to be prepared to do so very quickly. In both these countries the most likely sources of the needed nuclear materials are uranium enrichment plants. Proliferation in these cases has been officially denied, and therefore can also be categorised as both secret and ambiguous.

In short, the patterns of actual or strongly-suspected horizontal proliferation have changed sharply since China's first nuclear test in 1964. They have been marked by ambiguity and secrecy, rather than publicly used as a nuclear deterrent. They have made use of nuclear weapon production facilities publicly claimed to be for peaceful purposes. It is also possible that some other countries have a few nuclear weapons or are in an advanced state of latent proliferation, but have managed to keep their efforts secret.

Nearly 50 nations have refused to sign the Non-Proliferation Treaty (NPT). These include India, Israel, Pakistan and South Africa, as well as two announced nuclear weapon states, France and China. Furthermore the United States and the Soviet Union have been in technical violation of Article VI of the treaty, which requires them 'to pursue negotiations in good faith on effective measures relating to cessation of the nuclear arms race at an early date and to nuclear disarmament'. It remains to be seen whether recent announcements of agreement in principle by these countries to eliminate all intermediate-range nuclear missiles, to be followed quickly by negotiations regarding a 50 per cent cut in strategic nuclear warheads and further restrictions on nuclear testing will be considered by the rest of the world as meeting these countries' obligations under the NPT. Whether or not this is the case, continued dependence on nuclear deterrence by any countries seems likely to continue to stimulate other countries to do the same.

A BRIEF OVERVIEW OF NUCLEAR WEAPON TECHNOLOGY

The most difficult stage in the production of fission weapons is the acquisition of the needed special nuclear materials – plutonium or highly-enriched uranium (which can include uranium-233 produced by neutron irradiation of thorium in reactors). The information necessary for designing such weapons has been public for many years. The non-nuclear components of fission weapons can be made with materials and equipment that are accessible world-wide.

Reliable fission weapons with yields in at least the kiloton range can be made using plutonium produced in nuclear power plants, although plutonium produced specifically for nuclear weapons is easier to use for higher-yield weapons. The reason is that plutonium made in power reactors tends to have a high content of plutonium-240. This isotope fissions spontaneously at a sufficient rate to release neutrons that can initiate an explosive nuclear chain reaction before the optimum time. Without sophisticated designs, the yields of nuclear weapons using plutonium made in power reactors tend to be unpredictable, but still with a lower limit in the vicinity of 1 kiloton. With more sophisticated designs, however, power reactor plutonium can be used for making reliable, light-weight nuclear weapons with yields greater than that of the bomb that destroyed Nagasaki. Such pre-initiation problems are not important in implosion-type nuclear weapons that use only highly-enriched uranium.

Unsophisticated nuclear weapons can be made with 5–10 kilogrammes of plutonium or with about 20 kilogrammes of highly enriched uranium. With design principles that have been extensively used by all five announced nuclear weapon states, however, nuclear weapons that require considerably smaller quantities of these materials can be made to explode with yields greater than 10 kilotons. (Complete fission of 1 kilogramme of plutonium or uranium releases 17 kilotons of energy.)

Tritium, the radioactive isotope of hydrogen with a half-life of 12·3 years, plays an important role in most modern nuclear weapons. Its fusion with deuterium produces about ten times as many neutrons, for the same energy release, as fission. Incorporation of gramme quantities of tritium in fission weapons can therefore 'boost' their yield considerably without adding significant weight. Boosted fission explosives can either be used alone as warheads, or as triggers for thermonuclear weapons. Tritium is an important component of neutron bombs. Very small quantities of tritium can also be used in initiators of the chain reactions of fission weapons.

Requirements for nuclear tests depend on the data available from previous tests, the types of nuclear weapons and specific design features, and the character of available non-nuclear testing methods and facilities. Unsophisticated fission weapons of the implosion type do not require nuclear tests if the contained special nuclear materials are close to critical before the implosion starts, and extensive use is made of non-nuclear tests of the implosion system, along with detailed performance calculations. It is therefore feasible that a country developing first-generation, relatively unsophisticated fission weapons might not require any nuclear tests to be confident of the approximate performance of stockpiles of such weapons. Confidence in more advanced fission weapons, without boosting, could be increased considerably by so-called 'zero yield' tests of implosion systems with smaller amounts of plutonium or uranium than needed for a nuclear weapon, yet still producing a fission chain reaction with observable radiation output at a very small yield that could not be detected at a significant distance. It is also possible that terrorists or other criminals that have acquired the needed quantities of plutonium or highly-enriched uranium by theft or purchase from a black market could feel reasonably confident that an untested, crude fission bomb would explode with a yield in the vicinity of at least a kiloton. Boosted fission weapons, on the other hand, require testing in the kiloton range to provide confidence in their performance. The need for testing is even greater when it comes to thermonuclear weapons, for which, unlike boosted weapons, a large fraction of the yield is directly accounted for by thermonuclear fusion. Horizontal proliferation of thermonuclear or boosted weapons thus requires tests, and is therefore much more difficult to keep secret than proliferation of first-generation fission weapons.

A major concern about continued vertical proliferation is the development of so-called 'third-generation' nuclear weapons. These can be generally characterised as nuclear explosives that make use of any of countless ways to enhance or suppress any of the many different forms of energy released in a nuclear explosion, while also achieving highly directional release of forms of energy selected to cause specific types of damage. Development of such weapons requires extensive nuclear testing, however. Furthermore deployment of such weapons, especially for offensive or defensive use in space or high altitude, would also require testing above ground. Many of the effects of such weapons cannot be confidently simulated in underground tests. Perhaps most importantly, pursuit of new nuclear weapon technologies by the superpowers could stimulate proliferation of nuclear weapons in other nations without large nuclear weapon stockpiles, but already in advanced states of latent proliferation.

PRESENT AND PROJECTED STOCKPILES OF SPECIAL NUCLEAR MATERIALS

The quantity of plutonium that is now in spent fuel from the world's civilian power reactors is greater than the quantity of plutonium in all the world's nuclear weapons – about 500 metric tons, against 250 metric tons.[2] By the year 2000 the quantity of plutonium in weapons is not expected to increase substantially, and may decrease dramatically if nuclear disarmament agreements lead to rapid dismantlement of the weapons and disposal or peaceful use of the weapons' plutonium. The total quantity of plutonium that will have been produced in power reactors by then, on the other hand, is expected to be greater than 1500 tons. As much as 350 tons of this plutonium is likely to have been extracted from spent fuel and recycled through thermal power reactors or used to start up and fuel breeder reactors.[3]

In sharp contrast, the world's stockpiles of highly-enriched uranium are overwhelmingly accounted for in nuclear weapons, somewhere in the vicinity of 1000 to 1500 tons.[4] This compares with the order of a few tons, at most, of highly enriched uranium for research reactors and a very small number of power reactors.

The total quantity of tritium used for nuclear weapons is estimated at roughly 200 kilogrammes.[5] Since tritium decays at a rate of about 5·5 per cent per year, at least 10 kilogrammes or so per year need to be produced to maintain present nuclear weapon stockpiles world-wide. This tritium is generally produced by neutron irradiation of lithium in military production reactors used also for producing plutonium.

USE OF NUCLEAR MATERIALS FROM DISMANTLED NUCLEAR WARHEADS AS FUEL FOR NON-MILITARY NUCLEAR POWER PLANTS

These brief overviews of past proliferation, nuclear weapon technology, and quantities of nuclear weapon materials in nuclear weapons and peaceful nuclear energy systems suggest several possible types of future coupling between nuclear disarmament and peaceful nuclear technology. One of these is briefly discussed in the remainder of this chapter.

Nuclear disarmament agreements that forbid the use of plutonium or highly-enriched uranium from dismantled warheads in other warheads require agreements about the disposition of these materials. A possibility that is receiving increasing attention is that of using these materials as fuel in non-military nuclear power plants. This would not eliminate the uranium-235 or plutonium in the warheads, since spent fuel from power reactors contains significant quantities of these materials – not all the uranium-235 is fissioned and new plutonium is produced. But it would render these materials no more accessible for use in weapons than the rapidly increasing stockpiles of stored spent fuel from reactors.

The fuel for most of the world's nuclear power plants is uranium-enriched to 3 per cent uranium-235. Since uranium at less than 6 per cent enrichment cannot sustain a fast neutron chain reaction, this uranium fuel cannot be used for making nuclear explosives. Appropriate dilution of highly-enriched uranium from nuclear warheads by mixing with natural or depleted uranium would therefore render it useless as a nuclear weapon material, but not as a reactor fuel. Using this approach would thus serve the double purpose of recovering much of the resource value of the weapon uranium, while making the uranium no more accessible for new weapons than the uranium now used in power reactors.

Plutonium has also been used for supplementing the fuel in a few of the light-water reactors that now account for most of the world's nuclear power. Plutonium from nuclear warheads could also be used for this purpose, but it would be easier to divert for military purposes than low-enrichment uranium. The reason is that chemical separation of plutonium metal from reactor fuel is much easier than uranium enrichment. This option would therefore require much more stringent safeguards against diversion or outright theft than weapon uranium after dilution. Present quantities of highly-enriched uranium and plutonium in the world's nuclear warheads have the potential to provide the nuclear fuel for all the world's present nuclear power plants for roughly four years.

Typical annual loadings of uranium-235 in low-enrichment uranium fuel for light water reactors, which account for about 90 per cent of the world's nuclear power plants, are about 1000 kilogrammes for every gigawatt (1000 megawatts) of electrical capacity. The world's currently installed nuclear electric-generating capacity is about 300 gigawatts.[6] Thus the present world's stockpile of roughly 1000 tons of highly-enriched uranium (more than 90 per cent uranium-235) in nuclear warheads could provide all the fuel needed by these power plants for about three years. This time is much shorter than a credible time for dismantling a majority of the world's nuclear warheads. Furthermore the world's installed nuclear electric-generating capacity is expected to increase substantially during the next decade. Therefore this potential 'market' for diluted uranium from dismantled warheads is very likely not to be saturated, at any credible rate of nuclear disarmament, as long as nuclear power continues to be a major source of energy.

Annual loadings of plutonium in mixed uranium–plutonium oxide fuel for these power plants could account for about 300 kilogrammes per gigawatt year, or roughly one-third the potential rate of use of uranium-235 in reactors fuelled only with uranium. Use of plutonium from dismantled weapons could therefore add another year to the time during which all the power plants could be fuelled entirely with nuclear materials from warheads. Using warhead-plutonium in power-reactor fuel, however, is more difficult than using diluted warhead-uranium for this purpose. Mixed oxide fuel costs much more to fabricate than uranium oxide fuel, partly because plutonium is such a hazardous material. Plutonium in all the chemical forms it would take if incorporated into fresh fuel for power plants is much easier to convert back to metal for use in nuclear warheads than re-enriching uranium after it has been diluted by natural or depleted uranium. Safeguards to prevent diversion or outright theft of plutonium would therefore have to be much more stringent than safeguards applied to uranium-235 after it has been diluted. It is therefore possible that these problems may make it more attractive, overall, to convert only the warhead-uranium to reactor fuel, and dispose of the warhead-plutonium directly, without using it as fuel. Any such disposal method, however, should be at least as effective in preventing later use of the plutonium in weapons as ultimate disposal of the plutonium remaining in spent fuel from present nuclear power plants, whether or not they use plutonium in their fresh fuel.

In spite of these problems, several countries – notably Japan, West Germany, Great Britain, and France – have started recycling plutonium from reprocessed spent reactor fuel. The projected world total amount of plutonium separated by the year 2000 is 300 to 400 tons, nearly twice the plutonium now in the world's nuclear warheads.[7] In any case means to verify that all the special nuclear materials in nuclear warheads is accounted for in the process of dismantlement, and none is diverted back to weapons, are likely to be called for in treaties for elimination of a large proportion of the existing nuclear weapons. Such verification must overcome several problems.

Continuing secrecy about nuclear weapon designs may require that the dismantlement be performed by nationals of the country owning the weapons, without detailed observation by inspectors from other countries. Verification that the devices to be dismantled are, in fact, operable nuclear warheads could depend partly on external on-site inspection of the removal of the warheads from missiles or other components of a nuclear weapon system. All the warheads of a given class might be externally inspected before dismantlement, using active or passive non-destructive assay techniques that depend on observations of neutron and gamma ray intensities measured outside the warhead. This can probably be done in such a way that the quantities of contained uranium or plutonium are not revealed, but accurate 'fingerprints' of all warheads of the same type could be compared to ensure that all configurations of the nuclear materials are the same. A more problematic possibility would be random underground testing of a very small fraction of the warheads of a particular type, allowing foreign inspectors to measure the yield.

All warheads taken to a dismantlement facility could be subject to 'containment'-type safeguards against withholding or diverting any of the contained special nuclear materials before the extracted materials could be analysed and measured by multinational or international inspectors. Such safeguards depend on batch processing inside an enclosure that can be thoroughly inspected before a new set of warheads is brought in, and kept under external observation to ensure that no materials leave the enclosure through unauthorised channels. If the quantities of plutonium or highly enriched uranium in specific warheads are considered secret, warheads of more than one type could be dismantled together in each batch, revealing only the aggregate quantities of these materials.

After dismantlement and determination of the quantities of each type of special nuclear material in each batch, the materials could then be placed under bilateral or international safeguards similar to those now used for safeguarding nuclear materials intended only for peaceful purposes. Safeguards would also need to include physical protection of the materials against theft by territorists or criminals. This is now a *national* responsibility, extending beyond bilaterial or international safeguards designed to *detect*, but not physically *prevent* material diversion.

CONCLUSION

It remains to be seen whether nuclear disarmament can reduce the risks of nuclear war sufficiently for the residual risks to be acceptable to a majority of the world's population, while at the same time vigorous growth in the world's dependence on nuclear energy for peaceful purposes continues. Initially, at least, use of nuclear materials from dismantled weapons as fuel for peaceful purposes may help progress to be made towards that goal, by stimulating considerable improvements in the effectiveness of arrangements for preventing diversion of the materials from peaceful to military purposes, while at the same time eliminating large numbers of nuclear weapons.

Notes

1.   *The Sunday Times*, London, 5 October 1986, pp. 1–3.
2.   David Albright, 'World Inventories of Plutonium', Princeton Univer-

sity Center for Energy and Environmental Studies Report no. 195 (rev. 1) (June 1987) pp. 19, 21, 24 and 99.

3. Ibid., p. 35.
4. Frank von Hippel, David Albright, and Barbara Levi, 'U.S. and Soviet Stockpiles of Fissile Materials', Princeton University Center for Energy and Environmental Studies Report no. 168 (1986).
5. Thomas B. Cochran, William M. Arkin, Robert S. Norris, and Milton M. Hoenig, *Nuclear Weapons Databook, Vol. 2: U.S. Nuclear Warhead Production* (Cambridge, Mass., 1987) p. 62. (World inventory of tritium was taken to be roughly twice that of the United States.)
6. A. De Volpi, 'Fissile Materials and Nuclear Weapons Proliferation', *Annual Review of Nuclear Particle Science*, vol. 36 (1986) 87.
7. Albright, 'World Inventories of Plutonium'.

12 Pakistan's Nuclear Weapons Programme and its Implications

Jorma K. Miettinen

Pakistan's long-running nuclear weapons programme evidently began to make striking progress in spring 1987. Statements by both international experts and Pakistani leaders, President Zia ul-Haq, and Dr. A. Q. Khan, the Director of the uranium-enrichment facility at Kahuta, served to confirm this, although some Pakistani officials have later attempted to reinterpret these statements.[1] Pakistan's nuclear programme has a long history. It started in 1955 on a modest scale but was accelerated between 1958 and 1977 when Zulfikar Ali Bhutto in different capacities acted as its chief promotor. He sent hundreds of young scientists to Europe and North America for training in Nuclear Sciences[2] and ordered from Canada a 125 MWe heavy water/natural uranium power plant called KANUPP, which started operation in 1972. He also concluded with France in 1976 an agreement on a nuclear reprocessing plant to be constructed at Chasma, an agreement which France terminated in 1978. Pakistan has, however, continued its construction and it is now said to be close to starting operations.[3]

After the loss of the 1971 war against India, Pakistan decided to start a crash programme to produce nuclear weapons.[4] The immediate plan was to process the KANUPP fuel but this was slowed down by the withdrawal of France from the project in 1978 and hampered by the International Atomic Energy Authority (IAEA) safeguards. A second route opened in 1975 after Khan, while working from 1972 to 1975 in the Netherlands at a physics laboratory in connection with the three-power ultracentrifuge uranium-enrichment plant in Almelo, got an opportunity to copy all the blueprints of the plant and a list of 100 subcontractors. Soon after Khan had left the theft was revealed.[5] Immediately after returning to Pakistan, Khan started construction of a gas centrifuge plant in Kahuta, near Islamabad, which in 1980 is reported to have approached an initial operational capability. In 1980 President Jimmy Carter cut down all military and other aid to

Pakistan on the basis of the Symington Amendment which forbids US aid to states which have a nuclear weapons programme. The Soviet occupation of Afghanistan and the change of Presidency in the United States changed this situation. In 1981 the Congress passed a special law which made aid possible if the US President confirms that the receiver country 'has no nuclear explosive device'. This President Ronald Reagan did, and from 1981 to the end of 1987 Pakistan received $3·2 billion in military and economic aid.

The operation of the Kahuta plant may have been slowed down in 1980 because little was heard about it between 1980 and 1983. In 1983–4 it was reported to be operating with 1000 tubes capacity. In 1985 Zia gave an assurance that '5 per cent enrichment level is not exceeded' in the Kahuta plant[6] and the US Congress agreed to continue military aid under the $3·2 billion package approved in 1981. The Reagan Administration even agreed to supply Pakistan with sophisticated air-to-air missiles to fend off Afghan aircraft violating Pakistan's air space.

Although emphasis has been on the uranium route it is evident that Pakistan has not given up the plutonium route, either. In 1982 it got into operation a new small reprocessing plant called New Labs in Rawalpindi, supplied by French and Belgian firms. It has a capacity to extract 10 to 20 kg plutonium per year. Pakistan may have a secret plutonium-producing reactor operational by now. In August 1984 it announced that it was producing high-quality graphite.[7] Its KANUPP reactor and the 5 MW research reactor PARR are both under IAEA safeguards and plutonium produced in them has to be placed under IAEA safeguards too.

In 1986 and 1987 the pace of statements and revelations increased. On 4 November 1986, the *Washington Post* quoted a classified US intelligence report that Pakistan had produced an unspecified amount of weapons-grade uranium.[8] In March 1987 Zia declared in an interview given to *Time* that 'Pakistan has the capability of building the [nuclear] bomb'. He added, however, that Pakistan does not intend to do so.[9] In April 1986 *Der Stern* stated that Kahuta has 14 000 centrifuge tubes.[10] David Albright believes that the lower figure of 1000 is more credible because of the reported material problems of the centrifuge tubes.[11] One technician working in Kahuta is reported to have told *Stern* that centrifuge tubes 'were exploding all over the place' because of wrong steel being used in the rotors.[12] Several Pakistani agents have been caught quite recently in attempts to purchase illegally from the United States so-called maraging steel containing nickel, cobalt and molybdenum and having a very high tensile strength. It would be the right material. The same agents ordered pure beryllium, a metal which has hardly any other use than as a neutron reflector in a nuclear bomb.[13] A Pakistani agent was also caught in the United States trying to smuggle 50 Krytrons, quick electronic switches used almost exclusively to ignite nuclear bombs.[14]

It takes 238 separative work units to produce 1 Kg 94 per cent U^{235} from natural uranium. If Kahuta has 1000 operative tubes each having a separative capacity of 5 units per year it could produce 21 kg of weapon-grade uranium (one bomb) per year. With 14 000 centrifuges it could produce 300 kg (15 bombs) per year, assuming around 20 kg are needed per bomb.

As already mentioned, the programme started in 1972 after the loss of the 1971 war with India. Since 1945 Pakistan has lost three wars to India, which is a much stronger military power. When India exploded its 'peaceful' nuclear device in May 1974 the then President Bhutto swore that Pakistan would build a bomb 'even though we had to eat grass'. According to V. D. Chopra, the Pakistani military establishment has given a rationale for the bomb programme to American experts: it would 'neutralize an assumed Indian nuclear umbrella under which Pakistan could re-open the Kashmir issue; a Pakistan nuclear capability paralyses not only the Indian nuclear decision but also Indian conventional forces and a brash, bold Pakistani strike to liberate Kashmir might go unchallenged if the Indian leadership was weak or indecisive'.[15] P. K. S. Namboodiri quotes the American professor Stephen Cohen in this matter:

> Pakistan belongs to a class of states whose very survival is uncertain, whose legitimacy is doubted and whose security-related resources are inadequate. Yet these states do not go away nor can they be ignored. Pakistan (like Taiwan, South Korea, Israel and South Africa) has the capacity to fight, to go nuclear, to influence the global strategic balance (if only by collapsing), and lastly, is in a strategic geographical location, surrounded by the three largest states in the world and adjacent to the mouth of the Persian Gulf . . .[16]

The outspoken Khan has given the following rationale for the Pakistani nuclear deterrent:

> The appreciation by the threshold countries that nuclear weapons confer enormous advantages upon the country having them and can affect the imbalance in manpower, natural resources, industrial

potential and military strength, is a very big incentive for nuclear proliferation and has contributed a lot to the desire of these countries to keep all options open. India's persistence [*sic*] and continuing quest for international recognition as more than a 'pawn on a global chessboard' forced it to detonate a nuclear device in 1974. Pakistan's future policy is to remain closely tied to Indian actions. If India openly starts a weapon programme, the deep-rooted Pakistani fears of India, especially after its active role in the dismemberment of Pakistan in 1971, would put tremendous pressure on Pakistan to take appropriate measures to avoid a nuclear Munich at India's hands in the event of an actual conflict, which many Pakistanis think very real.[17]

As mentioned earlier, the Soviet forces in Afghanistan and the Reagan Doctrine led to the recommencement of the provision of US military aid to Pakistan, which had been stopped by Carter in 1980 because of Pakistan's bomb programme. Pakistan channels weapons to the Mujaheddins, gives them sanctuary and thus ties down Soviet forces in Afghanistan which is economically, politically and even in human lives costly to the Soviet Union. Economically the Afghan war is estimated to be costing the Soviet Union $12 billion annually. It has lost the support of the Third World, and particularly the Arab nations, and their protests continue. As for human lives, according to a US source, 'The US State Department estimates that about 15 000 Soviet troops have been killed in Afghanistan. Soviet officials have privately informed visiting Americans that the figure stood at about 25 000'.[18] So important, then has been Pakistan's support of the Mujaheddin that the Reagan Administration has continued to give generous military aid to Pakistan in spite of it almost overt bomb programme. In addition, it gave direct weapons aid to the Afghan guerrillas worth $680 million in 1987.

Pakistan is asking Washington for $4·02 billion for the next six years (1988–93) for a gigantic weapons package including 20 ultra-modern F-16C multipurpose fighter planes in addition to the 40 F-16s it received in 1986. The new request includes also three E-3A AWACS planes (or three small E-2C Hawkeye Planes). Because of financial considerations the Pakistanis probably could not, however, purchase them, particularly the AWACS planes which cost about $800 million each. Hence they may ask for a rental arrangement. Furthermore the planes evidently have to be accompanied by American pilots and technicians.

The Pakistani request to Washington is justified by the Afghan air strikes on the refugee camps and towns near the Afghan border, but the planes would also give Pakistan parity if not superiority against India (which has 1000 war planes against Pakistan's 300). The list also contains the most modern US battle tanks, M1s (or older, modernised M-60s), Harpoon and Tow-2 missiles and other modern weapons. Such a package of superweapons coupled with the vigorous nuclear programme would certainly induce an Indian nuclear weapons programme in response. And India would probably go much further than Pakistan; it is about 15 years ahead in nuclear weapons technology – it exploded a device in 1974, which Pakistan has still not yet done – and its nuclear technological base is much larger. India, moreover, had already increased its military budget for 1987 by 43 per cent.

Zia is in a difficult position. He wants quite visibly both the aid *and* the bomb; and it is questionable whether the Reagan Administration will be able to continue to give the aid, for there is strong opposition in Congress, particularly that of the influential Senator John Glenn.[19] And if the approaching *détente* helps the Soviet Union to withdraw its forces from Afghanistan it would be difficult for Zia to justify the weapons requirement any more. But Zia needs economic aid to feed the three million refugees. Pakistan's internal order has also badly deteriorated. Amounts of heroin, automated weapons in large numbers in the hands of terrorists and criminals (said to be supported by the Afghan secret police), and ethnic disturbances have grown tremendously in the last few years.[20]

It seems likely, then, that a new arms race is emerging in the Indian Peninsula in spite of the fact that both Pakistan and India badly need a lessening of tensions in order to be able to devote their resources to domestic problems. Let us hope that the emerging superpower *détente* will also bring regional *détente* to this over-populated and troubled region.

Notes

1.   Leonard S. Spector, *Going Nuclear* (Cambridge, Mass., 1987) pp. 104–24; David Albright, 'Pakistan's Bomb-making Capacity', *Bulletin of the Atomic Scientists* (June 1987); and *Newsweek*, 16 March 1987.
2.   P. K. S. Namboodiri, 'Pakistan's Nuclear Posture' in K. Subrahmanyam (ed.), *Nuclear Myths and Realities* (New Delhi, 1984).

3. V. D. Chopra 'Pakistan's Nuclear Project' in V. D. Chopra and R. Gupta (eds), *Nuclear Bomb and Pakistan: External and Internal Factors* (New Delhi, 1986).
4. Namboodiri, 'Pakistan's Nuclear Posture', p. 143.
5. The Government of the Netherlands, *Report of the Inter-ministerial Working Party Responsible for Investigating the 'Khan Affair'* (The Hague, 1979); and Namboodiri, 'Pakistan's Nuclear Posture', pp. 164–94.
6. Leonard S. Spector, *The Nuclear Nations* (New York, 1985) p. 118.
7. *Nucleonics Week*, 2 August 1984.
8. *Washington Post*, 4 November 1986.
9. *Time*, 30 March 1987.
10. *Der Stern*, 30 April 1986.
11. Albright, 'Pakistan's Bomb-making Capacity'.
12. *Der Stern*, 30 April 1986.
13. M. Holderness, 'Special Steel Evidence of Pakistan's Nuclear Plans', *New Scientist*, 30 July 1987.
14. Chopra, 'Pakistan's Nuclear Project', pp. 15–22.
15. Ibid, p. 22.
16. Namboodiri, 'Pakistan's Nuclear Posture', pp. 162–3.
17. A. Q. Khan, 'The Spread of Nuclear Weapons Among Nations: Militarization or Development' in Aga Khan (ed.), *Nuclear War, Nuclear Proliferation and Their Consequences* (Oxford, 1986) p. 423.
18. Marin Strmecki, 'Gorbachev's New Strategy in Afghanistan', *Strategic Review* (Summer 1987) pp. 31–42.
19. M. Crawford, 'Glenn Asks Reagan to Halt Pakistan Aid Pending Review of Nuclear Programs', *Science*, 13 March 1987, p. 1321.
20. Strmecki, 'Gorbachev's New Strategy in Afghanistan'; and M. Isphani, 'The Perils of Pakistan', *The New Republic*, 16 March 1987, pp. 19–25.

Part IV
Problems of Command and Control

13 Routine Nuclear Operations

Paul Bracken

Two arms competitions exist between the United States and the Soviet Union. The first is the classical arms race where each side accumulates nuclear missiles, bombers and submarines. This competition has been studied extensively. It focuses on questions of deterrence, perceptions, mutual interactions, and arms control programmes to eliminate or restrain *numbers* of weapons. The START negotiations, the Intermediate Nuclear Forces (INF) negotiations and the Anti-ballistic Missile (ABM) Treaty are all structured to take account of this dimension of the arms race. There is, however, another more obscure competition that goes on between the two nations. That is the *operation* of military forces. A virtually continual game of feint, bluff, and stand-down goes on, involving the airplanes, submarines, ships and other forces of the two sides. This game, which has received relatively little attention, has become such an ingrained feature of the superpower nuclear relationship that for all practical purposes it is widely seen as 'routine' by those inside the respective national security establishments. Why are submarines brought in close to any enemy coast, or bombers flown up to his air defences? The answer, for the Soviet and American security and intelligence bureaucracies, is that they have always done it.

It is time to start thinking about this second, operational, arms competition because it needs to be the focus of more political attention. Perhaps it too needs restraint through arms control. Yet until some of its broad features are understood and debated serious policy discussion is impossible. It is premature to propose sweeping changes and restrictions to current practices, but it is not too soon to lay the groundwork for imposing arms control on this set of activities. Conceptually, operational arms control is about twenty years behind ordinary numerical arms control. The subject needs a lot more attention and debate.

A distinction between routine and non-routine activities is often used in the analysis of complex operations. It describes the differences

219

between normal practised activity and that employed in unusual situations. Since there has never been a nuclear war between the United States and the Soviet Union, nor even a serious bilateral nuclear alert, it is fair to say that a serious nuclear crisis would be an abnormal, unique situation in which many new and unanticipated phenomena would arise. Non-routine operations would be an extension of routine rehearsed operations, but they would involve much more as well. Indeed, non-routine behaviours are likely to be important intelligence tip-offs of attack, and therefore any venturing into this territory could reveal serious instabilities in a nation's ability to remain at high alert for an extended period. However, operations during an intense crisis or in actual wartime would involve added special strategic and political considerations quite different from routine peacetime operations. For this reason, the discussion here will concentrate on the more frequent and familiar activities of nuclear forces when there is no threat of war.

Pieces of the nuclear operations landscape are familiar to most students of the strategic situation. The United States until 1968 maintained an airborne alert of some of its nuclear bombers to insure against a Soviet surprise attack. Large-scale exercises were undertaken to test the capabilities of the bomber forces. For example, bombers were rapidly launched with live weapons aboard and sent towards Soviet airspace. Upon reaching the Soviet radar line, still outside actual Soviet territory, the planes would veer away at the last possible minute. To a Soviet radar operator or commander in some underground bunker the thought had to occur that this nerve-wracking bluff might just be the real thing: perhaps this was a surprise attack or a malfunction that had taken place, so that the bombers would not be called off.

On the Soviet side, different but analogous operations have been used. For example, the Soviets patrol some of their nuclear weapon-firing submarines off the Atlantic and Pacific coasts of the United States. In these positions they could launch strikes on Washington and interior bomber bases in less than 15 minutes. By sailing in closer they can decrease US reaction time even further.

Several generalisations about these routine nuclear operations are worth making. The first is that few individuals outside a small cell within government understand the totality of what is going on. For the public, it is rare for anything more than the isolated incident to be reported in the mass media. A Soviet submarine collides with a US carrier in the Sea of Japan, a Soviet reconnaissance aircraft is fired on as it is about to enter Soviet airspace, or a US intelligence aircraft is confused with a commercial airline which is then fired upon by Soviet air defences. Information on these operations comes from sensitive intelligence collection and, if possible, there is an understandable bureaucratic tendency to conceal the sources of the collection.

Within government, surprisingly, the picture for most officials is not much better. Many routine operations are so highly classified that a strict 'need to know' rule is applied. This cuts out a lot of people, most especially in the foreign ministries. Even relatively high officials in the Soviet and American foreign ministries may not know of or understand the nature of most of these operations because they do not have clearance to gain access to their details. What is known is garnered from public sources, or second-hand information from friends closer to the operation. Yet even here a restraint applies, because a diplomatic official will not enhance his or her career prospects by poking highly secret compartmented military operations. This should not really surprise us because history shows many comparable or analogous examples. In 1914, for example, the German Foreign Office had essentially no information about the nature of German operational plans.

One might surmise that intelligence agencies or military services themselves would be informed about these operations. But even here compartmentation reigns. There is a traditional wall that separates intelligence from operations, and this is one reason for the high success rate of surprise attacks throughout history. Intelligence personnel frequently have low status in military organisations, and since they command nothing in terms of military forces, they too have no need to know about operations and plans of their own side. As for military organisations, they will tend to have a great deal of knowledge about very narrow operations. The Soviet Navy, for instance, will obviously have people who understand how their intelligence vessels come in close to the US coast, or how they hug the US fleet. However, these individuals will know nothing about Soviet exercises for launching the Strategic Rocket Forces on warning, or about signals intelligence in Europe. In the United States similar compartmentation is also the rule rather than the exception.

If the foreign ministries and intelligence agencies of the two sides are shut out from the total picture of routine nuclear operations, who then does have an overview of them? At least in the United States special groups are formed within the Joint Chiefs of Staff Organization and the political leadership to ensure that legitimate oversight is represented in decisions on such matters. In the Soviet Union, one assumes (or hopes?) a comparable set-up exists between the Politburo and the General Staff. But since relatively small numbers of people have an accurate picture of the totality of even routine operations, the circumstance of who happens to be in charge matters. A national leader might be ill-prepared to understand what is going on, or what he is approving. Or he may be sick or incapacitated. The central question concerning the informational features of routine nuclear operations, then, is one familiar to political scientists. The problem of perceived images has long been studied in this field: how images and perceptions, rather than actual events in the world, shape behaviour and decisions. But the subject would be better described as one of 'compartmented perceived images'. Only a dozen people or less might understand the extent of routine intelligence and military operations. The small numbers create opportunities for interpretations that are not reviewed by any broader body of people.

Another feature of routine operations is that they depend on the reactions of the competitor. It is necessary to delegate routine authority to local commanders because there may not be time available to request guidance from higher military or political commanders if something goes wrong. The South Korean airliner incident of 1983 is a case of this. No one could have foreseen the complications arising from routine US intelligence collection flights near Soviet territory in the Far East. There was no internal lobby to ask critical questions about what could go wrong in a confused situation, or even to ask *beforehand* whether Soviet air defence fighters could talk to a commercial airliner if the need arose.

A country could, of course, so severely restrict the authority of its field commanders as to prevent them from taking any significant action whatsoever in this kind of situation. In so doing, however, new vulnerabilities are opened up, especially if the other side finds out that authority is severely restricted. It does not seem that the classic political mechanism for control of military operations, the use of rules of engagement, is an adequate administrative solution for modern quick-reacting forces. We are relying on 200-year-old management methods to govern late twentieth-century forces, some of which carry nuclear weapons, and some of which have as their mission the probing of the opponent's nuclear warning and alerting systems. A relatively small number of people are in adequate possession of enough information to understand the overall picture, and tens of thousands of people in the command and control organisation are given rules of engagement and restraints on operations without the slightest information about the overall context of what they are doing.

What motivation encourages a nation to undertake routine nuclear operations in the first place? Nuclear forces can be used for a variety of purposes in a peacetime non-crisis environment. A great deal of experience, both Soviet and American, suggests that these operations have mundane military rationales behind them. Radar frequencies are determined, and tactical responses estimated. But the more interesting motivation is political and strategic. Routine operations can be used for subtle intimidation, as a demonstration of bluster, to create uncertainties and anxieties in an opponent, or simply to remind the other superpower that nuclear weapons exist and must always be a factor in his decisions. These political and strategic aspects of routine operations can be grouped under the heading of large-scale 'impression management'.

A nation may simultaneously engage in publicly improving political relations while operating its nuclear and other military forces in such a way as to remind the leadership of another nation that it possesses large military forces that could be triggered into action during a crisis. Let us consider a purely hypothetical example of an illustration of this type of behaviour. During an arms control negotiation a country suddenly flushes its nuclear-firing submarines from port. They put to sea in a way that evades detection by the intelligence sensors of the other side. This in itself demonstrates a keen knowledge of how the other's sensors operate, and shows that in an actual crisis the submarines could simply 'disappear'. It is meant to shock the other side into understanding how primitive its defences really are. A short time later the submarines are co-ordinated suddenly to appear on the opponent's sensors, all at once, undergoing provocative firing preparations. All of this, of course, would be 'practice'. The public would never know any of this, nor would the diplomats of either side. The only people who would be aware are a handful of highly-placed intelligence officials, some senior military advisers, and the national leaders.

Why would anyone do this sort of thing? A complex of reasons account for the motivation. Every senior decision-maker understands the benefit of keeping an opponent off-guard, guessing about intentions and about what game is being played. In order to do this, he must create a sense of the unexpected in peacetime, if he is to have any hope of convincing his opponent not to test him with an actual crisis or war. Provocative moves with strategic forces, such as simultaneously having a bomber exercise suddenly overlaid on top of ordinary exercises can show that one is not weak in an arms control negotiation, something that might be especially important if one is actually making concessions, or if a leader is weakened domestically on the home front. It has been reported, for example, that during the Watergate years in the United States in the 1970s, nuclear forces were operated in some particularly unusual 'routine' ways. Such actions would never see public or diplomatic scrutiny, outside the narrowest of compartmented channels.

Another strategic reason for undertaking such operations would arise from attempts at intimidation regardless of the context. A nation might feel that military force was the currency of international politics. In addition, internal bureaucratic politics could influence these types of operations. A leader might be drawn in two different directions at once, and might try to solve the problem by conducting carrot-and-stick approaches to negotiation, safe in the knowledge that a warlike image would not arise because of the secrecy organic to these operations. However, the leader for reasons of personality or national style might emphasise the carrot more than the stick.

Here the Soviet Union's behaviour seems noteworthy. At the cost of tremendous political setback to its publicly-stated political interests of better relations with countries on its borders, Moscow has engaged in extremely provocative operations directed at Sweden and Japan. Against Sweden, the Soviet Union has launched dozens of submarine penetrations of territorial waters, in full knowledge that Stockholm detects many of these intrusions. However, so far Sweden has sought to play down these incidents, although the public's reaction has been most negative against the Soviet Union. In Japan, a continual stream of Soviet air and sea sorties around Japanese waters has contributed to a steady worsening of relations between these two nations. What political gain Moscow gets from these is difficult to discern. Their overall effect, however, is to create opinions in the West that the Soviet Union is not willing to harness its routine military operations, and until this happens, concern will continue to exist.

It seems that, if the superpowers can sit down to talk about the numbers of nuclear weapons, and conflicts in particular nations, then it is entirely plausible that they might one day discuss routine nuclear operations as well. Doing so might have a number of positive effects, although a great deal more consideration needs to be given to this subject. Restrictions on routine nuclear force operations might minimise the chances of loss of control in crisis by separating the forces of the two sides. At a more political level, such restraints could further delegitimate nuclear weapons from the currency of international relations by placing an even larger taboo on using these systems to achieve political objectives. This is something that should be approached very carefully, however, especially by the United States. The United States still sees advantages in retaining the threat of nuclear weapon use in some circumstances. Nuclear weapons, in many people's minds, have kept the peace for four decades. Yet at some point we may want to match cuts in nuclear arms with restraints on routine operations. Arms control restraints on the number of forces have not seemed to damage either side's security, and limitations on operations could contribute to the positive easing of tensions by restricting the ability to engage in secret threats and diplomacy. During the next few years the groundwork for thinking through these delicate issues needs to begin, so that our conceptual machinery and vocabulary are adequate to the task of negotiating these important problems.

14 Arms, Decision-Making and Information Technology

Allan M. Din

ARMS AND INFORMATION TECHNOLOGY

Historically, the technological arms race has often been claimed to be the main reason for the world finding itself in a permanent condition of extreme over-armament accompanied by a constant threat of cataclysmic nuclear confrontation. While there is certainly some truth in this claim, it has had the unfortunate result of attaching a distinctly negative connotation to technology when measures attempting to control and curtail the arms race are discussed. Some technologies have a very narrow military field of application and may therefore justifiably be subjected to criticism within the context of arms control. Other technologies, especially nuclear weapon and nuclear energy technologies, require a more elaborate arms control assessment. Finally, there are technologies which have both important weapon system applications *and* a potential for furthering arms control and treaty verification.

To avoid the impression that arms control efforts might be too strongly associated with a negative attitude towards technology, it is essential that more attention be focused on technological developments which may contribute positively to the cause of arms control. In the present chapter, it will be argued that information technology is an important, but hitherto insufficiently appreciated, factor in a multitude of modern weapon developments which fuel the technological arms race; at the same time, however, it also has a potential as a dedicated arms control technology which merits more emphasis.

In the 1980s, the technological arms race is reaching new levels of technical sophistication and complexity. The most recent example of this is the effort which is currently being expended on research and development for a panoply of star wars weaponry developed within the US Strategic Defense Initiative (SDI) programme. The debate, in

226

this regard, has mostly turned around the feasibility or non-feasibility of developing particular types of weapons, such as laser and particle beam weapons. Also in other areas of the technological arms race, interest often focuses mainly on the individual features of weapon types. Thus, for example, there has been much discussion on the technical capabilities of the new MX and Midgetman missiles, the B1 and the Stealth bomber. Unfortunately, only comparatively little public attention has been paid to more intangible issues, such as command and control, strategy and doctrine, which are central to any assessment of the effect on international security which the new technologies might have.

The overall performance of weapons sytems may be subject to several shortcomings due to large-scale system features which add to the basic problems of making the individual weapons work as planned. Information technology, including computer hardware and software as well as sensors and communications, is a key ingredient in the workings of complex systems and therefore its role in modern weapon systems requires a careful, but unfortunately often neglected, analysis. Information technology is an essential element in many weapon components: computer chips are ever-present in sensor and guidance parts of 'smart' weapons and on-board microprocessors are commonplace in aircraft, missiles and setellites. These uses of information technology have led to highly improved weapon precision and lethality but also to new types of failure modes caused by the fragility of electronic components, especially in hostile military evironments like the one generated by the Electromagnetic Pulse (EMP).

Much military research and development is orientated towards an upgrading of the computer hardware basis of information technology. Projects concerning Very High Speed Integrated Circuits and Very Large Scale Integration are deemed essential for fielding new miniature weapons with high information processing rates. For this purpose, computer chips with submicron features are being developed and new materials, like gallium arsenide, make their appearance. Novel computer architectures based, for example, on parallel processing may play a decisive role in future computer applications where massive data flows and high-speed processing are required.

Though developments in computer hardware are, and will remain, essential for advanced weapon technology, it is increasingly being realised that software development is a very critical, and often neglected, element. This has prompted much recent alarmism about the crisis in military software which threatens the performance of a variety of existing and planned weapon systems. In many cases software development and maintenance are responsible for almost 90 per cent of total weapon costs, but software quality still leaves much to be desired. It is important to note that these shortcomings exist, and persist, within rather traditional applications of information technology and the situation is not likely to improve with the development of even more sophisticated software applications involving, for example, artificial intelligence languages.

The virtues and problems of information technology in the context of the command and control of complex weapon systems are difficult to analyse and assess because the technology is rather new and many developments are taking place in military programmes on which technical details are scantily available. One such programme is the Strategic Computing Initiative (SCI) started in 1983 by the US Defense Advanced Research Projects Agency (DARPA); its most important project is precisely concerned with battle management and Command, Control, Communications and Intelligence (C^3I) for use within the SDI. A general assessment of the potential and problems of information technology applications is possible, however, despite the secrecy shrouding much of the research in the overlap areas of SCI and SDI.

In the context of armaments, the interest in information technology is not limited to its use in the modern weapon systems which contribute to the arms competition. There are also important arms control applications in areas such as seismic monitoring, satellite verification and modelling for strategy and negotiations. These applications are discussed below.

DECISION-MAKING AND ARTIFICIAL INTELLIGENCE

A large body of problems associated with the use of weapons and the control of weapons systems are intimately related to the question of efficient decision-making. This circumstance supports the assertion made above that information technology is of fundamental importance for weapon assessments. In fact, at the level of individual weapons such as, for example, a fighter aircraft, the amount of information about malfunctioning, guidance, external threats and fire control is substantial and both evaluations and decisions have to be made rapidly. The basic problem is one of fusing incoming data to initiate an appropriate response. At the level of large-scale command

and control of weapon systems, the problem is the same even if the incoming information and the subsequent decisions have a very different character. On both levels, the development of modern weapons has been so rapid that one is faced with an information explosion which implies that decision-makers, working under the stress of crisis or conflict, will be called upon to deal with extremely complex choices vastly exceeding the processing capacity of the human mind.

One simple solution for avoiding the information explosion is information compression: some pieces of information are cut away or kept for later retrieval while other, presumably more relevant, data are transmitted directly to the decision-maker. The risk is, of course, that some minor data under certain circumstances turn out to be highly important, but nevertheless remain concealed with potentially serious consequences. One example of this kind happened in connection with the US Space Shuttle accident in 1986 when information on the fatal fuel leak was not displayed directly in the control room but could only be recovered after some time from piles of data tapes. In this example it might not really have mattered if the pertinent information had been readily available, but one may not in general be sure that this would never be the case. While there can be no absolute guarantee that crucial data about elements of complex systems are not overlooked or discarded, it may still be possible to devise intelligent ways of retrieving, selecting and presenting relevant information as a basis for making decisions in situations where the human brain for various reasons is unable to cope. The relevant aspect of information technology for such endeavours is artificial intelligence.

Artificial intelligence (AI) has many connotations and is notoriously difficult to define precisely but, for the purpose of the present discussion, it will be sufficient to think of AI as synonymous with computer-aided decision-making. Various kinds of AI developments have been and are being pursued with different kinds of ambition level, from mimicking the human brain to more pragmatic expert systems, the latter being the most promising for near-term applications. The essential ingredients in an expert system are a base of knowledge and data together with a so-called inference engine which uses input information and the base in a computerised evaluation so as to present conclusions and/or multiple choices through a user-friendly interface to a human decision-maker. The expert system fulfils the function of sorting and compressing large amounts of information in an 'intelligent' way and in so short a time span that the

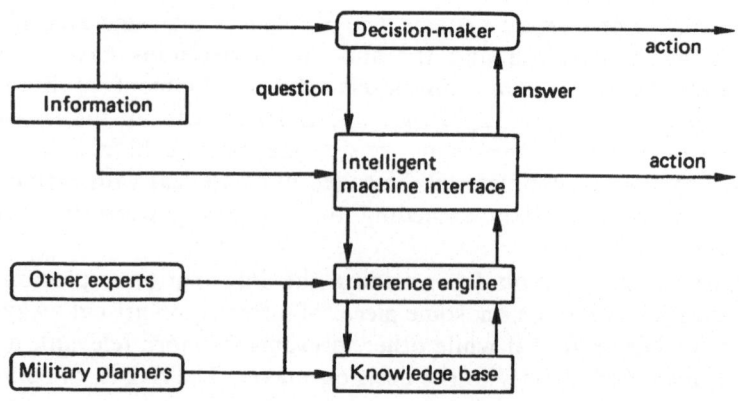

Figure 14.1   Outline of an expert system

human mind is unable to compete with it (see Figure 14.1). The level of human intervention in the functioning of an expert system may vary. At one extreme, the operator is in full control by making a long chain of proper decisions upon presentation of multiple choices; at the other extreme, the operator may be completely excluded from the loop when the computer is devised to make 'intelligent' decisions at all levels and, at the end of the process, to carry out a specific action. The latter option might be invoked in very complex situations under heavy time pressure.

INTELLIGENT WEAPON SYSTEMS

Some modern weapons contain electronics and/or computer chips to the extent that it is customary to term them 'smart' weapons. They are, however, 'dumb' insofar as they function with only a few sensors and simple operating instructions and follow a rather predictable pattern of action. The objective of the DARPA programme for strategic computing is precisely that this situation should change and that a new generation of more 'intelligent' weapon systems will soon see the light of the day. Some extracts from the 1983 SCI document illustrate the point:

As a result of a series of advances in artificial intelligence, computer science and microelectronics, we stand at the threshold of a new generation of computing technology having unprecedented capabilities. The United States stands to profit greatly both in national security and in economic strength by its determination and ability to exploit this new technology.

Computing technology already plays an important role in defence technologies such as guided missiles and munitions, avionics and C^3I. If the new generation of technology evolves as we now expect, there will be unique new opportunities for military applications of computing. For example, instead of fielding simple guided missiles or remotely piloted vehicles, we might launch completely autonomous land, sea and air vehicles capable of complex, far-ranging reconnaissance and attack missions. The possibilities are quite startling, and suggest that new generation computing could fundamentally change the nature of future conflicts.

In contrast with previous computers, the new generation will exhibit human-like 'intelligent' capabilities for planning and reasoning. The computers will also have capabilities that enable direct, natural interactions with their users and their environment as, for example, through vision and speech.

Using this new technology, machines will perform complex tasks with little human invervention, or even with complete autonomy. Our citizens will have machines that are 'capable associates', which can greatly augment each person's ability to perform tasks that require specialized expertise. Our leaders will employ intelligent computers as active assistants in the management of complex enterprises. As a result the attention of human beings will increasingly be available to define objectives and to render judgments on the compelling aspects of the moment.

As is often the case when the discussion is about new technology, it appears difficult to draw a clear distinction between fact and fiction, between what can be done in special situations and what may be accomplished for large-scale systems, and between near-term and long-term applications. It is an inescapable reality, however, that an ambitious information technology programme has been initiated and that computerisation is encroaching at the higher levels of military decision-making. The initial SCI programme was oriented in three different directions. First, the pilot's associate project is to develop an ensemble of integrated aircraft systems which as a key ingredient will involve an expert system with advanced speech input/output and real-time computer technology; this will allow the pilot to concentrate on only a few control functions and, eventually, under severe threats and time constraints, let the computer system take automatic action. Under a second project title, the technology for an autonomous vehicle will be developed with the objective of producing a moving machine capable of autonomous navigation and tactical decision-making. It will combine the use of laser range scanners and colour TV cameras with high-speed computing technology and advanced image understanding. This project may also be seen as part of wider-scale US Army-inspired efforts to develop sophisticated battlefield robotics which have generated exaggerated claims about radical changes in future tactical warfare. Thirdly, the project on navy battle management is to develop machine intelligence technology for US naval commanders which will handle, more or less automatically, the complex threat, planning and battle management situations of big naval carriers. This project has a command and control scope which goes far beyond the navy application area; in fact, the main interest in the battle management potential of the SCI now lies in the strategic defence field.

The realization that Battle Management/Command, Control and Communications (BM/C³) are central to a successful development of SDI was for some time obscured by disputes about various weapon elements. At present, the importance of this part of SDI is underscored by the effort going into the so-called National Test Bed (NTB), which will be a huge computer facility for running simulations of BM/C³. This facility is bound to address the crucial problems of how decision-making can be implemented by a distributed computer system, how far automation must be driven and to what extent human beings must be left out of the decision loop. A more recent extension to the SCI programme has been the project which deals specifically with the information technology and architecture needed for implementation of AirLand Battle (ALB) concepts on the European tactical battlefield.

COMMAND AND CONTROL PROBLEMS

Many of the developments in information technology discussed above are likely to fuse with other emerging technologies into weapons systems for a modern tactical battlefield which will have a high degree of automation. The military dream of an automated battlefield was spelled out, for example, in 1969 by General William Westmoreland and partly implemented during the Vietnam war, albeit without much success. In the 1980s, however, the relevant technologies are seen by some military planners as having matured to a degree which warrants their inclusion in both strategy and doctrine. The most prominent examples of this trend may be found in the emergence of the new war-fighting doctrines of the North Atlantic Treaty Organisation (NATO) involving the concepts of ALB and Follow-On-Forces-Attack (FOFA). These concepts emphasise deep strikes into Warsaw Pact territory which for their execution require a very high level of computerisation and automation to achieve effective battle management.

The command and control problems involved in ALB and FOFA are staggering. Huge amounts of information from elaborate systems of ground-, air- and space-based sensors about potential threats and targets will be collected by an 'electronic hilltop'. Military operations are, however, likely to evolve during a very short time span, so that centralised, and therefore vulnerable, command and control cannot be relied upon and, as a result, distributed decision-making becomes a necessity. Automation like, for example, the pilot's associate is therefore a natural part of the picture. The short decision times involved in ALB and FOFA large-scale operations and the trade-offs between centralised and distributed control create several types of risks. In particular, the escalation risk is not negligible since some local threat assessments may easily give rise to action which is out of proportion to the more global threat. At the extreme, with deep strike action control running astray, an initial conventional weapon conflict could become a nuclear one.

The command and control problems inherent in strategic defence systems are different from those of the automated tactical battlefield but are not likely to be less formidable. The main difference lies in the global scope of missile defence with its muliple tiers, including boost, mid-course and terminal phases, with the reliance on extensive satellite support, as well as with the coupling to the use of offensive nuclear weapons. The problem of trade-offs between centralised and distributed decision-making remains and is even further exacerbated because of the characteristics of strategic defence weapon systems involving, for example, laser and particle beams. The engagement time for such sophisticated weapons is measured in fractions of a second, a circumstance which implies prompt local threat assessment and action to prevent the early loss of space assets of importance to the defensive shields. At present, C^3I related to offensive nuclear weapons relies heavily on satellite-based sensors and communication systems. Strategic defence systems, with their weapon platforms and new types of sensors, will enhance immensely the already important role of space; at the same time, the anti-satellite capability which invariably accompanies anti-missile systems, will increase the vulnerability of command and control functions. Concomitantly, the risk of accidental nuclear war is also likely to increase.

The system issues raised by SDI, of which command and control are but elements, will be analysed in connection with the NTB-simulation facility referred to above. It is, however, an open question whether this kind of exercise will ever produce a plausible architecture for strategic defence systems which would appear sufficiently reliable to satisfy both military and political decision-makers. One very important problem is software reliability and validation; the millions of lines of computer code required for the 'Star Wars' computer system are likely to contain hundreds and thousands of 'bugs' which no simulation, necessarily vastly different from the realities of actual conflict, could possibly identify and remove.

VERIFICATION TECHNOLOGY AND ARMS CONTROL MODELLING

In the context of arms control, information technology and its novel 'intelligent' elements have a number of interesting applications which may be developed further than has been the case hitherto. This is manifest in connection with verification technologies where large amounts of data must be processed, for example to certify compliance with arms control treaties.

In the field of satellite monitoring, the superpowers have so far had a virtual monopoly, commonly referred to as the National Technical Means of verification, which has played an important role in the provisions of a number of treaties, like the Limited Test Ban Treaty, the Anti-ballistic Missile Treaty and the SALT Agreements. A variety of reconnaissance satellites have, over the years, provided each of the superpowers with detailed scrutiny of the military installations of the other side and have thus made possible verification of compliance as well as detection of other features of strategic importance. To carry out such a detailed monitoring activity requires a sizeable technical equipment for image processing and a large staff which scans and analyses the data. Currently, satellite monitoring and advanced image processing is developing in an interesting way outside superpower control. For example, the SPOT satellite launched by France in 1986 now makes it possible to have a close look at almost any part of the earth's surface with a resolution of ten metres. While this is no match for the resolution of some superpower satellites of down to a few tens of centimetres, it is still sufficient to open new perspectives for arms control verification, for example in connection with European security, and for crisis monitoring. In particular, the idea of an International Satellite Monitoring Agency (ISMA) acquires a new actuality from a technological point of view.

Even with a relatively low resolution of ten metres, an arms control verification satellite will provide very large amounts of data which must be analysed within an acceptable time and cost framework. It is on this point that new information technology for rapid image processing and pattern recognition could provide new approaches that make the ISMA proposal more feasible than was previously the case. The analysis and interpretation of satellite pictures may be facilitated to a large extent by 'intelligent' computer processing, although some human analysts will still be required for the final processing of the information. Another verification area where advanced information processing is of great importance is in seismic monitoring of underground nuclear tests. This is already the case for the present seismic arrays and will be even more so if a very low threshold, or a complete, nuclear test ban is concluded. A world-wide system based on regional arrays and number of 'black' boxes near potential test sites will require a complex communication and analysis of the data which could only be done by some automatic filtering during the information flow. Expert systems for such kinds of application have already been studied in some detail.

In arms control modelling, there is an interesting potential for novel applications of information technology, including artificial intelligence techniques. The latter have already been applied with interesting results to knowledge-based simulation of nuclear strategy issues, such as deterrence, escalation control and war termination. A similar methodology could be used in an arms control context for studying stability questions in connection with future drastic reductions in nuclear stockpiles, for example.

Information technology also offers the possibility of developing some novel analytical tools for negotiations on arms control in various areas. Given the complexity of weapon characteristics and deployments, the use of computerised databases is a necessity for negotiators. It is, however, in principle possible to go a step further by integrating these databases with a rule base embodying such matters as doctrines, strategy, fall-back positions, historical lessons and psychological perceptions so as to produce an expert system which could be a useful tool for arms control analysts. Attempting to set up such an expert system would be useful in itself because it would help unravel the decision-making processes involved in negotiations.

15 Controlling Theatre Nuclear War

Desmond Ball

INTRODUCTION

Over the past decade, both official defence establishments and independent strategic analysts have devoted increasing attention to the Command, Control, Communications and Intelligence (C^3I) systems which support the US and Soviet strategic nuclear forces, and to the role of these systems in crises and in strategic nuclear war-fighting. In particular, specific consideration has been given to such critical issues as the extent to which current strategic C^3I systems enhance crisis stability or instability, and whether or not they would serve to control escalation in the event of a strategic nuclear exchange or, because of their vulnerabilities, would in fact contribute to the dynamics of the escalation process.

Several publications have now directly addressed the central issues concerning strategic command and control systems, capabilities and processes. Paul Bracken has argued persuasively that the dynamic interaction of US and Soviet C^3I systems in a crisis could well serve to exacerbate the situation and, indeed, as the systems are generated to increasingly higher alert levels, could lead to irresistible pressures for pre-emption.[1] Bruce Blair[2] and John Steinbruner[3] have demonstrated that, historically, the US strategic C^3I system has been much more vulnerable than the strategic forces which it supports, and hence has provided an extremely lucrative target set – with unfortunate consequences for crisis stability. Indeed, as Steinbruner has argued, the vulnerability of the US C^3I system offers the 'only imaginable route' by which the Soviets might hope to achieve 'decisive victory in nuclear war'.[4]

The present writer's work has been concerned with the functioning of C^3I systems *during* a strategic nuclear exchange and, in particular, with the ability of such systems to provide continued, informed and responsive control over the strategic nuclear forces beyond the very early stages of a strategic nuclear exchange. It would be extremely

237

difficult to maintain control over such an exchange beyond several hours or several tens of nuclear detonations. Limited nuclear war or controlled nuclear war-fighting is more a chimera than the basis of a sound national strategic policy.[5]

Daniel Ford offers a further perspective. He argues, at least implicitly, that the inability of the US strategic C^3I system to survive any first strike against it, or to provide any subsequent responsive control over the strategic nuclear forces, derives from conscious or more likely subconscious design – namely, that neither the US Air Force nor the Navy have ever accepted the basic US national strategic policy of flexible response and controlled escalation; rather, the Air Force continues to adhere to a preferred policy of massive pre-emption while the Navy remains committed to massive retaliation.[6]

No similar studies have been published with respect to theatre C^3I systems. There have, of course, been numerous official (classified) analyses of the vulnerability of the various theatre C^3I systems to nuclear effects and, more recently, to attacks by Soviet special forces (Spetsnaz operations). However, such critical issues as the extent to which these theatre systems are designed to support crisis management and to function across the spectrum from peacetime operations through various levels of conflict to the large-scale employment of tactical and theatre nuclear weapons, and whether in fact their inability to perform as planned would actually exacerbate the dynamics of the escalation process, have not been sufficiently addressed.[7]

This is especially remarkable since the likelihood of theatre nuclear war is somewhat greater than that of strategic nuclear war. Indeed, it is difficult to envisage any strategic nuclear exchange which has not been preceded by very extensive use of nuclear weapons in one or more theatres. Further, theatre C^3I systems are even more vulnerable than their strategic counterparts and, moreover, it is likely that theatre nuclear wars would be even more demanding on C^3I systems than would strategic nuclear exchanges. Hence, as General Richard C. Ellis, former Commander-in-Chief of the Strategic Air Command (CINCSAC) noted in 1982, 'If you think we're in trouble in strategic connectivity, we're in terrible trouble in the battlefield theater area.'[8]

The various factors impacting upon the design and operation of theatre and tactical C^3 architectures were described by Donald C. Latham, then the US Deputy Under Secretary of Defense for Command, Control, Communications and Intelligence (C^3I) in testimony in March 1982 as follows:

The thing we are faced with [concerning theatre and tactical C³ systems] is a very wide range of potential conflicts that really stress the system in terms of the geography you have to fight over, the types of forces you might meet, and so on.

So, the architecture, or how we would construct the system, is really driven by some quite unique theater aspects. For example, and these are just examples in Europe . . . high threat; NATO and interoperability; and nuclear weapons. . . . In the Pacific of course it is somewhat different—there is very expanded geography, there are a lot of joint operations. . . . The North Koreans have a very formidable capability. Going in there and reinforcing and coping with that threat drives a given kind of [unique] force structure and supporting C³.

In the Third World so far perhaps a different set of crises, but also demanding in the way you would be able to handle the forces.

If you look at the influences, then, on this, there are some very unique service interests. There are joint operations aspects, and of course the allied views – interoperability is key here. IFF [Interrogation Friend or Foe] . . . how do you tell a German aircraft from a British, from an American, and talk to each other?[9]

In summary, theatre C³I systems contain added complexities introduced by the necessity to co-ordinate operations with allies and other second parties, many of whom do not speak English and lack interoperable C³ systems. In some theatres, such as South-west Asia, Africa and South America, the C³I architectures are comparatively rudimentary and unpractised. For example, an assessment by the Department of Defense of the US invasion of Grenada in 1983 found that the US Navy and Army used incompatible UHF communications systems, and that shipborne satellite communications equipment was not designed for manoeuvring operations:

Poor communications was the single most glaring deficiency of the entire operations.

Communications between JTF-120 and Army ground forces were practically non-existent. All ground operations were conducted using the same single net, since JTF-120 had no other communications capability which could link with Army ground forces; although both Army and Navy forces have UHF, the frequencies are incompatible and therefore, UHF could not be used.

Every time a ship would turn, communication was lost until the antenna could be manually readjusted. The net result was lack of communication during critical stages of the operations.[10]

As General John A. Wickham, Chief of Staff of the Army, explained in Congressional testimony on 25 February 1986,

There were problems of communications. There will be problems in the future . . .
The satellite antennas on the top of Navy ships had problems in lock-on. We hadn't thought about ships steaming around as opposed to steady on a certain course.[11]

The geographical areas of some theatres are 'very expanded' compared to the continental United States (CONUS) or Europe, while the distances from Washington and the major CONUS C³I centres further increase the communications difficulties. In the case of the naval theatre, the C³I architecture is not only highly vulnerable; there are also grave deficiencies in the ability of National Command Authorities to exercise negative control, let alone precise and timely control, over significant aspects of naval operations involving nuclear forces.[12]

This chapter addresses certain aspects of command and control in the European theatre – which, notwithstanding the magnitude of the threat, has a more developed, extensive, comprehensive, sophisticated and practised C³I architecture than that of any other theatre. The primary concern is the implications of the complexity and vulnerabilities of the theatre C³I system for the control of escalation during a major conflict in Europe. It also explores some connections between the theatre and global/strategic C³I systems which suggest that a major war in Europe would greatly impede the transition to any limited or controlled strategic nuclear exchange between the United States and the Soviet Union.

THE COMPLEXITY OF THE EUROPEAN THEATRE C³I ARCHITECTURE

It is difficult to envisage a decision-making process or a C³I system as complex, or one less capable of functioning in a timely and responsive manner, than that of the North Atlantic Treaty Organisation (NATO). The Organisation consists of 16 sovereign nations, speaking a dozen different languages, and each with different military capabilities, national interests and threat perceptions. Three countries have independent national nuclear capabilities, while another six are involved in joint control of US nuclear weapon systems. There are far more nuclear weapons deployed in this theatre than in any other, and these weapons vary greatly in terms of their ranges, yields, purposes and effects – and hence in terms of their C^3I requirements. The picture is further complicated by the existence of several dual-capable weapons systems.

Indeed, the decision-making process is so complex that the most likely outcomes are chaos and paralysis interspersed with precipitous activity. This is recognised within NATO itself. There is a popular joke at NATO headquarters, recently related by Daniel Charles, as follows:

> One of the most difficult decisions confronting Soviet military planners in the event that they attack Western Europe will be whether to bomb the Alliance's headquarters or not. They probably won't, so as to ensure a maximum degree of confusion within NATO.[13]

The Alliance is structured so that there are essentially separate channels of political decision-making and military command and control.[14] The highest political authority is the North Atlantic Council, which consists of 'permanent representatives' from each of the 16 NATO member countries, and which is located at NATO headquarters in Evere, north-east of the centre of Brussels. The Defence Planning Committee (DPC), which is composed of representatives of all the member countries except France, deals with matters specifically related to defence. The DPC provides the forum in which the member Governments would express their views on the political and military objectives of any proposed use of nuclear weapons, as well as discuss such particular – and extremely critical – issues as alert measures (LERTCON levels) and dispersal of the NATO nuclear forces. The views of the six countries which host NATO nuclear forces (West Germany, Belgium, the Netherlands, Italy, Greece and Turkey), in addition to those of the United States and Great Britain, would be particularly important in these deliberations. Although France withdrew its personnel from the NATO integrated Military Headquarters and terminated the assignment of its forces to the international commands in 1966, it is reasonable to assume that it would also be consulted on these critical decisions in the event of war.

On the military side, there are innumerable headquarters and commands – including those of NATO itself, those of the US forces in Europe, and those of the European countries. Many of these commands are dual-hatted. Others are peacetime structures and would devolve into other arrangements in wartime situations.

The highest military authority in NATO is the Military Committee, which is responsible 'for making recommendations to the [North Atlantic] Council and the Defence Planning Committee on those measures considered necessary for the common defence of the NATO area, and for supplying guidance on military matters to the Major NATO Commanders'.[15] In peacetime, the armed forces of the various member countries remain under national command structures, but in wartime they would transfer (or 'chop') to an integrated NATO command structure consisting of three Major Commands: Allied Command Europe (ACE), commanded by the Supreme Allied Commander Europe (SACEUR), whose headquarters are the Supreme Headquarters Allied Powers Europe (SHAPE) at Casteau, on the outskirts of Mons, about 60 kilometres south-west of Brussels; Allied Command Atlantic (ACLANT), headed by the Supreme Allied Commander Atlantic (SACLANT), headquartered in Norfolk, Virginia; and Allied Command Channel (ACCHAN), headed by the Allied Commander-in-Chief Channel (CINCHAN), headquartered in Northwood, Great Britain. These are each two-hatted commands: SACEUR is a 4-star US General who in peacetime is the US Commander-in-Chief Europe (USCINCEUR); SACLANT is a US Admiral who in peacetime is Commander in-Chief Atlantic (CINCLANT); and CINCHAN is a British Admiral who as Commander-in-Chief of the Fleet is Britain's most senior naval commander and who in wartime would also serve as the Commander-in-Chief of the Eastern Atlantic (CINCEASTLANT) – which is, however, subordinate to SACLANT![16] (See Figures 15.1, 15.2 and 15.3).

The highest US command in Europe is the US European Command (USEUCOM), headquartered at Vaihingen, an outer suburb of Stuttgart in West Germany. USEUCOM has three components: US Air Forces Europe (USAFE), headquartered at Ramstein, about 60 kilometres south of Frankfurt; US Army Europe (USAEUR), headquartered in Heidelberg; and US Navy Europe (USNAVEUR), headquartered in Grosvenor Square, London. In wartime, the USEUCOM staff at Vaihingen (Stuttgart) would split, with some personnel transferring to SACEUR's wartime headquarters at Casteau (Mons),

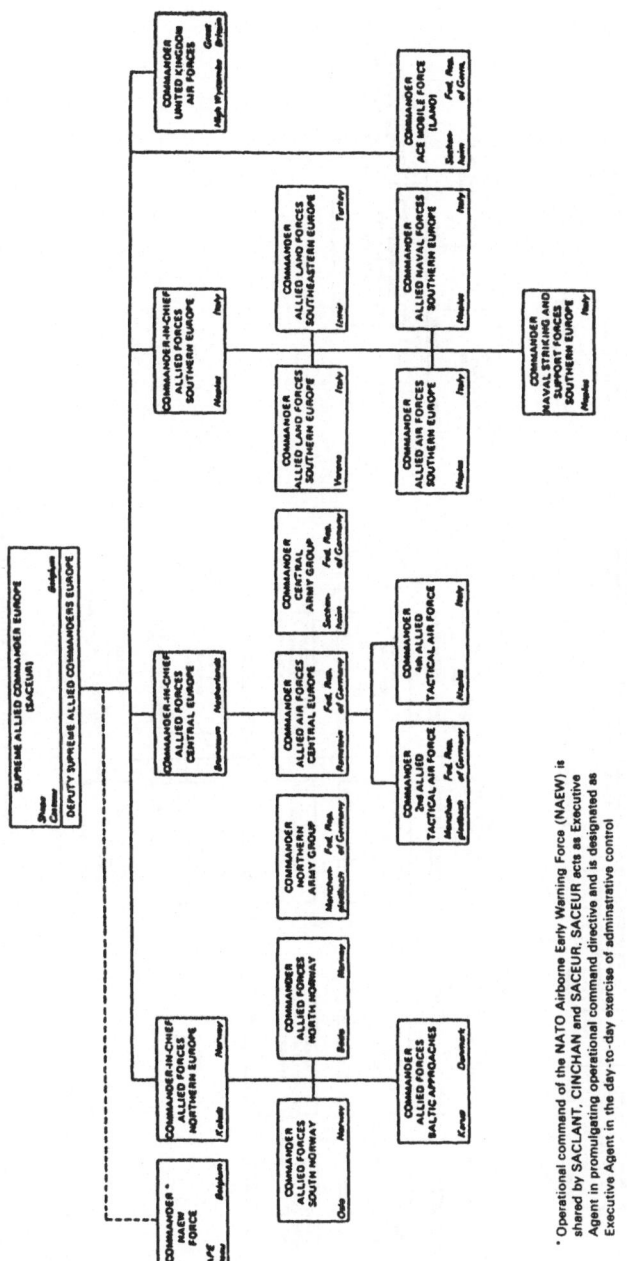

* Operational command of the NATO Airborne Early Warning Force (NAEW) is
shared by SACLANT, CINCHAN and SACEUR. SACEUR acts as Executive
Agent in promulgating operational command directive and is designated as
Executive Agent in the day-to-day exercise of administrative control

Source: *The North Atlantic Treaty Organisation: Facts and Figures.*

Figure 15.1   Allied Command Europe

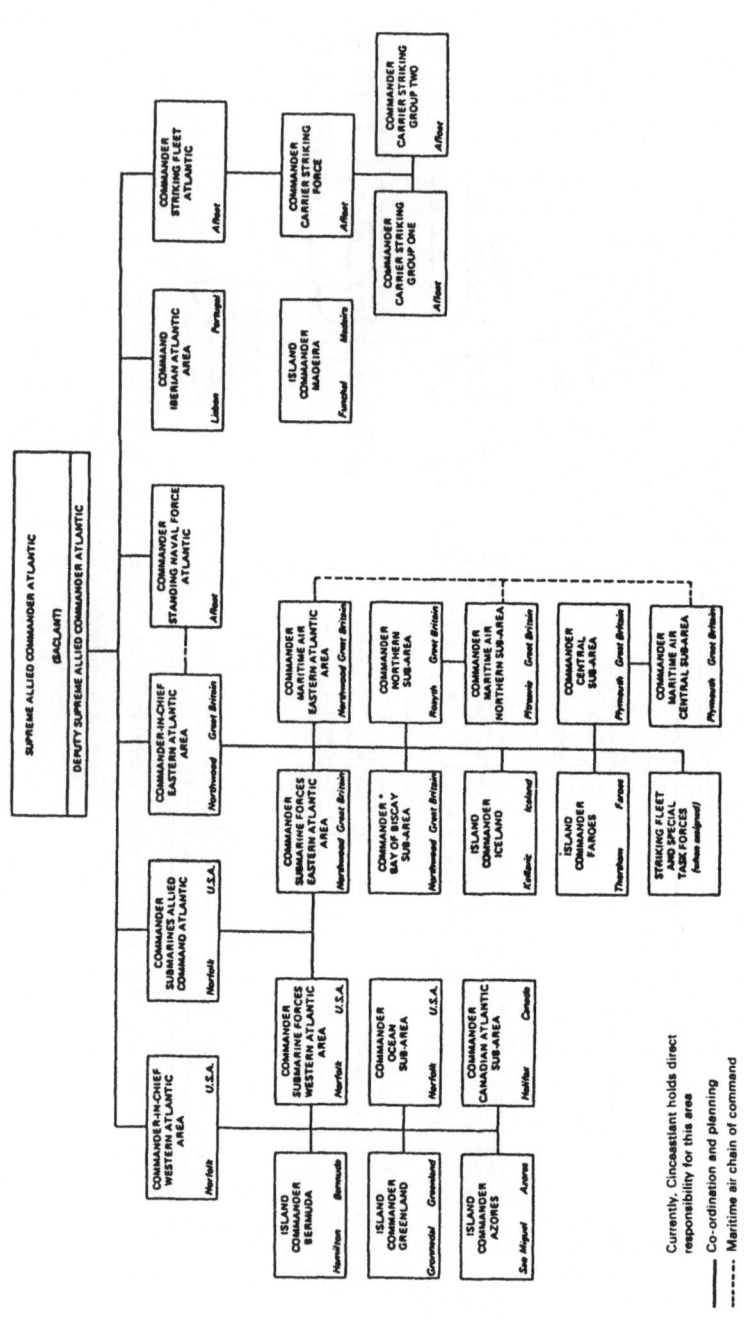

Source: *The North Atlantic Treaty Organisation: Facts and Figures.*

Figure 15.2 Allied Command Atlantic

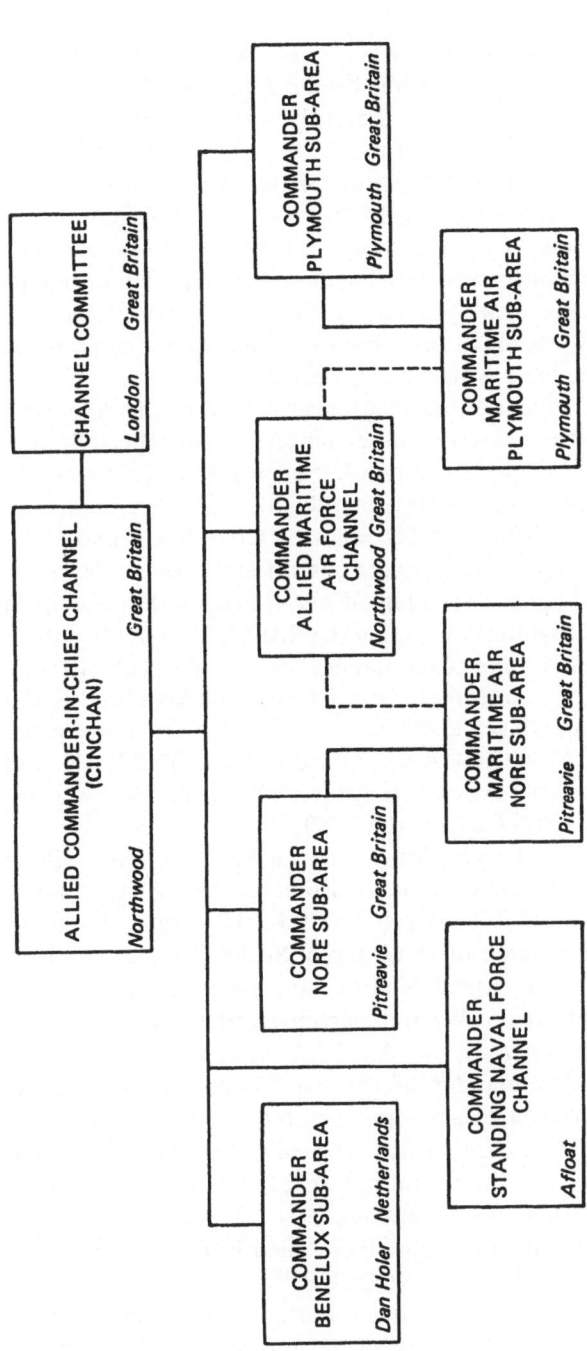

Source: *The North Atlantic Treaty Organisation: Facts and Figures.*

Figure 15.3 Allied Command Channel

forming the US element of the international staff of SACEUR, and the others moving to the new wartime headquarters of USEUCOM at High Wycombe, Great Britain.[17] Similarly, the USAEUR staff in Heidelberg would also split in wartime, with some transferring to the headquarters of the ACE Central Army Group at Seckenheim, West Germany, and the others moving to the wartime USAEUR head-quarters in Massweiler, West Germany. It is doubtful whether such organisational rearrangements and command and staff movements would be conducive to orderly planning and timely operations – yet they would probably occur at precisely that point in a conflict when orderliness and timeliness would be at a premium!

The complexity of the multinational peacetime/wartime command arrangements involves numerous other officers – and especially British and West German officers – in two-hatted positions. In the case of the Royal Navy (RN), for example, not only does the Commander-in-Chief of the Fleet hold the NATO posts of CINCHAN and CIN-CEASTLANT but his two subordinate operational commanders each also hold two NATO posts: the Flag Officer at Plymouth is also both the Commander of Plymouth Channel (COMPLYMCHAN), subordinate to CINCHAN, and Commander of the Central Atlantic (COMCENTLANT), subordinate to CINCEASTLANT and SAC-LANT; and the Flag Officer at Pitreavie is both Commander of the North Channel (COMNORCHAN), subordinate to CINCHAN, and Commander of the North Atlantic (COMNORLANT), also subordinate to CINCEASTLANT and SACLANT.[18]

In the case of the Royal Air Force (RAF), the Air Officer Commanding (AOC) Strike Command also holds the NATO position of Commander-in-Chief UK Air (CINCUKAIR), subordinate to SACEUR. The Commander of 18 Group of Strike Command (which is equipped with nuclear-armed Nimrod and Buccaneer aircraft for Anti-submarine Warfare (ASW) and maritime strike missions) is also the [NATO] Commander of the Air East Atlantic (COMMAIR-EASTLANT) and Commander of the Air Channel (COMMAIR-CHAN), both of which are subordinate to SACLANT. Finally, the commander of RAF Germany, which is based at Mönchen-Gladbach and which is part of NATO's multinational Second Allied Tactical Air Force (2 ATAF), is also Commander of 2 ATAF and hence subordinate to the Commander-in-Chief Allied Forces Central Europe (CINCAFCE) and thence to SACEUR.[19]

An obvious although quite impolitic question concerns the ultimate loyalties of these two-hatted commanders.[20] It is reasonable to assume that the senior NATO commanders would maintain some form of communication with their own national authorities.[21] How would CINCHAN act if directed to one action by NATO and another by the British political authority? Although British forces assigned to SACEUR require a specific order from him to use British nuclear weapons, would the Commander 2 ATAF go against a negative SACEUR decision if he was instructed by the British Prime Minister to use a Jaguar or Tornado or Harrier with nuclear weapons to save an element of the British Army on the Rhine (BAOR)? Alternatively, would the CINCEASTLANT refuse to pass on an order from SACLANT to use a British nuclear depth bomb to defend a US/NATO Poseidon submarine if the British Government remained opposed to any use of nuclear weapons?

This picture is further complicated by the existence of national US, British and French nuclear forces and associated C^3I systems which are independent of the NATO arrangements – such as Tomahawk Sea-Launched Cruise Missiles (SLCMs) aboard US attack submarines (SSNs), and the British and French Submarine-launched Ballistic Missiles (SLBMs).

This complexity extends from the command arrangements to the supporting communications and battlefield Command and Control Information Systems (CCIS). As a result of a NATO directive which permits each member nation to provide its own communications and CCIS systems up to Corps level, there is now a multiplicity of such systems with little interoperability. According to one review, there are at least six NATO battlefield communications systems: Ptarmigan (Great Britain), RITA (Belgium and the United States), Autokonetz (West Germany), Zodiac (The Netherlands), Tri-tak (United States) and Mobile Subscriber Equipment (United States). Only Ptarmigan and Zodiac are directly interoperable with one another, and none is directly interoperable with NATO's Integrated Communications System (NICS).[22] In the case of battlefield CCIS, there are the US STOMA/SIGMA, the West German HEROS, and the British WAVELL systems. A similar situation obtains with regard to artillery fire control systems and many other important electronic systems.[23] In the case of IFF systems, US aircraft are equipped with systems which operate in the D-band (formerly known as the L-band) while West German systems operate in the E/F-band (formerly known as the S-band).[24] (In some exercises, as many as half of the participating NATO aircraft have been 'shot down' because of the lack of IFF commonality).[25]

One authoritative study of theatre communications and CCIS systems has concluded as follows:

> For the typical senior commander, allied or U.S., whose forces must use these systems, they represent the largely unplanned splicing together of ill-fitting components which have been delivered to his forces by relatively independent parties far away who have coordinated adequately neither with him and his staff nor with each other. And they neither exploit the present capabilities of technology nor does the system for their development adequately provide that future systems will.[26]

More succinctly, despite the enormous investments and the technical sophistication of the equipments involved, it remains the case that NATO 'cannot accomplish the major aims of a modern C^3 system'.[27]

THE VULNERABILITY OF THE EUROPEAN THEATRE C^3I ARCHITECTURE

C^3I systems are invariably more vulnerable than the forces they control and support. Many of the measures which can be taken to enhance the survivability of the weapons systems are not applicable to C^3I systems. Weapons systems can be hardened, but it is difficult to harden satellite ground stations, early warning radars, large High Frequency (HF) or Very Low Frequency (VLF) communication systems, or signals intelligence (SIGINT) antennae. Some weapons systems can be camouflaged, but satellite control stations, radars and communications stations produce strong and readily identifiable electromagnetic emissions. Weapons systems can be dispersed or deployed in some mobile mode, but it has proved difficult to design C^3I systems with any substantial degree of mobility, and the requirements of centralised command place practical constraints on the extent to which some critical elements can be dispersed. Moreover C^3I systems are generally more vulnerable not just to all the threats to which the forces are subject but also to others more peculiar to command structures and telecommunications systems.[28]

The multiplicity of NATO and national command centres and communication networks provides more aim points for Soviet military planners, but it does little to enhance the redundancy and survivability of the theatre C^3I architecture overall. Virtually all of the command centres and communications systems are soft, and there is very little interoperability.

The C³I facilities in the European theatre are vulnerable to a wide range of Soviet actions – including nuclear strikes, conventional and chemical weapons attacks, Spetsnaz operations, and radio-electronic combat (REC) activities. Listed in the Appendix are some 60 critical C³I points in the European theatre which, if destroyed or incapacitated, would completely nullify any capacity for controlled, informed, or responsive military operations within the theatre. The list is not meant to be definitive or to imply anything about Soviet/Warsaw Pact targeting doctrine or plans. Apart from the NATO headquarters in Evere in Brussels, the USNAVEUR headquarters in London, the HQ Commander Iberian Atlantic Area in Lisbon, and the HQ Commander Allied Forces South Norway in Oslo, it does not include the national capitals. (There are good and obvious reasons why the Soviet Union might wish to exclude the national capitals from any initial counter-C³I strike.) In fact, it would clearly be unnecessary to destroy all of the sites listed. Many of them are subordinate headquarters which would be unable to operate following the destruction of their superior commands. The continued operation of many of the early warning, intelligence and communications stations would be futile in the absence of survivable centralised processing command and relay centres. Indeed, it is likely that the destruction or incapacitation of less than 20 of these 60 sites would totally immobilise command and control across the whole European theatre.

Very few of the European C³I facilities have any significant degree of protection against nuclear blast effects or even sustained assault by conventional ordnance. The only hardened underground command centres are the following: 'Pindar' bunker beneath the Ministry of Defence in Whitehall, London;[29] the alternative British governmental headquarters at Hawthorn and the associated wartime headquarters of the Commanders-in-Chief Committee (CISC) at Corsham, in Wiltshire;[30] the West German 'emergency government' complex in the Eifel Mountains near Marienthal, about 20 kilometres south of Bonn;[31] the new underground wartime SHAPE headquarters at Casteau;[32] the alternative wartime headquarters of NATO's Fourth Allied Tactical Air Force (4 ATAF) at Kinsbach, West Germany;[33] the RN/CINCHAN underground headquarters at Northwood, Middlesex;[34] the underground RAF/Strike Command/CINCUKAIR headquarters at High Wycombe, Buckinghamshire, which has recently been extended to include a new underground wartime headquarters for USEUCOM;[35] the underground wartime headquarters of NATO's Commander-in-Chief Allied Forces Southern Europe in Naples;[36] the French National Military Command Centre at Taverny, Val-d'Oise; the alternate French underground national military command centre at Mount Verdun (Lyon), Rhône; and the underground command centre for the French SSBN force at Houilles, Yvelines.[37] None of these underground centres could withstand blast overpressures of more than 1000 psi, hence they would be unable to survive deliberate nuclear strikes.

The decision to construct the alternative underground USEUCOM headquarters at High Wycombe, which was taken by the US Department of Defense and approved by the US Congress and the British Government in 1982, was the subject of considerable controversy. On 29 March 1982 Secretary of Defense Caspar W. Weinberger issued the *Fiscal Year 1984–1988 Defense Guidance* which instructed the Joint Chiefs of Staff

> to establish a European Command war headquarters in the United Kingdom by 1986 as part of a survivable European command and control system.
>
> The command and control system must ... survive enemy attack, including biological warfare and chemical warfare agents, nuclear effects, and acts of sabotage.[38]

The requirement for the underground war headquarters at High Wycombe was justified to Congress as follows:

> Until 1967, HQ USEUCOM was located near Paris. However, when France directed all foreign forces to leave her soil, USEUCOM moved to Patch Barracks at [Vaihingen] Stuttgart, Germany. The selection of Patch was an expedient. At the time, Patch was the only available location with facilities which could accommodate HQ USEUCOM. From a military standpoint, Patch has never been considered an acceptable wartime location. It is too close to potential attacking forces, and its [*deleted*] provide nothing in the way of protection. ...
>
> *Current Situation* – HQ USEUCOM currently occupies soft, above-ground facilities. They are malpositioned, extremely vulnerable and completely unsuitable for a wartime headquarters.
>
> *Impact If Not Provided* – If this facility is not provided HQ USEUCOM will not have a survivable facility from which to conduct its wartime mission.[39]

Although the threat of Soviet chemical weapons attacks against command centres has been recognised within NATO for more than a decade, it remains the case that only a handful have been protected against such a contingency. One is the new semi-hardened alternate war headquarters for USAFE. According to testimony presented to Congress in 1984,

> This facility is a chemically and biologically (CB) protected command post with reinforced concrete walls, roof and floor slab. It includes communication and computer centers, decontamination areas, stand-by power plant, vault and staff areas. Exterior walls and roof are constructed of 65 centimeters (25·5 inches) of reinforced concrete and armor plate doors.[40]

In fact, the Soviet Union could easily destroy all but about a dozen sites in the whole European C^3I architecture with special forces using only conventional ordnance. The Soviet military literature frequently alludes to the value of sabotage or 'diversionary' operations against command, communications, early warning, navigation and electric power facilities, and some 380 Spetsnaz teams are reported to have been specifically organised for operations against NATO C^3 facilities.[41] This is sufficient to dedicate at least one team to every command centre, satellite ground station, ACE HIGH communications station and SIGINT collection facility in Western Europe!

Numerous communications networks have been deployed in Europe over the past three decades to support both NATO and national defence forces. The most important of these[42] are: the ACE HIGH tropospheric scatter/microwave system, which consists of 49 tropospheric scatter links and 40 line-of-sight microwave links, extending from northern Norway through western Europe to Turkey, which has 570 voice, 260 telegraph and 60 data circuits, and which provides the minimum essential transmission system for early warning, LERT-CON and implementation signals; the NATO satellite communications (SATCOM) system, which has 21 fixed ground terminals, including one in almost all of the NATO countries; the British SKYNET SATCOM system; the US Defense Satellite Communications System (DSCS), which has some two dozen ground terminals in Western Europe; the US Cemetery Net HF system for the transmission of Emergency Action Messages (EAMs) to '[nuclear] field storage sites, delivery units and mobile and fixed command headquarters';[43] the NATO Last Talk HF system;[44] the Cross Fox HF system for communications between shore-based and fleet-based

NATO commands; the European Military Telephone Network for the US armed forces in West Germany; the US Automatic Digital Network (AUTODIN) and Automatic Voice Network (AUTOVON) systems; the US Air Force's North Atlantic Relay System (NARS), which provides a tropospheric scatter radio relay link between the United States and Great Britain across Canada, Greenland and Iceland;[45] and the Initial Voice Switched Network (IVSN) and Telegraph Automatic Relay Equipment (TARE) voice and message switching facilities.

The total number of facilities in these systems amounts to about 200. However, they have poor survivability and endurance, and are generally vulnerable to Soviet radio-electronic combat (REC) activities. Most of the facilities are soft, and what little efforts at hardening have been made have been inconsistent. For example, some microwave relay towers have been hardened, but 'we ... have routed the waveguides into these towers out in the open, where they can easily be destroyed'.[46] Some sub-systems (such as NATO's new 'shelterised' transportable HF equipment) are hardened against electromagnetic pulse (EMP) 'but we have not yet an EMP-protected network'.[47]

The various systems also have relatively few nodal points. Although the ACE HIGH network has 89 links, there are so few nodes that, according to a NATO official, 'about a dozen Russian satchel charges could take the whole thing out'.[48] The NATO SAT-COM network has only two master control stations (Kester in Belgium and Naples in Italy), although it is likely that the British SKYNET master control station at Oakhanger, Hampshire, and the US DSCS Network Control Facility (NCF) at Landstuhl, West Germany, could also control the NATO satellites. There are only three AUTODIN Switching Centers in Europe (Croughton in Great Britain, Pirmasens in West Germany, and Coltano in Italy), while all AUTOVON connections between the United States and Europe pass through Hillingdon in Great Britain.[49] Pirmasens is also the headquarters of the Cemetery Net system.[50]

These communications systems are also vulnerable to Soviet jamming. For example, on one occasion in the late 1970s when extensive use of electronic counter-measures (ECM) was permitted during an exercise in Europe, the exercise ground to a halt after only some 30 minutes!

The national PTT networks carry about 50 per cent of the NATO telecommunications circuitry.[51] In some countries, the PTT system is more efficient, redundant, hardened and secure than the US and/or

NATO systems. For example, the West German PTT system, which extends into several other NATO countries, has some 7000 'entry points' and its buried Grundnet is relatively hardened and secure. Indeed, some commentators have argued that the German PTT system is better than the dedicated US telecommunication systems in Europe.[52]

On the other hand, the national PTT systems contain some quite debilitating weaknesses. Even the West German PTT system has significant vulnerabilities. In Berlin, for example, there are several points where the underground cables cross into the Eastern sector and are known to be tapped by both Soviet and East German security/intelligence services. None of the national PTT systems are hardened against EMP. They are all vulnerable to Soviet SIGINT operations. And even though some national systems have relatively good redundancy and survivability, 'the cross-border connectivity is very often poor, relying heavily on a few very soft choke points'.[53]

Further, while some national PTT systems are quite efficient, others – and particularly those on the southern flank – are rudimentary. According to General Richard Ellis for example,

> You might be amused by some of the exposure I had to NATO command and control . . . In 1971 I was sitting happily in Wiesbaden, Germany, as vice commander in chief of U.S. Forces Europe, and unexpectedly I was sent to Ismir, Turkey, as commander of the Sixth Allied Tactical Air Force. That is the easternmost projection of NATO's air power . . . My communications – when I walked into my office, I'll never forget the terrible shock. The phone looked like a World War I instrument. I picked it up, finally somebody answered, and he sounded like he was on the other side of the world. And I said 'Who is this?' and his voice said, 'I'm your secretary.' He was right outside the door.[54]

With regards to the vulnerability of the NATO communications system to Soviet SIGINT activities, Major General Joachim M. Sochaczewski, former Assistant Chief of Staff Communications and Electronics at SHAPE, has noted that 'all NATO communications are indeed on a daily basis easily interceptable'.[55]

The Soviet Union maintains the most extensive and comprehensive SIGINT capability in the world. Some 350 000 Soviet personnel are engaged in SIGINT activities. There are more than 500 SIGINT ground stations located within the Soviet Union, Eastern Europe, and elsewhere around the world. Other SIGINT systems have been deployed on an extraordinary range of platforms, including submarines, surface ships, aircraft, space satellites, and various trucks and other vehicles. SIGINT systems have also been installed in Soviet diplomatic establishments in some 60 countries around the world.[56]

Much of this SIGINT activity is directed against NATO communications and other signals. For example, a large station at Riga is primarily concerned with monitoring Scandinavian signals; a station at Kaliningrad provides coverage of Scandinavian and West German signals; and a station at Lvov is primarily concerned with monitoring military communications emanating from West Germany. The Soviet Union maintains more than 150 SIGINT stations in Eastern Europe, with major concentrations in the German Democratic Republic (GDR) (more than 50 stations), Poland, Czechoslovakia, Hungary and Bulgaria. Some particularly important stations in the GDR are those at Zossen-Wuensdorf, Karlshorst (in East Berlin), Neuruppin, Eberswalde, Stahnsdorf, Prignitz and Brocken. The SIGINT stations in Poland are primarily directed towards West German signals; those in Czechoslovakia cover West German and Austrian signals; those in Hungary also cover Austria; while stations in Bulgaria cover signals emanating from Greece and Turkey.[57]

The Soviet Union also operates an extensive variety of SIGINT aircraft designed to monitor NATO signals. Particular attention is accorded the Baltic Sea and the German border. In the case of the Baltic Sea, airborne SIGINT operations are conducted from bases at Riga, Palanga, Bryusterport, Kaliningrad and Baltiysk (Pillau) in the Soviet Union, and Slupsk and Kolobrzeg in Poland. MiG-25R/Foxbat D aircraft, used for high-altitude, high-speed ELINT operations, are based at Brzeg in Poland, Kolomniya in the Soviet Union, and Werneuchen, northeast of Berlin. Other bases for ELINT aircraft in the GDR include Schonerfeld, Stendal, Altenburg and Welzow. MiG-25R/Foxbat D and Il-18 Coot A ELINT aircraft from Werneuchen and Schonerfeld make daily flights along the West German border.[58] (During the Soviet invasion of Czechoslovakia in 1968, the Soviet Union greatly increased its airborne SIGINT operations in order to closely monitor the Western response.) Soviet SIGINT vessels, formally known as AGIs (Auxilliary General Intelligence) are also routinely stationed in the immediate vicinity of the submarine bases at Holy Loch in Scotland and Brest in France, as well as in the North Cape area, the Skagarrak Strait linking the North Sea and the Baltic Sea, the Strait of Gibraltar, and the Strait of Sicily.[59]

SIGINT systems are maintained in Soviet diplomatic establishments in some 18 countries in Western Europe. In many of these countries (such as Great Britain, France, Belgium, Sweden, Switzerland and West Germany) more than one such establishment is used.[60] In Belgium, for example, both the Soviet Embassy and the Soviet Trade Mission are used for SIGINT operations. According to a report by HQ USEUCOM in December 1973,

Human intelligence (HUMINT) sources have recently confirmed that ... SIGINT intercept operations have been conducted [by both the GRU and KGB] from the SOVIET Embassy and SOVIET Trade Mission in BRUSSELS, BELGIUM, for a minimum of eleven years. These operations include, but [are] not necessarily limited to, direction-finding, plaintext analysis of voice, teletype, and Morse communications and analysis of the externals of encoded or enciphered messages. Among primary intercept targets are the strategic/tactical air and ground forces of the US, NATO and Western European countries. Some examples of exploited information are: combat readiness checks (e.g., Emergency Action Messages), planning/progress of major exercises/ operations, real time status of deploying forces, flight activity by airborne command posts, travel of VIPs, etc.

These same HUMINT sources confirmed that the GRU participates in real situations by providing the following information: During the SOVIET Exercise OKEAN 1970, the task of the SOVIET Technical Service (SIGINT) Group in BRUSSELS was to intercept US military communications traffic pertaining to the exercise. This intercept activity enabled the SOVIETS to determine what the US and its Western Allies learned about OKEAN, how much they know, how much they were getting, etc.[61]

The Soviet and other Warsaw Pact SIGINT agencies accord special attention to the circuits which would be used for the extensive international discussions that would precede any decisions to increase the NATO alert levels (LERTCONS) or disperse the nuclear forces; to the channels which would be used to communicate such decisions to the various commands and units; and to the systems which would transmit the nuclear release authorisations.

The communications security (COMSEC) assessment conducted by HQ USEUCOM following the Yom Kippur War in 1973 showed that the GRU and KGB SIGINT agencies were able to monitor combat readiness checks from the USCINCEUR Airborne Command Post; the real-time status of deploying forces, such as the departure of the

Fleet Ballistic Missile (FBM) submarines *Kamehameha* (SSBN-642) and *Simon Bolivar* (SSBN-641) from Rota following the declaration of DEFCON THREE on 25 October 1973; discussions relating to 'contingency planning operations and potential task force compositions'; traffic concerning aircraft departures from the continental United States; the movement of 'war material' from the ports of Nordham and Bremerhaven in West Germany; numerous communications concerning the movement of fuel and other logistic activity in the Mediterranean region; and other related air and naval activities.[62]

According to a staff member of the House Armed Services Committee, the Soviet Union demonstrated its ability to intercept communications relating to the proposed 'use' of nuclear weapons during the NATO Reforger exercise in 1976: 'In fact, to embarrass us, the Soviets, who listen in on all our communications over there, announced two hours before we had, that is, before our troops had gotten approval, that NATO was going nuclear during that exercise.'[63] And according to another account,

> The top level NATO C³ system is not just vulnerable; elements of it are so insecure that during the last major NATO exercise under Alexander Haig's tenure as Supreme Allied Commander, Europe, even the highly encoded traffic sent over a special communications net to NATO's nuclear units was intercepted so quickly by Soviet communications intelligence [COMINT] posts that the Russians broadcast a message in the *clear* on an open broadcast channel to the effect that, 'NATO's going nuke'.[64]

NATO ALERT LEVELS, NUCLEAR FORCE DISPERSAL AND NUCLEAR RELEASE ORDERS

The NATO doctrine of flexible response ideally requires a C³I system which provides the political and military authorities with continuous, real-time information from a variety of authoritative sources, and the ability to control the forces through a spectrum of contingencies ranging from conventional conflict through the limited use of battlefield nuclear weapons to a large-scale theatre nuclear exchange, at all times responding to political and strategic demands in precise and timely fashion. It should be able to endure for at least as long as the forces themselves, and to communicate with the Warsaw Pact authorities, in order to support war termination. It should be able to move readily and smoothly from a negative to a positive control mode, and vice versa. And it should not have vulnerabilities or procedural choke-points which invite precipitous action by the adversary.

The present NATO C³I system is very far from this ideal. Its prevailing characteristics are more likely to be paralysis interjected with paroxysms. It is obviously impossible to predict how the system will in fact function – it will depend on the nature of the contingency, and there are no precedents for any of the situations of concern here. However, the system is so structured as to inhibit clear, coherent and timely decision-making at the outset of conflict, and then rapidly to generate escalation after a few days of intensive conventional conflict, followed very soon by a loss of any meaningful centralised control.

The most escalatory points in any theatre scenarios are when the decisions are implemented to raise the alert level from LERTCOM 5 (peacetime) or 4 ('military vigilance') to LERTCOM 3 or 2; to disperse the nuclear forces; and to issue the nuclear release authorisations. In practice, it is quite likely that these moves will be simultaneous – a major and ominous step.

In peacetime, the nuclear weapons in Europe are stored in a relatively few storage sites (some 121, including 13 GLCM and Pershing II bases). They are well-known and fairly vulnerable.[65] In peacetime all of the land-based weapons in Europe are locked with Permissive Action Links (PALs) which prevent accidental or unauthorised detonation. And in peacetime the NATO command system remains skeletal, with the forces remaining under national control while awaiting 'the chop' to transfer to NATO command.

The most compelling reason for unlocking the PALs and authorising the release of the weapons at the same time as dispersing them is that subsequent communications between the command authorities and field units would be too difficult and the process of attempting to match particular PAL codes to their corresponding warheads impractical. There would also be the fear that a central headquarters holding the PAL codes would be attacked and the codes destroyed if the Soviets were to use the dispersal decision as a trigger to pre-empt.[66] Hence, as Bracken has observed, 'practically speaking, a strong pressure exists to release any needed codes at the same time that the weapons are dispersed from their storage sites'.[67] And Charles has noted that, during the NATO WINTEX Command Post Exercises (CPXs), when the release procedures are practised, there is evidently no distinction made between the decision to distribute the warheads to their delivery systems and the nuclear release order.[68]

The NATO command structure is not conducive to timely or expedient decision-making. The complexities of the command arrangements and the profound and excruciating dilemmas which would face the political and military command authorities would probably mean that it would take two or three days to decide to disperse the nuclear forces and authorise nuclear release.

The central dilemma, which derives from the very nature of NATO's peacetime posture and C³I structure, is 'the conundrum between vulnerability and provocation' – that is, the measures required to enhance survivability of the forces 'might be apprehended and trigger preemption'.[69] A second dilemma derives from the known fact that NATO procedures generally require many hours and perhaps days to work through, and therefore for decisions to be relevant to emerging situations they would have to be taken considerably in advance – when the available intelligence is unlikely to sustain them and when NATO leaders would be very concerned not to feed the very developments they would be hoping to avoid.

These dilemmas would be compounded by the fact that the most critical actions would be unprecedented and NATO leaders would be fearful of unleashing unforeseen consequences. NATO has never moved beyond LERTCOM 4.[70] (In 1968, when the NATO Council met to consider various alert measures in connection with the movement of Soviet troops into Czechoslovakia, the 'deliberations gradually ground to a halt. Not a single one of the alert measures was implemented'.[71]) Similarly, NATO has never dispersed the nuclear warheads. Indeed, dispersal has never even been realistically practised, even during the WINTEX CPXs, apparently because the subject is too sensitive for the participants to address.[72]

It is likely that the process of authorising nuclear release would take from 24 to 72 hours. The US Army's Field Manual FM 100-5, *Operations*, issued in 1976, contains a chart depicting the 'bottom up' release process, beginning with a request from corps level for the release of nuclear 'packages', and moving through the subsequent requests and release sequence to the receipt of release authorisation by the fire-support element some 24 hours later.[73] (See Figure 15.4.) According to Catherine Kelleher, other estimates, supported by evidence from the WINTEX CPXs, suggest that 'perhaps as many as sixty hours would be required under the most favourable conditions'.[74] In the particular case of nuclear depth bombs, the exercise evidence is that the release sequence typically takes between 48 and 72 hours – by which time, of course, the target submarine is likely to have

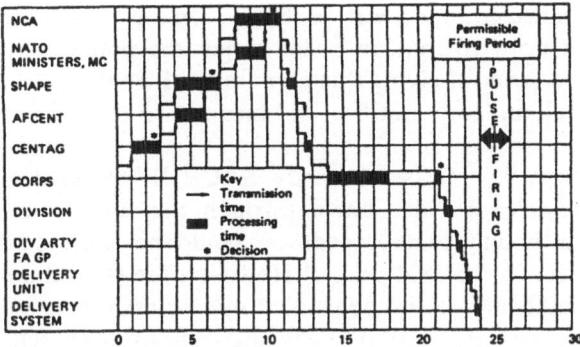

NCA NATO Foreign Ministers, Military Committee²
MINISTERS, MC NATO Foreign Ministers, Military Committee²
SHAPE Supreme Headquarters, Allied Powers Europe
AFCENT Allied Forces Central Europe
CENTAG Central Army Group
DIV ARTY Division Artillery
FA GP Field Artillery Group

¹ US and British NCA authorise release of nuclear weapons to using units after:
² NATO Ministers, consulting Military Committee, approve their use. French release procedures are separate.
Strictly US channels from corps to theatre to JCS to NCA and back, then same as above to delivery system in any theatre except Europe.

Source: Adapted from Army FM 100-5, *Operations,* 1 July 1976.

Figure 15.4 NATO theatre nuclear release channels

long gone! Hence, as an account of US Army doctrinal developments prepared by the US Army's Training and Doctrine Command (TRADOC) has noted, 'by the time a commander could clearly demonstrate the time to be right for nuclear weapons, it would already be too late'.[75]

There are, therefore, alternatives to the 'bottom up' process. In the case of the GLCM and Pershing II intermediate-range weapons, and probably also for the short-range Army systems, there is the 'top-down' process whereby SACEUR initiates the release sequence.[76] In principle, SACEUR would still have to secure prior release authority from the US and British National Command Authorities, but this could be expedited by 'early notification'. General Bernard Rogers, then SACEUR/US CINCEUR, testified in 1982:

The system that is used . . . is that I go to the political authorities at NATO headquarters with the request. I go also to the Ministers of Defense of all the nations and I go also to the two nuclear powers [the United States and Great Britain] simultaneously with my request for release. But prior to that time there would have been a warning message that I was probably coming to ask for release. And even prior to that, in order to get the political authorities thinking in terms of giving this permission, I would have sent what I would call an 'early notification' message to them.[77]

Having secured this release authority, SACEUR would then have to transmit the release order down through the successive levels of command, although some expedition could be achieved by leap-frogging direct to the field units.[78] However, the procedure remains somewhat cumbersome and dependent upon vulnerable communications systems, and unsuited to timely and responsive action. It is also possible that some pre-delegation has been given to SACEUR, as US CINCEUR, in some strictly-defined and contingent circumstances and with respect to US nuclear weapons in Europe. Such resort to national expedients, however, would probably be at the expense of further efforts at co-ordinated control across the theatre.

Whatever the mechanisms and time periods involved in taking and communicating the dispersal and release decisions, the consequences are likely to be two-fold. First, the distribution of unlocked nuclear weapons to numerous and widely-dispersed field units involved in many different tactical situations is likely to render futile the subsequent efforts to control and co-ordinate their deployment and employment. Further, the transition from negative to so-called positive control is unlikely to be reversible in any well-managed fashion. Secondly, it is likely that, through SIGINT activities and other means, the Soviet authorities would be able to keep abreast of NATO alert, dispersal and release developments. The difficulty for Soviet intelligence would be determining whether these developments were merely precautionary, or whether they were preparatory to launching a nuclear attack. There would certainly be great pressure to strike the storage sites and the supporting C^3 systems before the decision to disperse and unlock the weapons could be implemented.

One of the functions of an ideal C^3I system is to support war termination. Indeed, this is an essential requirement for doctrines of flexible response and escalation control. It means that some form of direct communication link should be maintained between the NATO and Warsaw Pact authorities, and that the NATO political and military authorities are able to exercise the strictest control over the forces at all levels at all times. The NATO C^3I system would seem to be unable to satisfy this latter condition. Given the vulnerability of

NATO communication systems, once some significant number of nuclear weapons had been used, it would be largely fortuitous if all the remaining weapons could be recalled from the hundreds of disparate field units and secured harmoniously. At best, there would probably be a few isolated detonations. At worst, operations would continue which would nullify an intra-war pause or armistice.

STRATEGIC C³I AND THEATRE WAR

Despite what might reasonably be inferred from some aspects of US strategic nuclear planning with respect to limited or controlled nuclear strikes against targets in the Soviet Union, it is most unlikely that any use of strategic nuclear weapons by either the United States or the Soviet Union against targets in the other's heartland would be the first move in any conflict between them. Rather, such use is likely to follow a period of large-scale military action, probably involving substantial use of battlefield and theatre nuclear weapons, in an area of vital interest to both adversaries and during which the dynamics of the escalation process would have already been set in motion. Military casualties on both sides, as well as collateral casualties among allied civilians, are already likely to be very high *before* any strategic nuclear exchange. In the European theatre, possible fatalities range from two to 20 million, assuming extensive use of nuclear weapons with some restraints, up to 100 million if there are no restraints at all.[79] Moreover it is almost certain that some of the command and control facilities, communications systems, early warning stations and intelligence posts that would be required to control a strategic nuclear exchange would have been destroyed or damaged in the conventional or theatre nuclear phases of a conflict.

Extensive segments of the US strategic C³I system are based in the European theatre. The US Defense Satellite Communication System (DSCS) satellites are used for communications with both the strategic and theatre forces, and the DSCS NCF at Landstuhl, as well as the stations with alternative control capability at Croughton and Oakhanger, together with many of the two dozen other DSCS terminals in Europe, would be likely targets for Soviet attention in a theatre conflict. The Classic Wizard/White Cloud ELINT ocean surveillance satellite ground station at Edzell in Scotland is one of only five ground stations in the system, and would again be likely to attract Soviet attention in a theatre conflict. The US Air Force's Satellite Control Facility (SCF) at Oakhanger is one of only eight such facilities. The

SCF network controls the great majority of US Air Force satellites, as well as CIA photographic surveillance and CIA and NSA electronic intelligence (ELINT) satellites. The ELINT satellites, together with the new CIA KH-12 photographic surveillance satellites, are becoming increasingly important sources of real-time theatre intelligence.

Early warning systems, such as the Ballistic Missile Early Warning System (BMEWS) station at Fylingdales and the US Defense Support Program (DSP) satellite early warning system, with one of its three ground stations at Kapaun, West Germany, are used to detect both ICBMs and shorter-range ballistic missiles. SOSUS and other Anti-submarine Warfare (ASW) systems would also be important in both theatre and strategic conflicts.

The likely destruction of some of these systems in a theatre conflict would generate considerable escalatory pressures. For example, any attempt by the Soviet Navy to move its SSBNs and/or surface fleet through the Greenland/Iceland/United Kingdom (GIUK) gap would doubtless be accompanied by attacks on the SOSUS, SIGINT and related facilities in Iceland, Norway and Great Britain. The expectation that these systems would soon be unavailable would cause many US Navy commanders to strike against any ships that were in their 'sights', be they Soviet SSBNs or whatever else, while they still had the opportunity.[80]

It also follows that the United States would find, at the outset of any strategic nuclear exchange with the Soviet Union, that its strategic C³I system was already impaired, with important geographical segments missing. The prospects for controlling the strategic exchange would be correspondingly weakened.

CONCLUSION

War by its very nature is not subject to precise control. There are inherent tendencies toward escalation. Information is obfuscated and distorted by Clausewitz's 'fog of war'. The horror, confusion and hatred produced by extreme violence generates irrationality. Outcomes are determined less by careful planning and management than by Machiavelli's 'fortuna'. The use of nuclear weapons greatly exacerbates both the human and physical problems of command and control.

Strategic policies and doctrines based on notions of escalation control therefore warrant the most careful scrutiny. This is particularly the case where their application is sought in theatres at some remove from Washington and Moscow. In the European theatre, assessments of the implications of C³I systems and arrangements for crisis stability, flexible response and escalation control must focus on the complexity and vulnerability of the C³I architecture.

There is probably, in practice, little scope for reducing this complexity and vulnerability. The complexity is, to some extent at least, inevitable in an alliance of 16 sovereign states, each with different capabilities, interests and threat perceptions. Although understandable, the decision-making structure and command and control arrangements spawned by this alliance are quite unsuited to the demands of rapid and timely decision and implementation that would pertain on the European battlefield. And the development of procedures to circumvent the present system do not resolve the fundamental problems. To resort to national machinery, to give commanders greater autonomy, to institute command and control structures in peacetime that would more closely reflect those needed for warfighting, to clarify the pre-delegation arrangements for SACEUR/ CINCEUR – these measures would serve only to undermine other aspects of stability and escalation control.

Many of the vulnerabilities are inherent in the nature of C³I systems, and would not be significantly alleviated by the expenditure of additional resources. Over the next ten years, it is likely that more than $50 billion will be spent on C³I systems in the NATO theatre, including more than $15 billion on the acquisition of new communications systems. As Cushman has noted, 'the obstacles are not financial ... but are primarily organizational and institutional'.[81] There is no centralised process for designing and acquiring a coherent and efficient theatre architecture, and it is reasonable to assume, given the minimal extent to which high-level interest and attention is devoted to the issue, that most of the present deficiencies will remain.

This is not to say that no improvements can be made. Concentration of effort on a few particular problems could achieve some progress. Three areas would be worthy of special attention. First, with respect to command and control facilities, consideration should be given to the design and deployment of a network of alternative ground-mobile command posts, which have increasingly been deployed in the US itself – the mobile command posts (MCPs) for the National Command Authority (NCA) in the Washington area; the SAC HERTS system; the NORAD RAPIER system; and the Mobile Ground Terminals (MGTs) for the DSP satellite early warning system. Secondly, much greater attention should be accorded to COMSEC and communications deception, in order to reduce the ability of Soviet SIGINT agencies to monitor particularly sensitive European communications and the confidence which the Soviet leadership might place in signals interception as a trigger for preemption. And, thirdly, consideration should be given to the strictest possible separation of strategic and theatre C^3I assets. Despite the additional financial costs which would be involved, there would be great merit in withdrawing critical strategic C^3I systems, wherever technically feasible, to the Soviet Union and the continental United States; in physically separating those strategic C^3I facilities which it is determined are necessarily required abroad from areas of potential theatre operations; and for dialogue between the United States and the Soviet Union to identify agreed remaining C^3I systems that would be exempted from such theatre operations.

In general, however, the problems which derive from the complexity and vulnerability of theatre C^3I systems can best be addressed by changes to the strategic policies and postures which the C^3I systems are intended to support. Perhaps the most important contribution to theatre rationalisation would be the removal of the short-range battlefield nuclear weapons. There is very little that can be said in favour of these weapons. They contribute little to deterrence. Their military efficacy lies in either early employment or particular tactical engagements – situations which would involve their detonation on West European territory, and which would require a more responsive decision-making process than NATO could possibly pretend. Insofar as some of them depend on dual-capable delivery systems, they tend to constrain the conventional commitment and hence contribute to the pressures for escalation.

At the same time, there must also be enhancement of NATO's conventional capabilities. This does not necessarily involve either increased conventional forces or improved offensive capabilities. But there is no way that the battlefield nuclear weapons can be removed and the deficiencies in the theatre C^3I architecture addressed without much greater attention being accorded to the conditions of viable conventional deterrence.

Notes

1. Paul Bracken, *The Command and Control of Nuclear Forces*, (New Haven, Connecticut, 1983).
2. Bruce G. Blair, *Strategic Command and Control: Redefining the Nuclear Threat*, (Washington, DC, 1985).
3. John Steinbruner, 'Nuclear Decapitation', *Foreign Policy*, no. 45, Winter 1981–2, pp. 16–28.
4. Ibid., p. 19.
5. Desmond Ball, *Can Nuclear War Be Controlled?*, Adelphi Paper no. 169 (London, 1981).
6. Daniel Ford, *The Button: The Nuclear Trigger – Does It Work?* (London, Boston, Sydney, 1986).
7. Some noteworthy exceptions to this observation are Shaun Gregory, *The Command and Control of British Nuclear Weapons* (Peace Research Report no. 13, School of Peace Research, University of Bradford, December 1986); Daniel Charles, *Nuclear Planning in NATO: Pitfalls of First Use*, (Cambridge, Massachusetts, 1987); and Catherine McArdle Kelleher, 'NATO Nuclear Operations', in Ashton B. Carter, John D. Steinbruner and Charles A. Zraket (eds), *Managing Nuclear Operations* (Washington, DC, 1987) pp. 445–69.
8. Richard C. Ellis in *Seminar on Command, Control, Communications and Intelligence* (Program on Information Resources Policy, Harvard University, 1982) p. 2.
9. US Congress, House of Representatives, Committee on Armed Services, *Hearings on Military Posture and H.R. 5968* (Washington, DC, 1982)
10. US Congress, House of Representatives, Committee on Appropriations, *Department of Defense Appropriations for 1987* (Washington, DC, 1986) Part 1, p. 159.
11. Ibid., p. 140.
12. See Desmond Ball, 'Nuclear War at Sea', *International Security*, vol. x, no. 3 (1985–6) pp. 3–31.
13. Charles, *Nuclear Planning in NATO*, p. 19.
14. *The North Atlantic Treaty Organisation: Facts and Figures* (NATO Information Services, Brussels, 1984), chapters 4 and 5; and *NATO Handbook* (NATO Information Service, Brussels, April 1986).
15. *NATO Handbook*, p. 35.
16. *The North Atlantic Treaty Organisation: Facts and Figures*, pp. 102–12.
17. Charles, *Nuclear Planning in NATO*, pp. 127–8.
18. Gregory, *The Command and Control of British Nuclear Weapons*, pp. 92–3.
19. Ibid., pp. 85–6.
20. Ibid., p. 86.
21. Charles, *Nuclear Planning in NATO*, p. 128.
22. Gregory, *The Command and Control of British Nuclear Weapons*, p. 99.
23. Ibid., p. 100.
24. Michael Feazel, 'NATO Debates Electronic Warfare Spending Levels', *Aviation Week and Space Technology*, 2 April 1984, pp. 62–5; 'Choice

of Common Frequency Emerges as Major IFF Issue', *Aviation Week and Space Technology*, 21 May 1984, p. 98; and 'U.S., NATO Agree in Principle on IFF Frequencies Approach', *Aviation Week and Space Technology*, 8 October 1984, p. 27.

25.   'IFF Frequency Debate Nearing Resolution?', *Defense Electronics*, October 1984, p. 21.

26.   Lt. Gen. John H. Cushman, *Command and Control of Theater Forces* (Program on Information Resources Policy, Harvard University, April 1983) p. ES-3.

27.   Franz-Joseph Schultz, 'C³ in NATO', in Werner Kaltefleiter and Ulricke Schumacher (eds), *Conflicts, Options, Strategies in a Threatened World*, (Kiel, 1982) p. 223.

28.   See Ball, *Can Nuclear War Be Controlled?*, pp. 9–14.

29.   Duncan Campbell and Patrick Forbes, 'New Whitehall Bunker', *New Statesman*, 26 July 1985, p. 6; and Paul Brown, 'MoD Digs Deep for Bunker', *Guardian*, 25 July 1985, p. 4.

30.   Duncan Campbell, 'Maggie's Bunker', *New Statesman*, 24 September 1982, pp. 6–8.

31.   'Preparing for a Nuclear War: W. Germany Tests an Emergency Government', *Philadelphia Inquirer*, 9 March 1985, p. 1.

32.   Duncan Campbell, *The Unsinkable Aircraft Carrier: American Military Power in Britain* (London, 1984) p. 187.

33.   Ibid., p. 74.

34.   Harold Jackson and David Fairhall, 'US To Pull Back War Centre to Britain', *Guardian*, 10 December 1982, pp. 1, 28.

35.   Ibid., p. 28.

36.   Ibid.; and Charles, *Nuclear Planning in NATO*, p. 128.

37.   William M. Arkin and Richard W. Fieldhouse, *Nuclear Battlefields: Global Links in the Arms Race* (Cambridge, Massachusetts, 1985) pp. 282–7.

38.   Richard Halloran, 'U.S. Plans War Headquarters in Britain', *New York Times*, 11 December 1982, p. 3.

39.   US Congress, House of Representatives, Committee on Appropriations, *Military Construction Appropriations for 1983* (Washington, DC, 1982) Part 1, p. 1406; and US Congress, House of Representatives, Committee on Appropriations, *Military Construction Appropriations for 1983* (Washington, DC, 1982) Part 5, p. 206.

40.   US Congress, House of Representatives, Committee on Appropriations, *Military Construction Appropriations for 1985* (Washington, DC, 1984) Part 6, p. 187.

41.   William Schneider, 'Trends in Soviet Frontal Aviation', *Air Force Magazine*, March 1979, p. 76.

42.   R. J. Raggett (ed.), *Jane's Military Communications 1987* (London, 1987) pp. 791–5.

43.   Defense Marketing Services, 'EUCOM C³', *DMS Market Intelligence Report* (Greenwich, Connecticut, March 1982).

44.   Campbell, *The Unsinkable Aircraft Carrier*, p. 184.

45.   Ibid., pp. 86–7, 226.

46. Major-General Joachim M. Sochaczewski, 'The Role of Communications in NATO', *Military Technology*, no. 6, 1984, p. 154. See also 'NATO Planning Enhanced C³ Over Decade', *Aviation Week and Space Technology*, 7 November 1983, p. 81.
47. Sochaczewski, 'The Role of Communications in NATO', p. 154.
48. Cited in Benjamin F. Schemmer, 'No NATO C³I "Check-out Counter" ', *Armed Forces Journal International*, December 1982, p. 92.
49. Campbell, *The Unsinkable Aircraft Carrier*, pp. 190–1.
50. Ibid., p. 184.
51. Sochaczewski, 'The Role of Communications in NATO', p. 154.
52. Schemmer, 'No NATO C³I "Check-out Counter" ', p. 92.
53. Sochaczewski, 'The Role of Communication in NATO', p. 154.
54. Ellis in *Seminar on Command, Control, Communications and Intelligence*, p. 2.
55. Sochaczewski, 'The Role of Communications in NATO', p. 154.
56. See Desmond Ball, 'Soviet Signals Intelligence', Bruce L. Gumble (Executive Editor), *The International Countermeasures Handbook* (Palo Alto, California, 1987) pp. 73–89.
57. Ibid., pp. 73–8.
58. Dick van der Aart, *Aerial Espionage: Secret Intelligence Flights by East and West*, (New York, 1986) pp. 83–6, 105–9; Bill Sweetman and Bill Gunston, *Soviet Air Power: An Illustrated Encyclopedia of the Warsaw Pact Air Forces Today* (London, 1978) p. 100; and information provided by William Arkin, Institute for Policy Studies, Washington, DC, 16 April 1984.
59. Charles C. Petersen, 'Trends in Soviet Naval Operations', in Bradford Dismukes and James M. McConnell (eds), *Soviet Naval Diplomacy* (New York, 1979) pp. 52–3.
60. See Ball, 'Soviet Signals Intelligence'.
61. Headquarters United States European Command, *COMSEC Assessment During October 1973 Mid-East Conflict* (HQ USEUCOM, Vaihingen, Germany, ECJ-A-73-0045-S, December 1973) pp. A1–A2.
62. Ibid.
63. US Congress, House of Representatives, Committee on Armed Services, *Department of Defense Authorization of Appropriations for Fiscal Year 1981* (Washington, DC, 1980) Part 4, p. 1946.
64. Schemmer, 'No NATO C³I "Check-out Counter" ', p. 92.
65. Frank Greve, 'Security Gaps Raise Fears About U.S. Warheads in Europe', *Philadelphia Inquirer*, 3 January 1983, p. 1.
66. Charles, *Nuclear Planning in NATO*, p. 55.
67. Bracken, *The Command and Control of Nuclear Forces*, p. 168.
68. Charles, *Nuclear Planning in NATO*, p. 50.
69. *Crisis Stability and Nuclear War* (A Report published under the auspices of the American Academy of Arts and Sciences, and the Cornell University Peace Studies Program, Ithaca, New York, January 1987) p. 40.
70. Ibid., p. 36.
71. Richard Betts, *Surprise Attack* (Washington, DC, 1982), p. 85.

72. Charles, *Nuclear Planning in NATO*, pp. 49–50.
73. US Department of the Army, *FM 100-5: Operations* (Washington, DC, 1976) p. 107.
74. Kelleher, 'NATO Nuclear Operations', p. 457.
75. John L. Romjue, *From Active Defense to Airland Battle: The Development of Army Doctrine 1973–82*, (TRADOC Historical Monograph Series, US Training and Doctrine Command, Washington, DC, 1984) p. 37.
76. Kelleher, 'NATO Nuclear Operations', p. 457.
77. US Congress, Senate, Committee on Armed Services, *Department of Defense Authorization for Appropriations for Fiscal Year 1983* (Washington, DC, 1982) Part 7, p. 4334.
78. Kelleher, 'NATO Nuclear Operations', p. 457.
79. Ball, *Can Nuclear War Be Controlled?*, p. 36.
80. Ball, 'Nuclear War at Sea', p. 22.
81. Cushman, *Command and Control of Theater Forces*, p. 96.

Appendix: Critical Points in the European Theatre C³I Architecture

COMMAND HEADQUARTERS:	
1 Apt-St Christol, Vaucluse, France	HQ, Brigade de Missiles Stratégiques (S-3 IRBMs).
2 Bodo, Norway	HQ Commander Allied Forces North Norway
3 Brussels (Evere), Belgium	NATO HQ.
4 Brunssum, The Netherlands	HQ Allied Forces Central Europe.
5 Bentley Priory, Great Britain	Alternative Air Defence Operations Centre, UK Air Defence Ground Environment (UKADGE).
6 Casteau (Mons), Belgium	Supreme Headquarters Allied Powers Europe (SHAPE). HQ Allied Command Europe (ACE). New underground HQ in same complex.
7 Hawthorn/Corsham, near Bath, Wiltshire, Great Britain	Alternative underground Cabinet and MOD HQ. HQ UK Defence Communication Network (DCN).

HQ Land Forces Army Command.
Wartime HQ UK Commanders-in-Chief
Committee (CICC).

8 Heidelberg, West
Germany

HQ US Army Europe (USAEUR).
HQ 4th Allied Tactical Air Force (4
ATAF).
HQ Allied Command Europe (ACE)
Mobile Force.

9 High Wycombe, Great
Britain

HQ Royal Air Force (RAF).
HQ RAF Strike Command.
HQ RAF Air Defence Operations Centre
(ADOC), UKADGE.
HQ NATO Commander-in-Chief UK Air
Forces (CINCUKAIR).
HQ Commander Air East Atlantic
(COMMAIREASTLANT).
HQ Commander Air Channel
(COMMAIRCHAN).
Underground Wartime HQ for
USEUCOM.

10 Houilles, Yvelines, France

HQ Force Océanique Stratégique (FOST),
SSBN Underground Command Centre.

11 Karup, Denmark

HQ Commander Allied Forces Baltic
Approaches.

12 Kinsbach, West Germany

Underground Wartime HQ, 4 ATAF.

13 Kolsaas, Norway

HQ Allied Forces Northern Europe.

14 Lisbon, Portugal

HQ Commander Iberian Atlantic Area.

15 Grosvenor Square,
London, Great Britain

HQ US Navy Europe (USNEUR).
US Navy Fleet Ocean Surveillance
Information Center Europe
(FOSICEUR).
HQ NSA Special Liaison Office (SUSLO)
and Naval Security Group (NSG)
Europe.

16 Marienthal, West
Germany

Underground West German 'emergency
government' headquarters, 20 km south
of Bonn.

17 Mildenhall, Great Britain

HQ 10th Airborne Command and Control Squadron (ACCS), with 4 EC-135 'Silk Purse' aircraft, provides alternative emergency airborne command post (ABCP) for USEUCOM.
HQ US Third Air Force.
Base for 2 SR-71 reconnaissance aircraft.
US Strategic Air Command (SAC) 'Giant Talk' HF communications station.

18 Mönchen-Gladback, West-Germany

HQ RAF Germany.

19 Mount Verdun (Lyons), Rhône, France

Underground alternate national military command centre, and alternate HQ for Commandement Air des Forces de Défense Aérienne (CADFA).

20 Naples, Italy

HQ Allied Forces Southern Europe.
HQ Allied Air Forces Southern Europe.
NSG Detachment.
Control Station for NATO Satellite Communications (SATCOM) System.
Control Station for US Navy Fleet Satellite Communications (FLTSATCOM) System.

21 Northwood, Great Britain

HQ NATO Supreme Allied Commander Channel (SACCHAN).
HQ Commander-in-Chief Eastern Atlantic (CINCEASTLANT).
HQ Royal Navy (RN), including HQ for Command and Control of UK SSBN/SLBM force.

22 Oslo, Norway

HQ Commander Allied Forces South Norway.

23 Pitreavie Castle, Great Britain

HQ Commander North Channel (COMNORCHAN).
HQ Commander North Atlantic (COMNORLANT).

24 Plymouth, Great Britain

HQ Commander Plymouth Channel (COMPLYMCHAN).
HQ Commander Central Atlantic (COMCENTLANT).

25 Ramstein, West Germany	HQ Allied Air Forces Central Europe. HQ US Strategic Air Command (SAC) Europe, and SAC 7th Air Division. HQ US Air Forces Europe (USAFE). HQ USAF Electronic Security Command (ESC) Europe.
26 Rheindahlen, West Germany	HQ 2nd Allied Tactical Air Force (2 ATAF). HQ British Army (BAOR).
27 Seckenheim, West Germany	HQ Commander Central Army Group (CENTAG).
28 Sirinyer, near Izmir, Turkey	HQ 6th Allied Tactical Air Force (6 ATAF).
29 Schwaebisch Gmuend (Bismark Kaserne), West Germany	HQ 56th Field Artillery Brigade (Pershing 11 IRBMs).
30 Taverny, Val-d'Oise, France	Underground National Military Command Centre (HQ for all French nuclear forces), and HQ Commandement Air des Forces de Défense Aérienne (CADFA).
31 Patch Barracks, Vaihingen, Stuttgart, West Germany	HQ US European Command (USEUCOM). HQ Special Operations Task Force Europe (SOTFE).
32 Verona, Italy	HQ Commander Allied Land Forces Southern Europe.
33 Vicenza, Italy	HQ 5th Allied Tactical Air Force (5 ATAF).
34 Whitehall, Great Britain	HQ UK MoD ('Pindar').

EARLY WARNING SITES:

35 Diyarbakir, Turkey	AN/FPS-17 and AN/FPS-79 early warning radars.
36 Fylingdales, Great Britain	Ballistic Missile Early Warning System (BMEWS).

| 37 | Kaupaun, Kaiserslautern, West Germany | Simplified Processing Station (SPS) for Defense Support Program (DSP) Satellite Early Warning System. |

INTELLIGENCE STATIONS:

38	Andoya, Norway	SOSUS Station and P-3 LRMP aircraft base.
39	Augsburg, West Germany	Largest communications intelligence (COMINT) complex in Europe. AN/FLR-9 circularly disposed antenna array (CDAA).
40	Brawdy, Great Britain	SOSUS Station.
41	Cheltenham, Great Britain	Government Communications Headquarters (GCHQ).
42	Chicksands, Great Britain	AN/FLR-9 CDAA.
43	Edzell, Great Britain	Classic Wizard/White Cloud ocean surveillance satellite ground station. Also AN/FRD-19 CDAA.
44	Frankfurt, West Germany	HQ NSA Europe.
45	Hof, West Germany	AN/FLR-12 SIGINT site.
46	Menwith Hill, Harrogate, Great Britain	Major US SIGINT satellite ground station.
47	Rota, Spain	SISS ZULU SIGINT station. HF/DF Net Control station for Mediterranean area. US Navy Fleet Ocean Surveillance Information Centre (FOSIC).
48	San Vito dei Normanni, Brindisi, Italy	AN/FLR-9 CDAA
49	Wurmburg, West Germany	La Faire Vite SIGINT Network Control Centre

COMMUNICATIONS CENTRES:

| 50 | Anthorn, Great Britain | VLF station. |

51 Coltano, Italy

US AUTODIN (Automatic Digital Network) Switching Center.
AN/GSC-39 DSCS terminal.
Main US communications station in Italy.

52 Criggion, Great Britain

VLF station.

53 Croughton, Great Britain

Principal US communications station in UK.
Three DSCS satellite ground terminals.
AUTODIN (Automatic Digital Network) Switching Center.
US SAC 'Giant Talk' HF communications station.
US 'Mystic Star' HF system for Presidential and VIP emergency communications.
SAC 'Scope Signal III'.
Emergency Action Message (EAM) transmissions.

54 Hillingdon, Great Britain

AUTOVON (Automatic Voice Network) Switching Centre, connects the US and European systems.

55 Kester, Belgium

NATO control terminal for US DSCS and NATO communications satellites.

56 Landstuhl, West Germany

Main communications station for US forces in West Germany.
Network Control Facility (NCF) for Indian Ocean DSCS satellite. Also control facility for Atlantic Ocean DSCS satellite.
Terminals include AN/MSC-46, AN/MSC-61, and AN/FSC-78 systems.

57 Machrihanish, Great Britain

Communications centre for transmission of nuclear weapons release orders to NATO maritime aircraft.

58 Oakhanger, near Borden, Hampshire, Great Britain

USAF Satellite Control Facility (SCF) for control of photographic intelligence (PHOTINT) satellites and other military satellites. Master control station for UK SKYNET military satellite communications system.
NATO satellite communications (SATCOM) terminal.
US DSCS terminal.

59 Pirmasens, Great Britain   DSCS terminal linked to Fort Detrick, Maryland.
AUTODIN Switching Center.
National Command Authority (NCA) communications centre in Europe.
HQ of European Command and Control Console System (ECCC-S) for peacetime communications control of US Army and NATO nuclear weapons.
HQ of Cemetery Net for US Emergency Action Message (EAM) transmissions.

60 Rugby, Great Britain   VLF Station.

Sources: Duncan Campbell, *The Unsinkable Aircraft Carrier: American Military Power in Britain* (London, 1984); Shaun Gregory, *The Command and Control of British Nuclear Weapons*, (School of Peace Studies, University of Bradford, Peace Research Report No. 13, December 1986); Jeffrey T. Richelson and Desmond Ball, *The Ties That Bind: Intelligence Cooperation Between the UKUSA Countries – United Kingdom, the United States of America, Canada, Australia and New Zealand* (Boston, London and Sydney, 1985); 'Allied Command Defends Vast Area', *Aviation Week and Space Technology*, 21 May 1984, p. 57; and William M. Arkin and Richard W. Fieldhouse, *Nuclear Battlefields: Global Links in the Arms Race* (Cambridge, Massachusetts, 1985).

16 Problems in European Command and Control

George W. Rathjens

Desmond Ball has highlighted the physical vulnerability of the Command, Control, Communications and Intelligence (C^3I) systems of NATO, the incompatibility of different service and national communications systems, and the mismatch between the cumbersome decision-making structure of NATO and the requirements for rapid communications and quick decisions.[1] In so doing he has made it clear that NATO forces would probably be unable to use nuclear weapons in the European theatre with any semblance of coherence in the event of a large-scale European war. Whether this should be viewed as a truly sad state of affairs depends on one's view about the role of nuclear weapons in the defence of Europe.

Above and beyond the question of the efficacy of nuclear weapons, perhaps the debate over the weaknesses of C^3I should be broadened to address the need to communicate effectively, early on, and to deal effectively with crises before hostilities escalate. This requires a shift in focus away from vulnerabilities in C^3I *per se*, towards vulnerabilities and tensions arising from third parties who might precipitate a conflict in Europe. But before going further with this line of argument, it is helpful to review the different schools of thought concerning the use of nuclear weapons.

Much planning – and training – has been based on the assumption that it would certainly be desirable, and might be possible, to use nuclear weapons with selectivity and precision so as to affect military operations favourably; and for those holding these views, the deficiencies in NATO capabilities have to be seen as serious, and not only that, but scandalous. But there are many others, especially in Western Europe, who have grave doubts about basing plans and force postures on the assumption that nuclear weapons might be used in the ways described. Some just do not believe that, however much it might be improved, effective command and control capability could survive the use of significant numbers of nuclear weapons. Secondly, however the West might feel and plan, keeping conflict limited must depend as well

275

on adversary actions. Thirdly, and perhaps most importantly, there is the widespread conviction throughout Western Europe, but especially in West Germany, that any large-scale war in Europe, conventional, limited nuclear, or unlimited, would be completely unacceptable because of the level of damage that would have to be expected. Those so convinced will hardly be concerned that NATO's C³I system will not be up to managing a war which they are convinced must not be fought. Rather, their interest – if any – in C³I will be in having a system that will contribute to a perception on the Warsaw Treaty side that any deliberate attack will evoke a NATO response that will escalate to the point of massive nuclear attacks against the Soviet Union in a relatively short period of time, certainly before Western Europe could be overrun or engulfed in a highly destructive conventional or limited nuclear war.

During the late 1950s and perhaps the early 1960s it was credible that after attack from the East decisions might be taken by the United States to respond in just such a fashion, given the limited capacity of the Soviet Union to wreak destruction on the United States; but by the mid-1960s this was an increasingly dubious proposition. By 1979 Henry Kissinger was simply stripping the last shreds from a pretty tattered veil when he remarked in his famous Brussels speech that '. . . the European allies should not keep asking us to multiply strategic assurances that we cannot possibly mean, or if we do mean, we should not want to execute because if we execute, we risk the destruction of civilization'.[2] Surely by then, if not at least a decade earlier, it had to be clear that deterrence based on the threat of massive nuclear destruction of the Soviet Union would have to depend for its credibility mainly on the possibility of events getting out of hand: on force postures being designed, or developing willy-nilly, so that control of the use of nuclear weapons would have become, if not automatic, unpredictable and beyond the possibility of recapture by competent political authorities.

This is the system we have in Europe: a congeries of nuclear capabilities and C³I systems with all of the technical 'connectivity' and decision-making deficiencies identified by Ball. It is a system which would permit the effective use of nuclear weapons, if at all, only if there were no effective adversary attack against the weapons or the C³I system *and if* there were an extensive, irrevocable pre-delegation of authority for allocation and use. What we have constructed is a very elaborate doomsday machine.

One can be critical of it on a number of grounds. It is inefficient and its sensitivity is beyond either our or our adversary's ability to predict. One can speculate that with some ingenuity one could design a much less costly and more mechanistic system that would produce an irrevocable signal to launch several thousand nuclear weapons, given some specified stimulus; and perhaps it could be arranged that this be done, not with certainty, but on a probabilistic basis. It is hard, however, to imagine this occurring in the real world, and it is relatively easy to explain why, instead, we have what we have.

Consider the requirements. The West must be able to cope with some low levels of conflict without resort to the destruction of the Soviet Union – the crossing of borders by company-size forces, brief interruption of traffic between West Germany and Berlin, the sinking of a ship or two, the downing of an aircraft – but beyond that consensus becomes elusive. Flexible response seems like a reasonable doctrine for those, perhaps mostly American, who believe that nuclear weapons can, and, given attack from the East, should, be used in limited and selective ways. Since no one can, of course, *prove* that limited use is an illusion, even many who believe that virtually uncontrollable escalation would probably follow the use of nuclear weapons can be expected to support a policy for their limited use simply on the grounds that their expectations might not be realised. Events might, after all, develop in ways that would permit de-escalation following limited use; and surely, it is argued, one would not want to build force postures, plans, and a C^3I system that would preclude taking advantage of such an opportunity should it present itself. Thus not only the believers in nuclear war-fighting but many who are sceptical as well can reasonably argue for rectification of C^3I problems and, in some cases, for improvements in nuclear capabilities, for example, enhanced radiation weapons, that might make nuclear weapons more 'usable'. Many, particularly if they are Americans, will also argue for more support for conventional forces from West Europeans. But can one imagine proponents of this school favouring a mechanistic 'doomsday machine'? Hardly: it would require radical restructuring of their thinking, including acceptance of the risk of automatic destruction of the United States, which they believe might otherwise perhaps be avoided even in the event of serious conflict with the Soviet Union. Rather, this group will have a great interest in prevention of escalation and in having a posture of escalation dominance for the West across the whole spectrum of possible conflict.

For those who believe conflict in Europe at any significant level to be totally unacceptable – either because the immediate damage would be intolerable or because they believe catastrophic escalation would be virtually inevitable – everything must depend on deterrence, assuming, of course, a belief that there is a threat from the East. What is necessary, in their view, is a perception in the Kremlin that a nuclear exchange involving the United States and the Soviet Union would be a reasonably likely consequence of any significant conflict in Europe. While a mechanistic 'doomsday machine' might serve the purpose, it is inconceivable that the Americans – or for that matter a lot of Europeans – could be won over by any transparent design. The structure we have would seem, however, to serve the purpose at least moderately well, although there are frequent alarms about the possibility that withdrawal of theatre nuclear weapons and/or American troops could weaken the 'coupling' of conflict in Europe to an intercontinental nuclear exchange.

Thus the C³I posture we have in NATO can be explained as almost inevitable, given the assumption of a Soviet threat and yet radically different and apparently irreconcilable views about the role of nuclear weapons in the defence of Europe. It does not, however, serve anyone's interests well.

First, as Ball has explained, the system is grossly inadequate for 'war-fighting'. Secondly, for those who are sceptical about nuclear war-fighting, but who nevertheless believe that some capability is desirable 'just in case it proves possible', there is the dual dilemma: the system as constituted does not provide much capability; yet some of its components and doctrines – for example, the increasing propensity on both sides to develop capabilities to carry the battle deep into adversary territory and the retention of dual-purpose delivery systems – may make escalation to large-scale nuclear war more, not less, likely. Thirdly, for those with no interest in 'nuclear war-fighting', but who are nevertheless convinced of the need for deterrence of threats from the East, there can be little basis for strong belief, and certainly none for consensus, that the response capabilities of the system as currently constituted are anywhere close to optimal. The question is whether, for different levels of provocation that can be envisaged, the probabilities of catastrophic escalation are about 'right'. Some would argue that they may be too low: that, given a crisis in Europe, Warsaw Treaty forces might be used in attack against the West or to interfere in the Balkans or Eastern Europe, believing that this could be done effectively and with acceptably low risks of escalation. But others would argue that the risks of escalation may be too high: that events, particularly those having their origins in actions by forces over which neither superpower might have much control, could occur which would lead to catastrophe. (Scenarios analogous to 1914 are often mentioned.)

In considering C^3I, doctrine and weapons deployment, a third set of considerations is important: there is a widespread belief in the NATO countries, particularly in establishment circles, that solidarity is all-important, and for those who hold this view divisive debates weaken the Alliance, which is held to be a 'good' in itself, and serve the Kremlin's purposes. Fortunately, from this third point of view, the very nature of C^3I is such as to discourage inquiry and debate. Firstly, with some exceptions – the Airborne Warning and Control System (AWACS) programme is an example – C^3I generally does not involve 'big ticket' hardware like the Tornado aircraft which may grab headlines and be sources of international and national debate; and secondly, many aspects of C^3I are highly technical and highly classified, and thus neither accessible nor attractive matters for politicians to get involved in. Ball has suggested that the explanation for the sorry state of NATO's C^3I system lies perhaps in high-level neglect. No doubt he is, to an extent, right. But why the neglect? Surely it is in part because of the intangible nature of C^3I, the secrecy and lack of glamour, and the related fact that to a great degree C^3I and the planning of war are seen as the province of the military, but it is also because to get into C^3I is to raise questions politicians would just as soon not talk about: levels of damage in the event of war, the possibility of deterrence failing, risks of accident, and, above all, the role of nuclear weapons in Europe.

But still the question arises, are there not improvements that might be made? It is hard to imagine any which would meet with the approval of all three interest groups within NATO: those who believe that having 'nuclear war-fighting' capabilities is important; those who do not want such a capability, or perhaps not even a strong conventional force with offensive capabilities; and those who do not want divisive debate. The criteria are especially demanding when it is recognised that some in the second group may want the risk of escalation increased while others may want it lowered. Perhaps there have been a few developments that could be viewed as positive by virtually everyone, or at least as positive by some and not negative by the others. The installation on all of the US nuclear weapons in Europe of permissive action links (PALs) – devices which preclude nuclear weapons being used without a code having been transmitted from higher headquarters – may seem at first glance to be an example. While there has been serious – and effective – opposition to installing such devices on submarine-launched weapons because of fear that difficulties in communications could preclude transmission of the codes, there *seems* to have been relatively little opposition in the case of land-based nuclear weapons in Europe. The presumption has been that the codes could, and would, be transmitted in the event of crisis, and assuming this, there would seem to be little first-order effect on the ability to use nuclear weapons in conflict and little effect on escalation probability. Meanwhile, in the absence of crisis, having such devices on weapons would seem to be an unalloyed 'good': it would make the unauthorised use of weapons by terrorists or others virtually impossible. But even this relatively uncontentious case is not so simple: there is *some* possibility that codes would not be transmitted in a crisis, perhaps because of failure of communications, but more likely to avoid exacerbating the crisis; and both possibilities would serve to hurt war-fighting capabilities and weaken deterrence. Moreover there is a second-order effect: without such devices, it is at least possible that we would not want to deploy nuclear weapons in some of the places where we now have them, and this could mean both poorer war-fighting and weaker deterrent postures.

One confronts similar trade-offs when one looks at other deployment and C³I decisions in the European theatre. The removal of the INF weapons will probably not have much real effect on offensive capabilities but, in providing targets for both conventional and nuclear weapons, such missiles may facilitate 'coupling', making escalation more likely. Their removal raises the opposite prospect – a degree of 'decoupling' – and this has been a matter of concern, especially in West European establishment circles. The reverse side of this coin is the development and deployment of precision-guided conventional munitions, which may offer the prospect of improvement in conventional war-fighting capabilities and increasing the prospects for escalation by threatening the destruction of nuclear systems. Clearly, with no consensus on beliefs about 'war-fighting' and whether to increase or reduce the prospects for nuclear escalation there can be no possibility of consensus on these issues.

One other possible change in force posture and doctrine perhaps merits special comment: that is, movement towards a conventional 'defensive' defence posture, that is, greater emphasis on the use of barriers, thinning out or withdrawing forces from forward areas in Central Europe and less emphasis on armour, deep interdiction, bridging equipment and the like. Apart from political problems relating to highlighting the concept of two Germanies, it is hard to see why movement in this direction would be especially troublesome to any of the interest groups we have identified. Moreover the C^3I requirements for such a posture would be much less demanding than those for 'nuclear war-fighting', and more consistent with what we actually have and can reasonably expect to develop. What is really indicated is simply an improvement in capabilities to carry out *conventional* operations effectively. As Ball, and others, have pointed out, removing shorter-range nuclear weapons and drawing sharp lines between strategic and conventional C^3I systems and between delivery systems for nuclear and conventional weapons could be helpful in this respect.

There is, though, a still greater need, so far given short shrift in this chapter and in many others dealing with the possibilities of war in Europe: that is the necessity of recognising that looking at the problem mainly as one of two-sided war may be a grave mistake. The devastation of the First and Second World Wars and the expectations of what is likely in the event of a Third World War make any deliberate, premeditated attack by either NATO or Warsaw Treaty forces seem utterly insane. Yet we must still fear this possibility of conflict in Europe and of its escalating, perhaps to the point where nuclear weapons would be used and where effective control would disintegrate. There is still unrest – and the potential for more – in Eastern Europe, the possibility of conflict between Greece and Turkey, and the unsettled situation in and around Kosovo. While it may seem unlikely that conflict could develop in these areas and escalate beyond control, it surely seems more likely than a 'bolt out of the blue' attack by the main forces of the two great alliances. Moreover, as we have suggested elsewhere, the opportunities for greatest leverage – perhaps even 'the last clear chance' for avoiding catastrophe – may be in preventing such conflicts or in dealing with them very early in the game: long before either the use of nuclear weapons or deterrence based on them can have much relevance; and long before we need be much concerned about the problems of C^3I vulnerability or 'connectivity' during combat.[3]

Whether concern about the grievances of and possible militant actions by third parties over whom the superpowers may have imperfect control – and perhaps even little direct influence – is properly a C^3I issue may be open to question, but it may be a more fruitful area for greater emphasis than the hardening of communications to an Electromagnetic Pulse (EMP) or the vulnerability of radars.

Notes

1. Desmond Ball, 'Controlling Theatre Nuclear War', above, pp. 237–74.
2. Henry A. Kissinger, 'NATO: The Next Thirty Years', *Survival*, vol. xxi (1979) 264.
3. George Rathjens, 'Inadvertent Nuclear War and Crisis Management', *Scientia*, vol. cxx (1985) 337–41.

Part V
Arms Control: Successes and Failures

17 The Purposes, Achievements and Priorities of Arms Control

Paul S. Brown

THE PURPOSES OF ARMS CONTROL

United States policy with respect to nuclear weapons is one of deterrence. There is a spectrum of views as to what the nature of deterrence should be. Some believe that a nation should maintain the minimum capability needed to inflict immeasurable harm in retaliating against an aggressor. Others believe that deterrence should work in a measured way to counter aggression at any level in order to limit the conflict to that level. US policy reflects the latter thinking and the US strategic doctrine for deterrence is called 'Countervailing Strategy'.[1] US national security depends on the maintenance of an effective, reliable and survivable nuclear force to deter potential adversaries from any act of aggression on the United States or its allies. Deterrence has worked for over 40 years, in that major armed conflict between the superpowers has been avoided. Arms control can make deterrence work better.

Arms control serves a number of purposes. By placing limitations on the numbers and types of weapons that are developed and deployed, arms control can strengthen the framework of deterrence and reduce the threat that nuclear weapons would ever be used. Of particular importance is controlling and limiting weapon deployments that may be destablising in time of crisis. By promoting mutual trust and understanding among the participating nations, arms control can also serve to reduce the dangers of attack and accidental nuclear war. If deterrence were to fail, arms control could work to make the conflict less destructive. Finally, by managing the arms competition between nations, arms control can reduce military spending, allowing more resources for the civilian economy.

The premise of arms control is that nations realise that the deployment of more weapons and qualitatively-different weapon systems will do little to enhance, and may even diminish, national security. Nations must also have confidence that their existing weapons are adequate to serve their national security needs. These realisations must be satisfied for nations to be willing to negotiate arms control limitations. Of course it is also important to realise that some qualitatively different weapon systems, such as the US small mobile Intercontinental Ballistic Missile (ICBM), or the Soviet SS-25 missile, can enhance security and crisis stability.

An important aspect to arms control is the process, which is of considerable value. It is certainly better to be sitting across from one another at the negotiating table than to be rattling sabres. As Winston Churchill once said, 'jaw-jaw is better than war-war'.[2] The negotiating process keeps the channels of communication open in case crises or other situations arise. The process has considerable appeal to other nations whose people may feel much more secure when the Soviets and Americans are negotiating, particularly when they are negotiating successfully.

There is a common perception that arms control is disarmament. Arms control may lead to disarmament at some future time. However we must be realistic about what arms control can achieve in today's world. As long as we rely upon armed deterrence amongst nations, then the best we can expect from arms control is to make deterrence work better, increase international stability, and reduce the risks of war. These alone would be worthy achievements. Again, Churchill once said it very well: 'It is the greatest mistake to mix up disarmament with peace. When you have peace, there will be disarmament.'[3]

THE ACHIEVEMENTS OF ARMS CONTROL

The argument is often made that arms control has accomplished little in the past 40 years. It is also claimed that pursuing further efforts at negotiating new agreements is either not in the best interests of the United States or is simply a waste of time. The present writer takes a different view. One only has to examine the historical record to see what arms control has accomplished. Even though more certainly could have been accomplished with arms control, a number of successful treaties and agreements have been negotiated. While some of the treaties are unratified by the United States for various reasons, all of the treaties have been effective as arms control measures. Let us examine these accomplishments.

The Limited Test Ban Treaty

The Limited Test ban Treaty (LTBT) was first signed and ratified in 1963. The Parties to the Treaty agree not to carry out any nuclear explosion in the atmosphere, in outer space, or under water, or in any other environment that would cause radioactive debris to be present outside the borders of the nation conducting the test. While the Treaty was negotiated by the United States, the Soviet Union and Great Britain, over 100 nations have since become parties to it.

During the eight years prior to 1963, the nuclear nations had sought unsuccessfully to negotiate a complete ban on nuclear weapon testing. There were difficulties in agreeing to the monitoring measures that would be necessary to verify such a ban. The Soviets considered some of those measures to be overly intrusive. Monitoring of the LTBT could be done by National Technical Means (NTM), and the Treaty has been considered by some to be a compromise to substitute for the elusive Comprehensive Test Ban (CTB). In fact, once the three nations decided to pursue the LTBT, the negotiations took only ten days, and the text of the treaty filled only two pages.

The LTBT has been quite beneficial for environmental reasons in virtually stopping pollution of the atmosphere by radioactive debris from nuclear tests. Even China and France, non–signatories to the Treaty, have stopped testing in the atmosphere. The United States has had cause for concern over the years with some Soviet tests where radioactive gases were released and crossed international borders. While annoying, because these releases appear legally to violate the LTBT and suggest a lack of effort by the Soviets to contain the radioactivity, the releases seem to pose a limited environmental concern. (It should be added that there has been some debate about the legal interpretation of what the treaty forbids. The English version of the LTBT prohibits 'radioactive debris' from crossing international borders and debris is interpreted to mean all forms of radioactive contamination. The Soviets on the other hand claim that debris refers to particulate and not gaseous matter.) In any case, the Soviet releases are certainly of orders of magnitude smaller than they would be from atmospheric tests. Because of serious concerns about domestic radioactive contamination, the United States goes to great lengths to guarantee that radioactivity from a nuclear test will be contained within the ground. It may be that the Soviets do not share these concerns about such radioactive contamination. Since 1970, when the United States experienced a major sudden release of radioactivity in an event called Baneberry, it has experienced only two minor unexpected releases of radioactivity and these were seepages that were confined to the borders of the Nevada Test site.

When the LTBT was being discussed, concern was expressed on the impact the treaty would have on weapon development, weapon effects research, and Anti-ballistic Missile (ABM) development. Since atmospheric testing ceased, and with it atmospheric sampling, a source of intelligence data on each side's weapon development capabilities also ceased. Nevertheless the treaty's limitations have had an impact, and we have learned to work within these bounds. The costs associated with these limits have been well worth the treaty's benefits. It is noteworthy that the LTBT was the first real achievement in US–Soviet arms control and it was accomplished just after the Cuban Missile Crisis, which was a particularly strained time in US–Soviet relations. The LTBT helped improve relations between the superpowers and it set the stage for further progress in arms control negotiations, particularly those leading to the Non-Proliferation Treaty (NPT).

The Non-Proliferation Treaty

The NPT was first signed in 1968 and went into effect in 1970. Today, over 120 nations have agreed to abide by its terms. The NPT is intended to prevent the spread of nuclear weapons and to ensure that the peaceful nuclear energy activities of the non-nuclear weapon states will not be diverted to weapon development. In return, the NPT calls for the sharing of technology beneficial to the development of peaceful nuclear energy. The NPT is unique in that it was negotiated in a multilateral forum, the Eighteen Nation Disarmament Committee, which was the predecessor to today's 40-nation Conference on Disarmament.

The goals of the NPT appear to have been achieved in the main. When the Treaty went into effect, the number of nuclear weapon states was predicted to climb to 30 in a decade and the number of *known* nuclear nations today is generally acknowledged to be only six – the United States, the Soviet Union, Great Britain, France and

China, which have nuclear arsenals, and India, which has at least detonated a nuclear explosion. There are a few other nations who, according to newspaper reports, are either close to having or who already have nuclear weapons and do not appear to have conducted nuclear weapon tests.

The NPT has benefits which the superpowers fully appreciate. The two countries have had little difficulty in agreeing on steps to enhance its viability. Perhaps a large factor that has supported the non–proliferation of nuclear weapons is the security experienced by some non–nuclear nations through their alliance with the superpowers. Some nations are more willing to abide by the NPT when they see potentially hostile neighbours also accepting the treaty's terms.

Most of the parties to the NPT have been particularly dissatisfied with one aspect of the NPT, that having to do with Article VI. Nuclear arsenals have grown consideraby since the NPT went into effect and hence these nations have reason to be concerned. Article VI of the NPT commits each of the signatories 'to pursue negotiations in good faith on effective measures relating to cessation of the nuclear arms race at an early date and to nuclear disarmament. . .'. In other words, Article VI addresses the issue of vertical proliferation as contrasted with horizontal proliferation. Many nations believe that the best way to satisfy the goals of Article VI would be for the superpowers to negotiate a cessation of nuclear testing. But the current proposals in Geneva for major reductions in nuclear arsenals are more effective ways of achieving the goals of Article VI and of directly attacking vertical proliferation. Regarding the role of nuclear test limitations, such limitations must be considered in the context of major arms reductions, and will be discussed in more detail below.

The Anti-ballistic Missile Treaty and SALT I

The Anti-ballistic Missile Treaty was first signed and ratified by the superpowers in 1972 and then modified in 1974 by a protocol effective in 1976. The ABM Treaty was negotiated in the first set of Strategic Arms Limitation Talks (SALT I), along with a five-year Interim Agreement to limit certain strategic arms. The ABM Treaty and its protocol limit each side to a single ABM site consisting of 100 launchers and 100 interceptor missiles. Even more importantly, both parties agreed to limit qualitative improvement of their ABM technology in a number of areas: interceptors, the capability for rapid reload, and Multiple Independently-targetable Re-entry Vehicles (MIRVs) equipped with interceptors. The ABM Treaty also placed restrictions on the development of exotic new technologies and on sea-based, air-based, space-based, and mobile land-based ABM systems. The five-year Interim Agreement was structured to complement the ABMT and limit the offensive arms competition until a permanent strategic arms limitation agreement could be negotiated. The Interim Agreement essentially froze the number of ballistic missile launchers on both sides, allowing for some trade-offs of older launchers for an agreed number of new submarine launchers. Both sides also agreed not to convert launchers for light ICBMs into launchers for heavy ICBMs. The Agreement allowed modernisation and replacements, without any significant changes in launcher dimensions (less than 10–15 per cent of existing dimensions).

The ABM Treaty has been an effective one. It has effectively limited defensive deployments, which in turn could have spurred a spiral of offensive–defensive counter-deployments. Unfortunately the goal at the time of the ABM Treaty and Interim Agreement to restrain offensive deployments *per se* was unfulfilled.[4]

There has been recent concern about the continued viability of the ABM Treaty. On the one hand, certain Soviet actions have brought into question their continued compliance with the ABM Treaty – their most noteworthy action being the construction of the Krasnayorsk radar. While the individual Soviet actions appear to constitute only a limited military threat, taken together with their extensive research efforts on defensive technologies, the total effect has been an undermining of US confidence regarding Soviet intensions. There has been particular concern that the Soviets may be moving towards a nationwide ABM system, which is strictly prohibited by the ABM Treaty.

On the other hand, the Soviets have expressed considerable concern abut the US Strategic Defense Initiative (SDI). They have demonstrated their concern in the positions they have taken in current arms control negotiations. The present writer's views regarding SDI research are available elsewhere.[5] To summarise, SDI must remain a research programme until it becomes apparent that something exists which is worthwhile deploying. All SDI research should be completed within the limits of the ABM Treaty and any research departures from the ABM Treaty should be negotiated according to the provisions of the Treaty. Further, any deployments beyond what is allowed by the ABMT must first meet the Nitze criteria of 'survivability and cost-effectiveness at the margin' and must be subject to negotiation.[6] In the last two years, issues have arisen regarding the interpretation of the ABM Treaty and early SDI deployments. The Treaty language could have been more precisely worded in places, and arguments can be made either way about the interpretation of some of the articles and agreed statements.[7] However from a consideration of the text of the Treaty and what has been written and said on the intent of the wording, particularly by those who were involved in the Treaty negotiations, it may be concluded that the strict, as opposed to the broad, interpretation of the Treaty is the correct one.

Regarding early SDI deployments, concern may be expressed about the impact such a move would have on the political future of the Treaty, in the absence of Soviet agreement to amend the Treaty to allow such deployments. Concern may also be felt about the technical merits of early deployments. Such moves would jeopardise the research nature of the SDI programme and could seriously undercut research progress on promising advanced technologies that will be necessary for truly effective defences.

A system of space-based rockets called kinetic kill vehicles (KKVs) has been proposed as a first-tier strategic defence that could be deployed in the 1990s. Analysis done at the Lawrence Livermore National Laboratory has shown that, while early deployments of space-based KKVs might provide an effective defence against the current Soviet ICBM force consisting mostly of SS-18 missiles, the defence would require an order of magnitude more KKVs in order to defend against a modernised Soviet force with significant numbers of SS-24 and SS-25 missiles. The analysis further shows that Soviet counter-measures employing faster burning boosters and more rapid post-boost deployment would require even greater numbers of KKVs. It was concluded that the Soviets would readily be able to implement cost-effective countermeasures to currently proposed early deployments within the same time-frame as those deployments.[8]

The ABM Treaty has been a successful treaty in preventing a spiralling competition between offensive and defensive counter-deployments. Both superpowers will have to proceed very cautiously in the future if we are to preserve the benefits that the Treaty has to offer.

SALT II

The SALT II Treaty took almost seven years to negotiate and was signed in 1979. The Treaty, which was to last until 1985, was never ratified. Initially, this was owing in part to US disillusionment with other Soviet actions in the world, and in part to an intense debate in the United States between the Treaty's proponents and opponents about the merits and flaws of the Treaty and whether it could be verified. Although the Senate never gave its consent to ratify the Treaty and President Jimmy Carter withdrew the Treaty from the Senate after the Soviets invaded Afghanistan, Carter, and President Ronald Reagan after him, agreed to abide by the provisions of the Treaty as long as the Soviets agreed to do so. The Soviets made similar expressions of intent. SALT II was to last until 1985, and both sides agreed to abide by its limits, as well as by the limits of the Salt I Interim Agreement, until a follow-on, permanent agreement could be achieved. The Reagan Administration called this policy with respect to the SALT II treaty 'Mutual Interim Restraint'.

SALT II was made possible because it was perceived at the time that parity in strategic strength existed between the superpowers, even though the forces of the two countries had different structures. SALT II placed quantitative and qualitative limits on a variety of strategic systems. Other than calling for small reduction in 1981 of the ceiling for strategic nuclear delivery vehicles (SNDVs), SALT II did not call for major arms reductions. Both sides did agree in the next stage of talks (assumed at the time to be SALT III) to seek substantial reductions in strategic arms along with further qualitative limits. The Strategic Arms Reduction Talks (START) have, in effect, become the next stage.

The United States has developed over the years considerable concern about Soviet compliance with the unratified SALT II treaty.[9] These concerns have focused on several issues. The United States has claimed: first, that the encryption of telemetry in Soviet missile tests has impeded US NTM capability for verifying compliance with the Treaty; secondly, that the SS-25 ICBM was a second new type of missile and therefore not allowed by the Treaty; and thirdly, that the Soviets have exceeded the number of SNDVs allowed by the Treaty. The Soviets have replied to these complaints. For example, they have claimed that the SS-25 was simply a modernisation of the older SS-13 ICBM. And, of course, the Soviets have expressed concerns of their own.

The United States also expressed concern about a number of other issues relating to Soviet compliance with other treaties, particularly the ABM Treaty. These concerns have all been vigorously discussed in the Standing Consultative Commission (SCC). The SCC had been established in SALT I and later extended to SALT II to provide a forum where the two sides would work out problems that they perceived impeded compliance with the SALT Treaties. The two sides have been unable to resolve many of their differences on the US concerns. The Reagan Administration announced that it would continue the policy of 'Mutual Interim Restraint' only if the Soviets stopped their non-compliant actions. After repeated efforts to resolve the issues failed, the Reagan Administration announced in May 1986 that it would no longer bind itself to the limits of SALT II. The United States said that it would reconsider this action if the Soviets stopped those activities which it considered to be not in compliance with the Treaty.

It can be argued that these developments were quite unfortunate. SALT II had its drawbacks, because it stopped short of large cuts in weapon systems. However, it did have significant benefits, because it set ceilings on strategic systems and thereby helped ensure that parity between the two countries would be maintained. At the same time, SALT II allowed both sides to proceed with necessary weapon modernisation programmes and to address issues relevant to the asymmetric structure of their forces. In the process of doing these things, SALT II helped set a degree of predictability in the strategic balance between the two countries.

SALT II set the stage for major reductions in the follow-on START talks. Without the lid placed by SALT II on strategic systems and without a successful START agreement in the near term to take its place, even larger numbers of strategic systems could be deployed in the future. Even though the United States had legitimate concerns about Soviet compliance with the treaty, and these concerns diminished US confidence in Soviet intentions, the political benefits for the United States of continuing to adhere to the Treaty should have outweighed the risks of the non-compliant actions. The United States might have continued to try and work out its concerns in the SCC, and the Soviets certainly could have been more accommodating in that forum. There is a more general question that relates to the SALT II compliance issue, as well as to other Treaties: how to treat technical violations which do not appear to outweigh the overall value of a treaty. It is a question that will continue to arise in arms control deliberations.

Even with these events, it could still be argued that the seven years of negotiations were an excellent learning process which paved the way for further progress in arms control. Many issues were identified.

The Treaty contained a considerable amount of technical detail and precise definitions that can be applied to current and future arms discussions. The United States could also learn from the difficulties experienced in the compliance area when addressing the verification requirements for future treaties.

The Threshold Test Ban Treaty and the Peaceful Nuclear Explosions Treaty

The Threshold Test Ban Treaty (TTBT) was negotiated and signed in 1974. Negotiations then began on the companion Peaceful Nuclear Explosions Treaty (PNET) which was signed in 1976. The TTBT sets an upper yield limit of 150 kt on underground explosions. The PNET sets the same limit on individual nuclear explosions for peaceful purposes, and makes other provisions for group explosions. The PNET also provides for the conduct of PNEs on the territory of other countries. It is important to note that the TTBT was negotiated in only five weeks. This reportedly was the result of the desire of President Richard Nixon and Secretary General Leonid Brezhnev to have a treaty in hand prior to their Summit meeting in 1974. The two Treaties have never been ratified. However both countries agreed to abide by the terms of the TTBT with its effective date in 1976.

A major reason why the two Treaties are especially noteworthy is that they marked a change in Soviet attitudes towards nuclear test verification. Prior to 1974, the Soviets had insisted that National Technical Means of verification were adequate for monitoring test bans. In the TTBT negotiations, the Soviets accepted co-operative measures to enhance NTMs in the form of exchange of data on geology, geography and calibration yields. In the PNET negotiations, they accepted the need for in-country verification and allowed for the presence of personnel from the monitoring nation (the United States in this case) to verify the conditions of certain explosions and to measure directly the yields of individual explosions detonated as part of a group in order to guarantee that no individual explosion exceeded the 150 kt limit. The PNET also established a Joint Consultative Commission to address various issues regarding implementation of the treaties. The PNET has a technically complex protocol which was the product of 18 months of extended negotiations. The success in completing such an effort was the direct result of the desire on the part of both sides to conclude an agreement and the willingness to negotiate compromises to that end.

Ten years ago the 150 kt threshold was considered to be militarily important because it restricted the development of very high-yield weapons. Such weapons were considered to pose a particular threat in a first strike. Since then, as the accuracies of delivery systems have increased, the yields of warheads have decreased. So, in general terms, the 150 kt limit has imposed only a partial restraint on the arms competition between the superpowers. In fact, the directors of the US weapons laboratories have testified before Congress that their development programmes can meet all current military requirements within the 150 kt limit.[10]

An issue has arisen, however, that has diminished US confidence in Soviet compliance with the TTBT limit. Verification of the TTBT is carried out by teleseismic means which has large inherent uncertainties. These uncertainties were recognised when the Treaty was negotiated. However the body of seismic data from Soviet explosions is such that some analyses of the data conclude Soviet non-compliance with the 150 kt limit and the Reagan Administration has reported to Congress that the Soviets are in probable violation of the TTBT. Based on Livermore's assessment of the relationship between yields and seismic magnitudes for the Soviet test sites and the patterns of Soviet testing, it has been concluded that the Soviets appear to be observing some yield limit. Livermore's best estimate of this yield limit, given the statistical nature of the problem, is that it is consistent with TTBT compliance.[11] However, because of the uncertainty of the teleseismic yield estimates and the uncertainty in extrapolating US test experience to Soviet test sites, the possibility cannot be ruled out that a few Soviet tests may have exceeded the limit. In short, it is impossible to state with certainty that the Soviets have or have not violated the TTBT.

Are the few Soviet tests that appear to have exceeded the limit significant? The opinion of weapon scientists at Livermore is that the US weapons programme would gain very little of design significance by testing at the levels associated with the largest Soviet events. In assessing what is significant to the Soviets we are forced to use a 'mirror image' approach because so little is known about Soviet design practices. It is then concluded that the Soviets would not gain much either. It must be emphasised, however, that caveats must be placed on this answer because 'mirror imaging' can involve faulty assumptions and be subject to great uncertainties.

Why then seek improved verification of the TTBT? We are dealing here with a fundamental issue of confidence-building in the long term in the face of changing environments and value systems. The precision of TTBT yield estimates is insufficient to make it possible to render definitive judgements about Soviet compliance with the 150 kt limit. There are some in the United States who attach more military (and political) significance to the possible Soviet non-compliant actions with respect to the TTBT limit than others. An inability to distinguish reliably between compliance and violations of significance can undermine confidence in the arms control process and heighten international tensions. Means are available to improve the TTBT monitoring capability, and thereby create a situation of greater confidence. Regarding confidence, it should be noted that the Soviets have on occasion expressed concern about US compliance with the TTBT limit. Seismic uncertainties are universal. Indeed improved TTBT verification is the first priority goal which the United States has been seeking in the current nuclear testing discussions with the Soviets in Geneva.[12] The Nuclear Test Experts' Meetings (NTEM), and the decision on 17 September 1987 to begin full-scale negotiations on nuclear testing, will be considered shortly.

Treaties Establishing Nuclear-Free Zones

There have been a number of treaties that have established nuclear-free zones. These include the Antarctic Treaty, the Outer Space Treaty, the Latin-American Nuclear-Free Zone Treaty, and the Seabed Arms Control Treaty. All of these Treaties have successfully limited the spread of nuclear weapons to areas that do not yet have such weapons. The Antarctic Treaty, adhered to buy some 23 nations, including the United States and the Soviet Union, has successfully ensured since 1959 that Antarctica is used for peaceful purposes only. The Outer Space Treaty was modelled on the Antarctic Treaty and since 1967 over 100 nations have agreed to abide by its limits. The Treaty keeps nuclear weapons and other weapons of mass destruction out of space and it ensures that the moon and other celestial bodies will be used exclusively for peaceful purposes. Similarly, the Seabed Arms Control Treaty, negotiated in 1971, has kept the ocean floor outside each nation's territorial waters (agreed to in the Treaty to be a 12-mile limit) free of nuclear weapons. About 100 nations have agreed to abide by the Treaty's provisions.

The Treaty for the Prohibition of Nuclear Weapons in Latin America, effective in 1968, is similar to the other nuclear-free zone

Treaties. It differs in the respect that it applies to a populated area of the world and in that regard it works to prevent the proliferation of nuclear weapons. Twenty-five Latin American countries have agreed to abide by the terms of the Treaty. In Protocol I of the treaty, the four nations (France, the Netherlands, Great Britain and the United States) who have territories in the zone have agreed to abide by the terms of the Treaty. In Protocol II, the five nuclear weapon states agreed to support the denuclearised nature of the Latin American zone.

The successes achieved in negotiating this and all the other regional treaties indicates what can be achieved in arms control once nations recognise the mutual benefits to be derived for improved security and well-being.

Agreements to Reduce the Risk of War

There have been a number of successful agreements that the super-powers have signed that have reduced the risks of war, and in particular nuclear war. These include the Hot Line Agreement, the Accident Measures Agreement and the Prevention of Nuclear War Agreement. Rather than being treaties requiring an involved ratification process, these are executive agreements that are signed by the leadership or high level representatives of the two countries. The prime purpose of such agreements is to minimise the possibility for misunderstanding and its consequences during times of crisis. Measures include information exchanges such as launch notifications, constraints on military operations, and provisions for enhanced communications during crises.[13]

● In June 1963, the US and Soviet representatives to the Eighteen Nation Disarmament Committee negotiated and signed the Hot Line Agreement which established a direct communications link between the two countries. The link showed its value during the Middle East conflicts in 1967 and 1973. Concerns about the vulnerability and survivability of the communications link in the event of war led to an agreement in 1971 to upgrade the Hot Line using the latest US and Soviet satellite communications technology. The 1971 negotiations to upgrade the Hot Line were carried out as part of the SALT I process. In 1984, the superpowers again agreed to improve the Hot Line by adding a direct facsimile link to allow transmission of figures, charts and other pictures.

• The Accident Measures Agreement was also concluded as part of SALT I. Both superpowers agreed to take unilateral steps to improve operational and technical measures to prevent the accidental and unauthorised use of nuclear weapons. The agreement also committed each side to make arrangements to notify the other side in case of accidents, unexplained incidents and unidentified objects picked up by early warning systems. Included in the agreement was a provision to notify the other side of missile test launches that could be interpreted as a threat. In 1985, the superpowers completed a Common Understanding at the SCC to identify more clearly the kind of incidents that are of concern.

• The Prevention of Nuclear War Agreement signed in 1973 provides for consultations between the superpowers in the event of situations that might arise in which there is danger of a nuclear confrontation.

These agreements are all aimed at building up confidence between the superpowers rather than at reducing or limiting weapons. They have been successful and progress continues in this area. For example, in May 1987 the superpowers negotiated an agreement to establish Risk Reduction Centres in each nation's capital. The centres will contain modern communication equipment and will be staffed with diplomatic, military and intelligence personnel. They will provide a way to exchange information in support of existing and future such agreements. The new agreement was signed on 15 September 1987 when the Soviet Foreign Minister, Edward Shevardnadze, visited Washington.

Considerable emphasis continues to be placed on developing confidence-building measures. Negotiations for such measures were an important aspect of SALT II and continue to form an integral part of the current negotiations in Geneva. In recent years, there has been considerable activity in the United States studying how better to manage crises and enhance confidence. For example, there are efforts like the Harvard Avoiding Nuclear War Project[14] and the Harvard Negotiating Project.[15] Stanford University has also been active in studies of crisis management and nuclear war prevention measures.[16]

The Conference on Confidence and Security-building Measures in Europe

While thus far the discussion has been concerned with arms control agreements which involve nuclear weapons, it is certainly worthwhile to consider an arms control agreement dealing with conventional forces that was concluded in 1996 at the 35-nation Conference on Confidence and Security-building Measures in Europe (CDE). The 35 nations were from NATO, the Warsaw Pact and neutral countries. After three years of negotiations, they agreed to adopt measures that would reduce the risks of military confrontation in Europe and to refrain from the use of force in their relations. The measures include prior notification of military activities involving forces above a certain size and a provision for observers from the parties to the agreement to be at manoeuvres above a certain size. The agreement sets time requirements for the advance reporting of military activities. Two years' advance notice is required for a very large troop manoeuvre to avoid what might otherwise be interpreted as a show of force.

A significant part of the agreement was that a state could request a challenge on-site inspection (OSI) of another state in order to determine compliance with the agreement, with the number of OSIs that a state would have to accept being limited to three. The OSIs could be conducted on the ground or in the air, where the inspected state provides the aircraft and the inspecting state can use its own monitoring equipment. The fact that the Soviets agreed to these OSIs may be a significant indicator of their future willingness to accept effective OSIs for other arms control agreements now under negotiation. The propects for progress in arms control are enhanced when all the parties see value in the agreed measures, rather than viewing the measures as one-sided concessions.

The successful conclusion of the CDE agreement illustrates what can be accomplished in arms control when the parties realise the advantages to be gained for their national security. Owing to the widely varying views that are possible amongst nations, it is also noteworthy that 35 nations were able to come to closure on the agreement. The CDE agreement should go far in building confidence in Europe and in reducing the risks of an outbreak of war.

THE PRIORITIES OF ARMS CONTROL

Such are the achievements of arms control over the past 25 years. Considering the large nuclear arsenals that the superpowers have amassed in that time, it is certain that a great deal remains to be accomplished. Considering that parity in strategic balance generally exists between the superpowers, conditions now seem ripe to make major progress in arms reductions at the negotiating table. The present writer is optimistic that the current Geneva negotiations will produce meaningful results. These negotiations actually consist of three separate negotiations involving strategic arms (START), mid-range weapons (INF), and defence and space weapons. For the past year, the Americans and the Soviets have also been meeting in the Nuclear Test Experts Meetings discussing a broad range of issues related to nuclear testing. When Foreign Minister Shevardnadze visited Washington in autumn 1987 it was agreed to turn these discussions into negotiations.

What should the priorities of arms control be? In an article written in 1985 entitled 'What Went Wrong with Arms Control?', Thomas Schelling defined the issues very well. He said that we have lost sight of what arms control should be about. In his view arms control should avoid developments which would result in a strategic situation on one side that threatens the other side's ability to retaliate after being attacked. Such a situation would be extremely destabilising in a crisis.[17] In the present writer's view negotiations should also seek to reduce the weapon systems which most threaten the survivability of either side's deterrent. The difficult task is then to find ways in which the two sides can work out compromises, where both sides agree to cut back or limit weapons which are at the same time destabilising to the other side and an important element in the strategic forces of the possessor. For example, the United States is particularly concerned about the Soviet SS-18 missiles which pose a first-strike threat to its land-based missile forces. And the Soviets are concerned about the US SDI programme which they view as a threat to their retaliatory capability.

It seems that we are on the verge of major progress in Geneva. Now that it has been decided to negotiate a global zero–zero limit on INF missiles, the problems of verifying an INF agreement have become more tractable. The value of such an agreement will lie more in setting the stage for further progress in Geneva on strategic arms than it will in having any major impact on the balance of power in Europe. Nuclear weapons are political weapons, and INF weapons are perhaps the most political of all. There are still other weapons to accomplish the same mission for which the INF missiles were intended. The biggest gains to be achieved in Geneva will be in the area of strategic arms and will follow the INF agreement.

There is a fundamental issue that relates to arms control, and it has to do with modernisation, new technologies and research. Some say that new technologies drive the arms race, and sometimes they are correct. Others say that new technologies help make deterrence work better, and sometimes they are correct. The difficulty is to determine which technologies are desirable and then agreeing on this in Geneva. A considerable amount of time is devoted in Geneva to discussions on the control of technology.

A major thrust of the Soviet position in the current Geneva negotiations is to limit research on SDI. Earlier, the present writer gave his opinion that SDI research should go on as long as it is consistent with the ABM Treaty. The Soviet position has been to place much more restrictive limits on such research. The United States has been unwilling to accept this and the differences became an issue at the pre-summit meeting in Reykjavik, Iceland, in October 1986. The two sides were very near to agreeing on some major changes in nuclear arms, when the Soviet insistence that the United States make major concessions in its SDI research programme got in the way of agreements on strategic and INF weapons. There appeared, however, to have been a softening of the Soviet position when Foreign Minister Shevardnadze visited Washington in the autumn of 1987. The Soviets are now willing to allow research pursued within the traditional interpretation of the ABM Treaty.[18]

There were other concepts introduced at Reykjavik that, in the present writer's view, would have established some dangerous steps in nuclear policy. The notion that we could completely eliminate ballistic missiles in a decade was particularly disturbing to many policy experts, such as Brent Scowcroft, John Deutch, and R. James Woolsey who said in all seriousness in the *New York Times* that 'SDI has already protected the United States by preventing an agreement at Reykjavik that would have destroyed the American ballistic-missile deterrent within a decade'.[19] Similar words were used by James Schlesinger.[20] Christoph Bertram presented a cautious European point of view on Reykjavik in an interview in *Arms Control Today*, where he said that 'there was a very real and justified concern in Europe over what I would call the Utopian objectives of Reykjavik'.[21] And an interesting perspective on the elimination of ballistic missiles was recently discussed by Thomas Schelling, who sees merits in eliminating land-based missiles because of their destabilising vulnerability to pre-emptive attack, but sees 'no comparably strong case for eliminating or reducing the number of missile-carrying submarines'.[22] In short, the complete elimination of ballistic missiles in the near future, in the absence of carefully thought-out and well-planned arms control progress in other nuclear and conventional areas, could seriously undermine the deterrent capabilities of the Western World.

Another area where there has always been pressure to restrict technology is nuclear testing. In the 1950s and early 1960s, a nuclear test ban was the primary focus of arms control. The early test ban negotiations have been referred to as 'the SALT of their day'.[23] The problems of verification stood in the way of agreement on a CTB. The LTBT was concluded in 1963 as a compromise, a readily verifiable step towards a CTB. The LTBT was followed by the NPT, an agreement which most non-nuclear nations were willing to accept so long as the superpowers continued progress toward a CTB. Then came the TTBT and PNET which have already been discussed and which have posed their own set of verification concerns. During 1977–80, the Carter Administration reopened full-scale CTB negotiations. Considerable progress was made in achieving agreement on in-country seismic networks and on procedures for voluntary on-site inspections. Carter temporarily put aside consideration of a CTB in order to pursue SALT II, and the Reagan Administration in 1982 withdrew consideration of a CTB indefinitely, saying that it was a long-term goal.

Recently there have been increasing pressures on the United States to undertake more restrictive test limitations. The Soviet unilateral moratorium from August 1985 to January 1987 focused world attention on the issue. Other developments have occurred. There is the joint effort between the Soviet Academy of Sciences and the Natural Resources Defense Council in the United States which sets up seismic stations around Soviet and US test sites. There was also a proposal by a group of six nations (Mexico, Greece, Tanzania, India, Argentina and Sweden), called 'the Delhi-6', for a global seismic monitoring capability.[24]

The US Congress has also considered test limitations. Last year and this year, the House of Representatives passed amendments that would eliminate funding for nuclear tests greater than one kiloton as long as the Soviets agree both to abide by the same limit and to accept effective verification measures.[25] The House has passed arms control legislation in a number of other areas, including SDI, SALT II, chemical weapons, and Anti-Satellite weapons. In October 1986, when Reagan and Gorbachev went to Reykjavik, the House agreed to withdraw its arms control legislation so as to give the President negotiating flexibility. The legislation has been introduced again. The Senate has also considered legislation on test limitations. The Senate legislation is less restrictive than that considered by the House. In the past three years, the Senate has passed two non-binding 'sense of the Senate' resolutions asking the President to submit the TTBT and PNET for ratification, as well as to resume CTBT negotiations.[26] Currently there is Senate legislation pending for a two-year one kiloton moratorium with one 15 kt test per year.[27] The legislation calls for bilateral verification provisions. The Livermore Laboratory has analysed these provisions and finds that they fall short of what is required for effective verification.

Clearly there is increased interest at many levels in further test limitations. It must be emphasised that the process of arms control in the United States is complex, owing to a considerable amount of interplay between and within the Executive and Legislative branches of the Government. Proposed arms control agreements must have technical, military and political merits that will stand the test of time with varying options in dominance. These prerequisites must be met in order to pave the way for general acceptance of an international agreement.

The Reagan Administration has formally responded to all these concerns at an international level. In 1986, the superpowers started a series of technical meetings in Geneva which have been called the Nuclear Test Experts' Meetings (NTEMs). The present writer was the Department of Energy Delegate for the first three sessions in July, September and November of 1986. The first US goal at NTEM has been to negotiate improved verification measures for the TTBT and PNET. In Reykjavik, Reagan made the following proposal:

> The US and Soviet Union will begin negotiations on nuclear testing. The agenda for these negotiations will first be to resolve remaining verification issues associated with existing treaties. With this resolved, the US and USSR will immediately proceed, in parallel with the reduction and elimination of nuclear weapons, to address further step-by-step limitations on testing, leading ultimately to the elimination of nuclear testing.

The main Soviet goal regarding nuclear testing is to achieve a CTB. They appear to be willing to agree to intermediate steps along the way to a CTB, perhaps even along the lines of Reagan's proposal in Iceland. At Reykjavik Gorbachev said in a press conference:

We submitted the following proposal to the US President: Let us agree to start talks on banning nuclear explosions immediately after the conclusion of our meeting in Reykjavik. At the meeting we proposed that this be a process over the course of which we could examine at some stage, perhaps even on a top-priority basis, also the question of thresholds, the nuclear blast yield, the number of nuclear explosions per year, and the fate of the 1974 and 1976 treaties, and that we would move further toward the elaboration of a comprehensive treaty banning all nuclear explosions.

Just before Reagan went to Reykjavik, as a condition to get the US Congress to withdraw its legislation on nuclear test limits, he agreed to submit the TTBT and PNET to the Senate for advice and consent to ratify the treaties, whether or not Gorbachev agreed to the US proposal for improved verification of the two treaties. If Gorbachev refused the proposal, the President agreed to submit the treaties with a reservation that they would not go into effect until the verification provisions were acceptable. Gorbachev did not accept the proposal and the President submitted the two treaties, with the above-mentioned reservation, in January 1987 to the Senate to seek their advice and consent for ratification. A complicated debate developed in the Senate, as to whether to provide that advice and consent.[28] Considerable concern was expressed about the reservation. Some Senators asked if the President would return to the Senate for a second vote of approval on the acceptability of improved verification measures yet to be negotiated with the Soviets. Other Senators suggested that advice and consent be given without conditions; they suggested that the President should certify on his own the acceptability of any improved verification measures agreed to by the Soviets and that he should decide on that basis when the treaties would go into effect. Still other Senators suggested that the President defer his request for advice and consent until he had successfully negotiated the desired verification improvements with the Soviets. The Senate was unable to decide the issue and, in effect, the last suggestion was essentially chosen by default.

At a meeting in Moscow in April 1987 which included Secretary of State George Shultz and Foreign Minister Shevardnadze, the Soviets suggested performing joint verification experiments at each other's test sites. The nature of these experiments was discussed at NTEM. At a session in July 1987 the Soviets tabled a proposal for the conduct of 'joint experiments designed to improve verification measures of the

TTBT and PNET'.[29] The Soviet proposal has received extensive review by the US Government.

A step-by-step approach to further nuclear test limitations in parallel with major arms reductions is the correct approach. A major breakthrough occurred during Shevardnadze's visit to Washington, when the superpowers agreed to begin full-scale negotiations on nuclear testing. The two sides issued a joint statement on 17 September 1987:

> U.S. and Soviet sides have agreed to begin before December 1 full-scale stage-by-stage negotiations which will be conducted in a single forum. In these negotiations, the sides as the first step will agree upon effective verification measures which will make it possible to ratify the US–USSR treaties of 1974 and 1976, and proceed to negotiating further intermediate limits on nuclear testing leading to the ulimate objective of the complete cessation of nuclear testing as part of an effective disarmament process. This process, among other things, would pursue, as the first priority, the goal of the reduction of nuclear weapons and ultimately their elimination. For the purpose of the elaboration of improved verification measures, for the US–USSR treaties of 1974 and 1976, the sides intend to design and conduct joint verification experiments at each other's test site. These verification measures will, to the extent appropriate, be used in future test limitation agreements which may subsequently be reached.

Nuclear test limitations are more an effort to control technology than to control arms. Test limitations will not make nuclear weapons go away. The only thing that will make nuclear weapons unnecessary is vastly improved relations in the world and alternatives to war for resolving conflicts. As long as we rely upon nuclear weapons for a deterrent, then, in the present writer's opinion, nuclear testing is necessary to keep that deterrent viable.

An important view on the role of nuclear testing in arms control was expressed by Andrei Sakharov at the Forum for a Nuclear Free World in Moscow in February 1987. *Time* magazine printed Sakharov's statement, in which he said:

> Thus the question of nuclear testing is not critical for the restraint of the arms race. The issue of nuclear testing, in my opinion, is of minor, secondary importance in comparison with the other mili-

tary, technical, political and diplomatic problems involved in preventing thermonuclear calamity.... As long as nuclear weapons exist and are not banned, the decision regarding underground testing is the internal sovereign affair of each nuclear power.

I believe that eliminating the issue of a comprehensive nuclear test ban will facilitate negotiations on more urgent problems of disarmament. I have deliberately omitted any discussion of the propaganda and psychological aspects of the test-ban issue.[30]

The global strategic balance is constantly changing. The deterrent relationship among nations is a dynamic one. Military systems age and become obsolete, and new technologies present alternatives which make for a safer, more secure, more survivable and more effective deterrent. With or without restrictive test limitations, nations will continue to respond with new developments. Even with test restrictions, there will be new developments in nuclear weapons. These developments will lack the system optimisation that now occurs and uncertainties will increase. In fact we experienced a *de facto* CTB during the nuclear test moratorium of 1958–61. Changes indeed were made to the US stockpile during that time, and later nuclear testing showed that some of these changes involved errors in scientific judgement.[31]

Nuclear weapon technology is empirically-based. Assessment of existing weapons and of the impact of new technologies developed by either side rests on scientific judgement, and that judgement is based on nuclear test experience. New technologies can be those which we deem beneficial to enhance the safety and survivability of our own weapons, and they can also be technologies which our adversaries might use that could threaten the viability of our weapons. Hence it is arguable the most significant impact of a CTB would be to diminish, asymmetrically *vis-à-vis* the Soviets, the skills of the relatively few US weapon scientists capable of making these judgements. Science and technology thrive on the relationship between theory and experiment. Take away the experiments, and you remove the credibility of the theory. The present writer has developed computer programmes to simulate the physical processes in nuclear weapons. These programmes frequently made incorrect predictions, and it took experiments to make the necessary corrections. Take away the experiments, and you reduce the soundness of scientific judgement. There are many who believe that this would be a good thing to do in the case of nuclear weapon technology. I believe that it would be a dangerous thing to do. The end result would be to introduce uncertainty and instability into our political and military relationships.[32]

Notes

1. Harold Brown, Report of the Secretary of Defense to the Congress on the FY82 Budget, 19 January 1981.
2. Winston Churchill speech, Washington, DC, 26 June 1954.
3. Winston Churchill comment made on the indefinite adjournment of the Standing Committee of the Disarmament Conference in Geneva, 1934.
4. Unilateral Statement A of the ABM Treaty. On 9 May 1972 Ambassador Gerard Smith made the following statement:

> The U.S. Delegation has stressed the importance the U.S. Government attaches to achieving agreement on more complete limitations on strategic offensive arms, following agreement on an ABM Treaty and on an Interim Agreement on certain measures with respect to the limitation of strategic offensive arms. The U.S. Delegation believes that an objective of the follow-on negotiations should be to constrain and reduce on a long-term basis threats to the survivability of our respective strategic retaliatory forces. The U.S.S.R. Delegation has also indicated that the objectives of SALT would remain unfulfilled without the achievement of an agreement providing for more complete limitations on strategic offensive arms. Both sides recognize that the initial agreements would be steps toward the achievement of more complete limitations on strategic arms. If an agreement providing for more complete strategic offensive arms limitations were not achieved within five years, U.S. supreme interests could be jeopardized. Should that occur, it would constitute a basis for withdrawal from the ABM Treaty. The U.S. does not wish to see such a situation occur, nor do we believe that the U.S.S.R. does. It is because we wish to prevent such a situation that we emphasize the importance the U.S. Government attaches to achievement of more complete limitations on strategic offensive arms...

5. Paul S. Brown, 'Why Research on Defensive Weapons is Important', *Scientia*, vol. cxx (1985) 349–59.
6. Paul H. Nitze, 'On the Road to a More Stable Peace', World Affairs Council of Philadelphia, 20 February 1985.
7. L. E. Irish, 'SDI and the ABMT: Problems of Negotiations and Interpretation', *Law Quadrangle Notes*, vol. xxx, no. 3 (1985–6) 28–37.
8. G. H. Miller, Testimony before US Senate Committee on Appropriations, Defense Subcommittee, 26 March 1987; and 'Kinetic Kill Vehicles', *Energy and Technology Review*, July 1987, pp. 16–17.

9.   See for example, US Arms Control and Disarmament Agency, 'Soviet Noncompliance', 1 February 1986; and 'Analysis of the President's Report on Soviet Noncompliance with Arms Control Agreements', *Arms Control Today*, April 1987, pp. 1A–12A.

10.   R. E. Batzel, Testimony before the Senate Armed Services Committee, 26 February 1987; and S. S. Hecker, Testimony before the Senate Armed Services Committee, 26 February 1987.

11.   Batzel, Testimony.

12.   The verification measures proposed by the United States to the Soviets involve the use of on-site yield-measuring equipment called CORRTEX. CORRTEX consists of a cable buried in a hole alongside the hole containing the nuclear explosive. As the explosive shock expands and crushes the cable, we measure its change in length electronically and thereby deduce the position of the shock wave as a function of time. With these data it is possible to determine the explosive yield to an accuracy of better than 30 per cent.

13.   M. M. May and J. R. Harvey, 'Nuclear Operations and Arms Control', in *Managing Nuclear Operations* (Washington, DC, 1987).

14.   G. T. Allison, A. Carnesale and J. S. Nye, *Hawks, Doves and Owls* (New York, 1985).

15.   W. L. Ury and R. Smoke, *Beyond the Hotline: Controlling a Nuclear Crisis* (Cambridge, Massachusetts, 1984).

16.   J. W. Lewis and C. D. Blacker, *Next Steps in the Creation of an Accidental Nuclear War Prevention Center* (Stanford, California, 1983); and A. L. George, *Managing U.S.–Soviet Rivalry: Problems of Crisis Prevention* (Boulder, Colorado, 1983).

17.   T. C. Schelling, 'What Went Wrong with Arms Control?', *Foreign Affairs*, vol. lxiv, no. 1 (1985–6) 219–33.

18.   Paul H. Nitze, 'Arms Control after Reykjavik', *Washington Post*, 9 November 1986.

19.   B. Scowcroft, J. Deutch, and R. James Woolsey, 'A Way Out of Reykjavik', *New York Times Sunday Magazine*, 25 January 1987.

20.   J. Schlesinger, 'Nuclear Deterrence, the Ultimate Reality', *The Washington Post*, 21 October 1986.

21.   C. Bertram, 'Why Reykjavik Upset the European Allies', an interview in *Arms Control Today*, January/February 1987, pp. 15–18.

22.   T. C. Schelling, 'Abolition of Ballistic Missiles', *International Security*, vol. xii, no. 1 (1983–4) 179–83.

23.   G. A. Greb and W. Heckrotte, 'The Long History: The Test Ban Debate', *Bulletin of the Atomic Scientists* (August–September 1983) pp. 36–42.

24.   Parliamentarians for Global Action, 'Ending the Deadlock', 1986; and 'Document Issued at the Mexico Summit on Verification Measures', 7 August 1986.

25.   Amendment offered by Congressman L. Aspen to House Appropriations Bill No. HR 5438, *Congressional Record*, pp. H5754–6, 8 August 1986; and Amendment offered by Congresswoman P. Schroeder, *Congressional Record*, pp. H5702–8, 19 May 1987.

26.   See, for example, the Kennedy–Mathias amendment (Senate Joint

Resolution 29) to the FY85 Omnibus Defense Authorization Act (S2723), June 1984.

27. Underground Nuclear Explosions Control Act of 1987, S1106, Senators M. Hatfield, E. Kennedy, *et al.*, 29 April 1987.

28. A good description of the debate is given in P. Towell, 'Verification Debate Stalls Nuclear Test Pacts', *Congressional Quarterly*, 29 January 1987, pp. 162–3.

29. Press Statement by US NTEM Delegation, 21 July 1987.

30. A. Sakharov, 'Of Arms and Reforms', *Time*, 16 March 1987.

31. Paul S. Brown, 'Nuclear Weapon R and D and the Role of Nuclear Testing', *Energy and Technology Review*, Lawrence Livermore National Laboratory, September 1987.

32. This chapter was prepared as an account of work sponsored by an agency of the United States Government. Neither the United States Government nor the University of California nor any of their employees makes any warranty, express or implied, or assumes any legal liability or responsibility for the accuracy, completeness, or usefulness of any information, apparatus, product, or process disclosed, or represents that its use would not infringe privately owned rights. Reference herein to any specific commercial products, process, or service by trade name, trademark, manufacturer, or otherwise, does not necessarily constitute or imply its endorsement, recommendation, or favouring by the United States Government or the University of California. The views and opinions of authors expressed herein do not necessarily state or reflect those of the United States Government thereof, and shall not be used for advertising or product-endorsement purposes.

18 Progress and Failure in Arms Control

Edy Korthals Altes

There are various ways to tackle this vast subject. One of them is to present in detail an exhaustive account of the progress and failure in arms control during the postwar period. This could be termed the *descriptive* method. At its very best we would find it interesting. Having said so we would continue with our work as usual. The description does not affect us personally. In view of the *existential* nature of the arms race and the urgency of stopping it, a different approach may be proposed, namely to concentrate on two underlying, closely interrelated questions: first, why did arms control fail to limit effectively the postwar arms race; and secondly, what should be done to stop and reverse the arms race?

As far as the first question is concerned, there can be no denying that, notwithstanding great efforts in the field of arms control, the arms race has shown a steadily increasing momentum since 1945. The recently announced US–Soviet agreement to eliminate land-based intermediate- and shorter-range nuclear weapons from Europe could, however, lead to a historic breakthrough. The essential condition for that is that this agreement in principle is not going to stand on its own but that it will soon be followed by further effective arms reductions. Until now positive effects of arms control agreements in one area did not prevent extra efforts to expand the arms production in other fields. The sad fact is that efforts at arms control lag far behind technological 'progress'. The arms race seems out of control. Arms expenditures in the world have now reached well over 1000 billion US dollars per annum. If the militarisation of space is not stopped this sum will be doubled.

Roughly 750 000 scientists are engaged in the arms race. Many outstanding minds are at this very moment employed in the research and development of new, ever more 'efficient' weapons. Some of them are dedicating themselves to the question of 'how to kill the largest number of people in the shortest possible time at the lowest possible cost'. Others devote their best talents to obtaining weapons that may outsmart the opponent in a decisive way. Still others are trying desperately to invent adequate 'defensive systems'. As action breeds reaction and every new weapon leads to a countermeasure, the insecurity increases. There is no hiding from the grim reality: we are confronted with a dynamic arms race not only of terrifying proportions, but also with an increasing number of destabilising weapons and weapons systems. This development, if unchecked, is leading us to disaster. As a professional diplomat for well over 35 years, the present writer became increasingly worried about the irresponsibility of the arms race and the lack of any substantial results in past decades from the negotiations on arms reduction. It led to the publication of personal thoughts in a leading Dutch newspaper and to subsequent resignation as Ambassador.

One of the main reasons for a lack of real progress in arms control is the absence of a common concept of security between East and West. The outdated concept of 'national security' still prevails. In the name of this concept governments ask for ever-larger sums of money for spending on arms in order to ensure the security of their citizens. But the fundamental fact to grasp is that a greater security can no longer be obtained by procuring more arms, however sophisticated they may be. Many people and even governments, who should know better, are still imprisoned in a line of thinking which would be more appropriate to the pre-nuclear world than to the world of today. They fail to recognise that one of the basic facts in our nuclear age is the vulnerability of everybody, anywhere, even if he belongs to the most powerful nation in the world. No amount of arms spending is going to eliminate that basic reality. On the contrary, there is ample evidence that the basic insecurity increases with the introduction of new destabilising weapon systems. The time has come for an agonising reappraisal of the present course to disaster. A clear restatement of the basic concept of security is of crucial importance.

Our second question is: what should be done to stop and reverse the arms race? Here again many elements are involved. Here attention will be given to just two elements which are of key importance: first, the recognition that security can no longer be guaranteed on a purely national basis; and secondly, the responsibility and role of scientists.

Security can no longer be guaranteed on a purely national basis. Nowadays it can only be realised *together* with the potential opponent. For this reason the Palme Commission formulated in 1982 the concept of 'common security'. Although the author entirely agrees with this concept, he wonders on the basis of today's grim reality whether we do not need a station in between, namely the concept of 'mutual assured security', a more modest, but perhaps more realistic step. We are after all still a long way from a 'common' approach based on that deep sense of belonging together that characterises the members of a family or an alliance. First, we should try to come to agreement on a substantial improvement of the overall security situation. This could be achieved if the superpowers would accept that the security of each of them depends on the other. That means 'I am secure if you are secure' and vice versa. This only exists if my position in its totality (weapons, strategy, deployment, logistics and so forth) is not perceived as threatening you in such a way that, under stress, you may consider a pre-emptive strike necessary. In mutual security there is therefore a clear element of reciprocity. It is in each other's interest to reduce those elements in posture that may be interpreted as menacing. This implies: firstly, accepting the other state whether we like his regime or not; and secondly, refraining from actions to impose our will or from seeking superiority through a decisive breakthrough in arms technology.

The replacement of the outdated concept of 'national security' by the concept of 'Mutual Assured Security' (MAS) would represent a Copernican change in the relations between the two blocs: MAS instead of MAD (Mutual Assured Destruction). For the citizens it would mean a gradual moving away from the untenable and unacceptable situation of living under the permanent threat of assured destruction towards a more stable situation characterised by assured security. The concept of MAS should be translated into concrete actions and should be directed towards halting the quantitative and qualitative arms race. If we accept that war between the two blocs is no longer 'winnable' and can no longer be considered as a tool of policy, there is no need any more to continue the arms race. The perfecting of nuclear arms is no longer a necessity. MAS would also involve a drastic reduction of existing nuclear weapons to a minimum level and the elimination of chemical and biological weapons. Finally, it would require the reduction and restructuring of conventional forces. This means an Intermediate Nuclear Forces (INF) agreement followed by agreement on short-range nuclear weapons and tactical nuclear arms. This would lead on to a drastic reduction of all strategic nuclear arms to a level of minimum deterrence. It would also mean that strict adherence to existing arms-limitation treaties was possible , especially SALT II, and strict respect for the narrow interpretation of the Anti-ballistic Missile (ABM) Treaty. Needless to say, such action would demand the non-militarisation of space and the halting of the Strategic Defense Initiative (SDI) programme as well as similar programmes in the Soviet Union. There would be further benefits, including a Comprehensive Test Ban Treaty, a no-first-use declaration respecting nuclear weapons from both sides, and a nuclear-weapon-free zone in Europe.

The elimination of those weapon systems that are considered as highly offensive by the other partner would be the main aim of this policy, and would result in the adoption of a general non-offensive posture in forces, deployment, weapons, strategy and logistics. Thus a gradual reduction instead of expansion of conventional forces would be achieved and, inevitably, their restructuring in a 'defensive' posture. This would mean, finally, a major reduction in military R and D expenditures. We could expect a return to and intensification of the Helsinki process in all fields (military, economic, technical, scientific, as well as in cultural co-operation and human rights).

It is clear that this far-reaching programme cannot be realised all at the same time, for it is a gradual process. It is also obvious that adequate verification is necessary. The essential thing, however, is to start the process on the basis of a solid mutual acknowledgement that the long-term relationship is going to be based on 'mutual security'. It should also be realised that there exists a close link between the various elements, for instance between conventional arms and no first use of nuclear arms. Reduction of conventional arms to a low common ceiling would facilitate considerably the adoption of a 'no-first-use' declaration by NATO.

With MAS accepted as the dominating concept, the long process of real *détente* could be set in motion. MAS should be considered as a powerful magnet which aligns all elements of security policy in the direction of mutual instead of purely national aims. The unavoidable problem of 'linkage' would therefore present itself in a different light. The frustrating experiences of *ad hoc* negotiations on separate issues would be considerably reduced. Concessions in one field could be compensated in another field, as in security everything is related. Some steps should be taken on the basis of an agreement, others on a unilateral basis. A critical analysis may prove that it is in our interest to take certain unilateral steps because this would lead to greater security as the other partner would feel less threatened. Unilateral steps which do not jeopardise security could be very effective in obtaining a momentum in the arms reduction process.

Once real progress has been made with implementing MAS, we can move on to the much wider concept of 'common security'. The pressure of world problems will be increasingly felt and impose the necessity to co-operate. This should already be evident. The starvation and utter misery of millions of human beings in this world, existing at the same time as we witness the colossal waste in resources in the arms race, is a poor testimony to the degree of civilisation we have reached. There exists a close connection between these questions. The threat to national security leads nations to considerable expenditures on arms. For a guarantee of the security of nations a restructured United Nations with an effective police power would be of great importance. Turning to the practical implementation of the concept of MAS, it cannot be claimed that the concept of mutual assured security could be implemented in today's Europe without taking into account the two alliances, the North Atlantic Treaty Organisation (NATO) and the Warsaw Pact. For the time being they do provide stability and it is important to keep this in mind. If the European nations would care to have a greater influence on the vital issues affecting their security a restructuring of NATO seems inevitable. In the author's opinion the Alliance should be built on the two pillars of the United States and Western Europe.

European security depends on the relations between the two superpowers. If these worsen we suffer; if they improve we benefit. There is no nuclear sanctuary called Europe today. But this does not mean that the smaller and medium-sized countries can do nothing for the implementation of 'common security' in Europe. On the contrary. Especially at this critical period in history the European countries could and should exercise a decisive, positive influence for a radical improvement in the relations between the two superpowers. They could encourage, instead of making more difficult, an agreement on INF and further nuclear and conventional arms reductions. The expressed fears of several European statesmen about the implementation of the 'near-agreement' in Reykjavik especially on INF were not – in a larger perspective – justified. It would have been most regrettable if this historic opportunity to come to an agreement had been thwarted by shortsightedness. Now an INF agreement together with an acceptance of the concept of mutual assured security could be the beginning of a new era. The Western Europeans could also make it abundantly clear that they as Europeans not only share a common history and culture but that they share for better or worse a common destiny. Europeans in East and West want to live in peace. Any idea of a conflict, however 'limited', is unacceptable to them.

They should accordingly insist with great perseverance, in their respective alliances, on the concept of 'common' or 'mutual' security. They should also

- Plead for a nuclear-free zone in Europe.
- Intensify the Helsinki process in all fields, including that of human rights.
- Adapt cultural agreements to the pressing need for more contacts between universities, schools, cultural organisations and the media.
- Stimulate common projects and co-operation between scientists, industries and local organisations.
- Refuse to contribute, either in the media or in the schools, to a black picture of the 'enemy'.
- Increase openness in our societies.

Finally, let us consider the responsibility and role of scientists. The moment has arrived to face the fact that a continuation of the arms race will lead us to disaster. The concept of common assured security and – as in an earlier stage – of mutual assured security provides us with an important instrument. Whether it will be used effectively depends to a large extent on wide sectors of the population and on the vision and sense of responsibility of the politicians. There should be a strong and consistent effort to stimulate this process of rethinking. Of decisive importance will be the stand of scientists and engineers who design weapons. They wield great power and on their participation depends the continuation of the arms race. A manifest refusal to participate in research and development of weapons of mass destruction could have a decisive impact on public opinion and governments. Carl-Friedrich von Weizsäcker, deeply convinced of the need for a personal commitment, calls for a supreme moral effort by many who hold responsible positions. The present situation and the complexity of the problems demand a far greater personal effort from many people both in the East and the West.

19 Steps to Constrain Arms Races

F. A. Long

INTRODUCTION

'Arms race' has become the accepted name for a continuing and competitive deployment of military forces by a pair of nations or groups of nations. 'Arms competition' is probably a more descriptive term for the varied competitive approaches represented by force increases, military research and development, weapons production and deployment, and acquisition of military technology by other means. However 'arms race' has become the term of art to encompass all the stages of competitive arms acquisition.

The United States and the Soviet Union are an obvious pair of competitors, but so are the groups of nations that make up the North Atlantic Treaty Organization and the Warsaw Pact. India and Pakistan constitute a pair of poorer nations that are involved in a costly mutual arms race in which arms acquisition involves Research and Development (R and D) and production of new weapons, purchase of new weapons, and gifts of weapons and other material by more wealthy allies. The objective that guides these military programmes is the goal of national security. Unfortunately military programmes for national security normally turn out to be competitive. If nation A undertakes military moves to increase its national security, for example, by a build-up of military forces, its opponent B suffers (or believes it suffers) a loss in security. A common response by B, therefore, is to increase the size of its military forces, hoping thereby to increase its own national security. A continuing competition of this sort is an arms race resulting from a search for an elusive national security.

There are other components of national security besides military forces: an efficient civilian economy, a rising standard of living, and social justice, including such basics as health, education and welfare. The difficulty is that these non-military components of security are not easy to quantify, even in monetary terms. In contrast, military

316

force levels can be counted and compared (although their effectiveness cannot). It can be argued persuasively that expenditures on military forces detract from funds available for civilian programmes. Unfortunately there is no accepted analytical procedure to ascertain what balance of military and non-military expenditures leads to maximum national security. Even when (as has happened in the United States in the past five years) rapidly increasing expenditures for the military are accompanied by rapidly increasing national deficits, it cannot be convincingly shown that the increased military expenditures are the sole or even the principal cause of the deficit. It *can* be argued, however, in a general way that large military expenditures do come at the cost of desired civilian programmes. In specific areas the costs are reasonably obvious. Thus military research and development, in addition to its monetary costs, unavoidably diverts scientists and engineers from work on civilian-related problems. This diversion can itself be economically costly.

The most central concerns about military arms races and the costs of military programmes are generally political rather than technical. Government policies, inter-governmental politics and the political impact of military programmes are usually involved. Perhaps the most troublesome and dangerous aspect of arms races is that they are almost always politically destabilising. There is an invariable and corroding uncertainty about opponents' motives when they deploy new weapons or expand their military forces in other ways. Are they merely trying to keep up in a competition, or are they contemplating (dread phrase) a military break-out?

Surprisingly, the deployment of nuclear weapons has not modified this destabilising impact of arms races very much. The extensive build-up of intercontinental weapons by the superpowers, and the consequent mutual nuclear deterrence, would seem to preclude a war between them, given the devastating consequences. And one might reasonably expect the nuclear deterrence to extend to alliances where the superpowers are on opposing sides. Yet the conventional military confrontation between the Warsaw Pact nations and NATO has all the characteristics of a pre-nuclear arms confrontation. The massive forward deployment of Soviet forces hard against the West German border, and the NATO response, are intrinsically destabilising. Even so, the likelihood of conventional war in Europe remains very small, and in some measure this must be due to concern about a possible nuclear exchange.

It is saddening that a similar scenario seems to be developing between two large and poor nations, India and Pakistan. With Bangladesh now an independent and uninvolved nation, and with no significant new crises between India and Pakistan, one might have hoped for a subsidence in their military confrontation, but this has not happened. Both have large conventional military forces, many of them deployed close to their common border. Both appear to be additionally involved in costly and destabilising development pro-grammes for nuclear weapons. The consequence is greatly increased military expenditures by both nations; for example, India in the past four years has increased its military expenditures by about 60 per cent. An arms race between the two nations has developed, and shows every indication of increasing in intensity.

This brief introductory analysis suggests that there are two some-what distinct areas of great concern. One is arms races in the nuclear field, something that tends to be discussed under the term 'nuclear proliferation'. The second is arms races of the more widespread sort, where conventional weapons are involved. It has recently become clear that the major powers that have nuclear weapons are increas-ingly doubtful of their military utility, particularly since a likely outcome of a full-blown nuclear exchange between the United States and the Soviet Union is the obliteration of both. Certainly the situation with the two superpowers is one of rough nuclear balance and a growing interest in decreasing the number of their nuclear arms.

However, there are worrying developments that carry the possi-bility of nuclear arms races in other parts of the world. The common situation is a nuclear-have nation and a nuclear-have-not in juxtapo-sition. The fact that Israel has nuclear weapons (something no longer challenged, even if not formally announced) means that the adjacent Arabian countries cannot fail to be interested in getting nuclear weapons. The potential for nuclear proliferation and an ultimate nuclear arms race is real. Similarly India is concerned not only with the possible development of nuclear weapons in Pakistan, but with the significant deployment of nuclear weapons that has already occurred in China. There continue to be stirrings along nuclear lines in Argentina and Brazil, but these have fortunately been more muted recently.

In spite of these nuclear possibilities, the major world-wide concern will continue to be the build-up of conventional arms, and this is a phenomenon of significance to rich and poor nations alike.

STEPS TO CONTAIN ARMS RACES

There are three straightforward procedures that nations and groups of nations can use to constrain arms races:

- Negotiating bilaterial and multilateral treaties and other arrangements for arms control and disarmament.
- Unilateral steps towards arms control and disarmament taken by individual nations.
- Steps taken towards arms control and disarmament by international bodies that have authority or wide influence.

In addition, there are other important steps that nations and groups of nations, and also individuals and groups of individuals, can take that either help set the stage for arms control or help establish conditions that make the attainment of arms control more likely. Before examining in more detail these various approaches to constraint, it is useful to identify some of the obstacles to constraining or eliminating arms races. Lack of understanding of the components of national security is one. Internal political rivalries are another. The sheer momentum of military programmes and the tendency to give first attention to military views on national security are others. Suspicion and mistrust of rival nations is a common problem. None of these is insuperable, and all are alleviated by better understanding. Nevertheless the obstacles are real and serious.

The past decades have seen a continuing series of international negotiations on arms control. The majority of these have taken place in Geneva, Switzerland, where the Eighteen Nation Committee on Disarmament of the UN, later called the Conference of the Committee on Disarmament, has served as host. Almost all serious negotiations have been among self-selected groups of nations. Negotiations between the superpowers have been prominent among these. A substantial body of bilateral and multilateral agreements has been reached. An observer of the scene can simultaneously point to important areas of progress – the Anti-ballistic Missile (ABM) Treaty, the Nuclear Non-Proliferation Treaty, the Biological Weapons Convention – and bemoan the fact that so little has been accomplished in controlling the major consequences of arms races. Increasing military budgets, the competition among nations in military technology, and the incipient militarisation of outer space are some of the many areas where arms races continue.

There is general agreement that the most lasting and reassuring agreements for arms control and disarmament are negotiated treaties that are formally signed and ratified by the nations that are party to the treaties. The major incentive for negotiating arms control agreements is that there can be important *mutual benefits*. Arms races are costly, diversionary, uncertain in their impact, and often destabilising, so that the potential for gaining mutual benefits by minimising them is high. But negotiating an agreement that is acceptable to the several constituencies of the two or more nations that are negotiating can be tedious and frustrating. The military establishments of nations are frequently opposed to arms control agreements, and their influence can be considerable. Opposition by the industries that produce military technology can be important. To counter the opposing constituencies requires strong support from the general public and a vigorous and committed political leadership.

Unilateral arms control initiatives taken by a single nation on its own offer some obvious advantages. No negotiations are needed, decisions can be made quickly, and initiatives can be withdrawn if they are not accomplishing their purposes. These are real virtues, and unilateral approaches, as will be discussed later, clearly deserve more attention.

Nevertheless major moves for arms control and disarmament are probably better served by joint negotiation between opposing nations and their allies, and by formal assent to an agreed treaty. There are several reasons. A treaty is a serious and continuing obligation not to be entered into lightly, nor lightly abandoned. The indefinite character of most treaties, and the formal obligations inherent in them, are vital if military forces or specific weapons are to be reduced or abandoned, and if new programmes are to be banned. And only as part of a treaty can nations expect to obtain formally agreed procedures whereby adherence to the terms of the treaty can be verified, for instance, by on-site inspection, and there are important arms control measures where ability to verify is crucial.

If negotiated agreements for arms control are to assume a larger and more consequential role in eliminating arms races and achieving more disarmament, the pace and substantial character of the negotiations must be increased considerably. This calls for improved procedures for the negotiation process itself. But of equal or greater importance is a deeper appreciation of the potential benefits of meaningful arms control and disarmament, and more understanding of what specific arms control measures are mutually desirable and feasible. More scholarly and political analysis is needed of the general role of arms control and of the desired components of specific measures. In the United States the Arms Control and Disarmament Agency of 1961 was envisaged as a central performer and supporter of these studies and a lead agency in negotiation. Its role and significance have diminished greatly in recent years, but one may fervently hope that it will reassume these responsibilities under a new administration. Indeed a strong case can be made for similar agencies in the other major nations. However, independent analyses by citizen groups are also essential, as are extensive discussions between representatives of the nations that wish to negotiate arms control. Put another way, if arms control negotiations are to produce effective and significant arrangements, the needs and objectives, as well as the modalities, must be analysed seriously and must have wide public support.

There are a variety of 'confidence-building measures' that can help increase understanding and decrease distrust between nations. These merit more extensive analysis and utilisation. Perhaps the most important needs are openness and accessibility. The extent to which concern about and mistrust of the Soviet Union has diminished since Mikhael Gorbachev established his policy of *glasnost* is impressive. Many of the measures that have developed from *glasnost*, such as the release of Andrei Sakharov from exile, have not had a direct impact on arms control negotiations, but their indirect effects have surely been beneficial.

Unilateral measures by nations that seek to eliminate arms races and obtain more arms control can be important in several ways. They can in themselves be measures of arms control. They can be confidence-building measures. They can help set the stage for formal arms control negotiations. The importance of unilateral measures has long been recognised. Scholars have studied conditions when they are likely to be effective and have stressed the desired cumulative effect of unilateral measures of arms control that are deliberately designed to elicit a comparable response from an opponent. Heads of state have frequently used unilateral measures to aid the attainment of arms control measures. President Dwight Eisenhower and Nikita Khrushchev separately established moratoria on the testing of nuclear weapons to help the negotiation of a treaty to ban such tests. Khrushchev and US Secretary of State Dean Rusk separately, but in co-ordination, announced reductions in their nations' military budgets. President John F. Kennedy, in calling for further negotiations on a nuclear test ban, pledged that the United States would forgo nuclear tests in the atmosphere as long as other nations did the same.

More recently Gorbachev announced two significant unilateral initiatives, both designed to accelerate arms control negotiations between the superpowers. One was a moratorium on nuclear weapons testing, where his objective was to persuade the United States to resume the test ban negotiations that had almost been completed during the Jimmy Carter presidency. The Soviet moratorium on testing was continued for about 18 months. It was eventually terminated when it became evident that the United States was not going to respond. The other Soviet initiative, also designed to produce bilateral negotiations, was to ban any further tests of a Soviet anti-satellite (ASAT) weapon. Although the Executive Branch of the US Government did not respond to this initiative, the US Congress did by banning further US expenditures on testing US ASAT weapons for as long as the Soviet ban continued. So far the moratorium has not produced negotiations between the superpowers to ban deployment of these weapons, but it has served to inhibit deployment and thus keep options open.

There is a wide variety of possible unilateral initiatives relating to arms control. A recent analysis listed the following: troop withdrawals, weapons withdrawal, weapons test moratoria, no-first-use of weapons, restraints in other military programmes, and phased unilateral arms control initiatives. One could also list unilateral decreases in military budgets and cut-offs in military production, for instance, of fissile materials for nuclear weapons.

Another kind of analysis that is usually taken unilaterally and that can help restrain military expenditures, focuses on the social and economic costs of military programmes. A number of US economists have been concerned with these costs and with the impact of military expenditures. Some were dissatisfied with the economic utility of the Gross National Product as conventionally defined, and devised another measure of economic progress called Measure of Economic Welfare which was defined so that certain economic activities were non-contributory to the measure, or even negative in their contribution to economic welfare. Military expenditures in their analysis were in the main non-contributory. Another economist labels most military expenditures 'distractive', indicating that they contribute negatively to economic welfare. These arguments have not yet been accepted by many economists, but the general approach is interesting and merits further study.

A specific area where there is a mostly negative contribution from military expenditures is expenditures for military R and D. In the United States these are now approaching 40 per cent of all US expenditures for R and D, public and private. Since only a small fraction of military R and D, say 20 per cent, contributes to technology problems of civilian interest, it would appear that roughly one-third of the US scientific and engineering manpower is being diverted from civilian to military problems. In these days of fierce technological competition with Japan and West Germany, this is a major loss.

National security is composed of many things: military security, economic welfare, technological competitiveness and social welfare. A national programme that gives a high priority to military security is likely to neglect other important contributions to national security, and an overall diminishment in national security can be the result. Arms races, responding as they do to programmes of competing nations rather than to internal analyses of overall national security needs, are unlikely to optimise a cost-effective security effort. This suggests first, that nations try to minimise participation in arms races, and second, recognising that since military programmes, including military R and D, make minimal contributions to economic and social security, strive to reduce military expenditures substantially by efficient planning and programme management, by decreasing costly lower-priority overseas expenditures, and most importantly by negotiating arms control and disarmament measures that decrease the need for large military forces and large military expenditures.

20 Progress and Failure in Disarmament

Maj Britt Theorin

In the late 1980s progress in disarmament is certainly more in our minds than is failure in disarmament. The world has been brought close to a much awaited breakthrough, showing that the nuclear arms race can be reversed. A US–Soviet agreement to eliminate land-based intermediate- and shorter-range nuclear weapons from Europe would, by clearly demonstrating this, be an event of historic significance. In the slightly longer perspective of the last few years we have, however, seen much of both progress and failure. Indeed the ups and downs of a truly dramatic negotiating process can best be characterised in the words of Charles Dickens:

> It was the best of times, it was the worst of times, it was the age of wisdom, it was the age of foolishness ... it was the spring of hope, it was the winter of despair, we had everything before us, we had nothing before us, we were all going direct to Heaven, we were all going direct the other way – in short, the period was so far like the present period, that some of its noisiest authorities insisted on its being received, for good or for evil, in the superlative degree of comparison only.

This has, in essence, been a *Tale of Two Cities*, of Moscow and Washington, with important happenings in Geneva and Reykjavik.

The last few years have, on the one hand, been characterised by a remarkable renewal of disarmament negotiations. On the other hand, concrete results have been meagre and talks have taken place under the threat of new, and perhaps exceedingly dangerous, rounds of a spiralling arms race. Never perhaps has such a broad spectrum of arms issues been covered by a negotiating process – be it bilateral, regional or global. Never perhaps have such comprehensive proposals been addressed. Of the 95 per cent of nuclear weapons in American or Soviet hands, most categories are subject to negotiation; main exceptions being those with the shortest range and some sea-based systems. The bilateral agenda includes nuclear testing and the prevention of an arms race in space. And, as has already been touched upon, a first agreement leading to the scrapping of two entire categories of nuclear weapons appears to be within close reach. And the most remarkable goals have been set for negotiations between the two major nuclear powers: not only to prevent an arms race in space and terminate it on earth, but even, ultimately, to eliminate all nuclear weapons, everywhere.

Furthermore it may not be long before the elimination of another category of weapons of mass destruction, chemical weapons, is agreed in the multilateral negotiations at the Geneva Conference on Disarmament. And there is a growing, in some respects a new, international understanding of the need to apply disarmament to conventional weapons, too. There is accordingly no lack either of substantive proposals or of suitable forums in which to negotiate. This auspicious picture does not end here. On one crucial and long-standing problem in all disarmament efforts – on how to verify compliance with agreements – we have been seeing both a remarkable convergence of the positions of the major powers and steady progress in technical capabilities. This is indeed a bright and auspicious picture – everything before us, as Dickens put it. But even in the advent of an accord on missiles in Europe, the promise remains to be fulfilled.

Weapons continue to be developed at this very moment. Nuclear testing continues by all five declared nuclear powers. And we are on the verge of a new and possibly particularly dangerous, arms race in space. Existing arms limitation agreements have built up a low and far from perfect shield to protect us against a savage arms race. Clearly there is a chance that this shield may be superseded by a higher and better one; but there is also a manifest risk that the little protection we have may in a short time be blown apart by a raging tempest of new weapons development, especially as a consequence of a race between offensive nuclear and so called defensive systems. The SALT framework is no longer upheld, even though its quantitative limits still may play a restraining role. The ABM Treaty is under fire. The deployment of technologies now under research would call into question treaties such as the Partial Test Ban Treaty and the Outer Space Treaty.

The times we are now living in may thus very well be decisive: purposeful steps must be taken towards radical disarmament or a new arms race may render all efforts useless; new and increasingly effective agreements must be concluded or those we have may shortly be swept away. At this juncture the recent agreement in principle between the two major nuclear powers is welcome. And when superpower negotiations produce failure, the world must not let itself be trapped in pessimism. It must find openings even when they are difficult to discern. It must remain hopeful even when there may seem to be few reasons for hope. Equally important, when the nuclear powers achieve progress, the world must avoid being lulled into a false sense of security. It must stay vigilant and look forward. In the past disagreements over arms limitation have embittered relations between the major world powers and precluded urgently needed co-operation between them. Agreements on the same issues should now be a motor of such co-operation on crucial contemporary problems that range from regional conflicts to world-wide poverty and the urgent need to preserve and restore the environment we share.

A major agreement on disarmament, not just on arms control or arms limitation, would signal a new trend, and, it is to be hoped, the beginning of a new era in international relations. It can be particularly important in bringing about further disarmament. Indeed it must – if it is to remain a turning-point and not just be an exception in a long history marked by failure.

True, the agreement we hope to see President Ronald Reagan and General Secretary Mikhael Gorbachev sign in 1987 will in any case leave some 97 per cent of existing nuclear arsenals untouched. But it is much connected to other agreements. The link to agreements on strategic nuclear arsenals is clear. The latter can now be reduced without fear that they will merely be replaced by land-based systems with a somewhat shorter range. Various provisions in a first treaty could probably be transferred, with or without changes, to a treaty on strategic weapons. Much of the laborious technical work on elaborating definitions, procedures, arrangements for verification, and so forth has thus already been prepared. What is perhaps most important in this process is the signing of an agreement on strategic weapons which would benefit from the political momentum and support mobilised for the idea of negotiated reductions of nuclear arsenals.

There is also a link to agreements concerning other nuclear systems. Advocates have already appeared urging the need to compensate – as they say – for reductions in nuclear arsenals. But little would be gained if, for instance, sea- and air-launched Cruise missiles were to replace the weapons to be eliminated. Such a step would have negative consequences for security and stability, *inter alia*, in the North European and North Atlantic area. Today every fourth nuclear weapon is earmarked for *naval* deployment. Such weapons bring the nuclear arms race to all parts of the world. With the prospect of agreements on land-based nuclear weapons, they acquire additional importance. A minimum short-term requirement must at least be that arsenals at sea are not allowed to increase. Serious negotiations on naval nuclear disarmament are also overdue. As naval forces are not independent of other military forces, they should be considered in their general military context. Limitations on sea-borne nuclear missiles should also be agreed, bilaterally between the major nuclear powers, or in other contexts. Tactical nuclear weapons should be brought ashore. The possibility of negotiating measures of restraint on navigation with vessels carrying nuclear weapons on board is another matter to explore. The continuing practice of nuclear powers neither to confirm nor to deny the presence of nuclear weapons on board any particular ship at any particular time ought to be abandoned.

The argument is inescapable, namely that all land-based tactical nuclear weapons must be subject to negotiations. Those with the shortest ranges will be increasingly important when agreements to eliminate others are concluded. The deployment of battlefield nuclear weapons close to the East/West divide in Central Europe adds to the risk of early and inadvertent escalation of any military conflict in the area. To reduce this risk is a rationale of the corridor proposal first made by the Olaf Palme Commission. Measures to halt particularly destabilising developments must be considered both multilaterally and in negotiations between the major nuclear powers. The rapid development of weapons performance – in terms of accuracy, speed and versatility, the use of computers and satellites – makes quality increasingly important. *Qualitative aspects* must be considered in agreements on numerical limitations.

A comprehensive test ban treaty remains the foremost measure among those aimed at curbing the qualitative development of nuclear warheads. A test ban would be a major political turning-point. And it would help to halt the further proliferation of nuclear weapons. The United States and the Soviet Union have now also agreed to begin negotiations on nuclear testing matters. Their approach is a step by step one: first, improved verification arrangements for the two bilateral threshold agreements, then further intermediate limitations on nuclear testing and, finally, the complete cessation of testing as part of an effective disarmament process. The start of superpower negotiations in this field is to be welcomed. Still, it should be emphasised that the time has come for a definite end to *all* nuclear weapons testing and any agreement that leaves room for continued testing is clearly insufficient. Agreements must include a clear commitment to reach a complete and comprehensive test ban at an early, specified date. They can be steps in the right direction only if reductions are militarily significant, imposing real constraints on the ability of the parties to develop nuclear weapons at will. In this connection it should be recalled that the Partial Test Ban Treaty of 1963 was hailed by everyone at that time as a success and a source of great satisfaction. Although it has been beneficial to the environment, it would be considered a failure as a disarmament treaty in the eyes of many people today; it became not a first step but the only step. Our assessments of what is progress and what is failure may thus change over time.

Nuclear disarmament is an absolute priority because of the absolute character of nuclear weapons. But although other weapons of mass destruction may not threaten to obliterate human civilisation, they could cause casualties fully comparable with many nuclear weapons if used in densely populated areas. Chemical weapons have not yet been relegated to history. Reports that they have been used recently increase our concern that they are instruments of the present and the future and not of the past. The Conference on Disarmament in Geneva has made steady progress towards a ban on all chemical weapons. Most of the elements of a convention are in place. However, some technically and politically complicated matters still remain to be resolved by the 40 negotiating parties. The major military powers have manifested a common interest in working out a fully verifiable and truly comprehensive convention. This common interest will be no less important in the final stage of the negotiations.

Conventional weapons consume some 80 per cent of world military expenditure and have been used to kill some 25 million people in the last four decades. Without conventional disarmament, all efforts for international and regional security may be jeopardized. The goal of nuclear disarmament is not likely to be realised in isolation. There is a close link between nuclear and conventional armaments. For to achieve a world safe for conventional warfare is certainly not an aim to be pursued.

From a military viewpoint, many tactical nuclear weapons are, to a high degree, connected with conventional forces. Some nuclear powers obviously consider nuclear weapons as a means whereby they can compensate for imbalances in conventional forces. Also, in order to create conditions conducive to nuclear disarmament, it is imperative that fears caused by conventional forces and their potential offensive use be allayed. In Europe, the most heavily over-armed of all continents, members of the two alliances seem committed to starting negotiations on conventional disarmament within the framework of the Conference on Security and Co-operation in Europe (CSCE).

The author's own country, Sweden, would much welcome such negotiations. It is essential that all 35 CSCE states take part in a continuous exchange of views on the matter, which concerns the security interests of all of them. The purpose of a conventional disarmament process in Europe is to strengthen security by establishing military stability and balance at a substantially lower level than at present. The military forces of the neutral and non-aligned states do not threaten the security of any state. As far as Sweden is concerned, there is, on the contrary, good reason to believe that all parties consider a strong Swedish defence to be a contribution to stability and peace in our part of Europe.

The arms race would probably suffer a decisive blow if states were finally to realise that there are no durable technological or scientific short-cuts to solving political security problems. The fundamental reasons for which nations arm are, and will remain, political. Nations may arm because they harbour the intention of using their armed forces for a variety of purposes that range from the pursuit of world dominance to the liberation of occupied territory. They may arm in order to gain influence and power through their ability to coerce others or even to gain the prestige which certain weapons – most notably nuclear weapons – are supposed to provide. They may arm without any such hostile designs but in genuine fear, ill- or well-founded, that others may oblige them to defend their own sovereignity and territorial integrity. The arms race as such aggravates fear and suspicion in international relations. Like metastatic tumours in a cancer the arms race spreads further and further into the body. They must be eliminated, otherwise the patient, the world, will perish. The political core of the disease also has to be eliminated. If not, new dangerous cancers will grow. The medicine must be strict observance of international law, just solutions to regional conflicts, freedom for all peoples and respect for human rights. Confidence-building measures, such as those negotiated in Stockholm and now successfully implemented, have a part to play. So will economic, cultural and other forms of co-operation between states.

No leap forward in weapons technology can substitute for disarmament and political measures to reduce tension and conflict. The millenia of military history show that new weapons, devices or strategies will ultimately be matched by counter-weapons, counter-devices and counter-strategies. The nuclear powers are now about to establish the confidence and co-operation necessary for a successful search for genuine solutions to security problems facing them and us. These efforts must not be allowed to be frustrated by the old dream of the technological 'quick fix' – be that dream expressed in the American Strategic Defense Initiative or in any other manner. Political security problems can only be resolved by political solutions. The recent achievements of the superpowers may be considered either a giant leap forward or just a small, hardly significant, step. Whatever view is taken, it is the direction taken that is more important than the length of the first step – provided of course that they continue along the path they have staked out.

Proposals already made point the way forward. They include:

● Negotiations on substantial reductions of *all* nuclear weapons.
● Measures to halt the qualitative development of nuclear weapons, in particular by a comprehensive test ban.
● The early conclusion of a multilateral treaty banning chemical weapons.
● Substantial reductions of conventional forces in order to promote stability and balance.

In short, what is now required is determination and ingenuity at the negotiating tables and restraint in weapon laboratories and on testing grounds. Progress or failure in disarmament is to a large extent determined by the United States and the Soviet Union. But they do not act in isolation. They too are dependent for their security on the rest of the world. They too have to take into account the positions of other states, and domestic and world public opinion. Nuclear powers, allied states, neutrals and non-aligned – all have their role to play in the efforts to bring about disarmament.

Disarmament cannot be realised against or without the nuclear powers. Nor is it likely to be realised if the rest of the world leaves the matter to them. The results in bilateral negotiations that are now clearly apparent are to a a large extent due to the perseverance of a determined and knowledgable international opinion. In the moulding of this opinion, states, organisations, peace movements, scholars and others interact. Disarmament will only become reality if it is demanded by a determined and well-informed public. And it is our *personal* responsibility to act. More and more people have stood up to plead the cause of reason, of restraint, of survival. They have seen that humanity is faced with a choice: to disarm or to face the risk of being disarmed by a global and total war. It is for them and for us to make our times, in the words of Dickens, an age of wisdom, and to let the present spring of hope turn into summer.

21 Technology, the Arms Race and Disarmament

Joseph Rotblat

Scientists and physicians have made in recent years a great effort to educate and finally convince the men at the helm that a nuclear war would be suicide for all. When the leaders of the superpowers met in Geneva in 1985 they formally acknowledged this when they agreed that 'A nuclear war cannot be won and must never be fought'. Implicit in this phrase is that nuclear weapons are of no use as military instruments for waging war. Nevertheless an important role is still assigned to them, namely to deter their use by the adversary. A sufficient number of nuclear weapons must be maintained in the arsenals, so that – if attacked – enough will survive to ensure inflicting unacceptable damage in a retaliatory attack. This is the strategy of deterrence, the corner-stone of the security policy of many states.

The adherents to this doctrine never tire of telling us that peace in Europe has been maintained for four decades thanks only to the nuclear deterrent. This is proclaimed as a fact by political and military leaders in the West, even though it is only a supposition that cannot be proved one way or the other. One could equally postulate – also without proof – that the nuclear deterrent is responsible for the state of Cold War that prevailed during that period. One can easily make out a case that the political climate in Europe would have been much easier, that there would have been much less tension, if it were not for the nuclear arms race. This is so because, in order to justify the enormous military spending, it is essential to generate an atmosphere of fear and mistrust, it is necessary to create the image of an enemy, who is depicted as evil, aggressive and waiting for any sign of slackening in our defence posture to pounce and overwhelm us. But even if we accept that the existence of nuclear weapons has contributed somewhat to the avoidance of war in Europe so far, can we rely on the nuclear deterrent for the indefinite future? The answer to this question hinges on the stability of the deterrent; it depends on the answer to two further questions: has the deterrent proved to be stable in the past, and have we reasons to believe that it will be stable in the future? In my view, both these questions must be answered in the negative. I arrived at this conclusion after an analysis of the history of the nuclear arms race. This can be summed up in one sentence: at no time has either of the superpowers been satisfied with its nuclear arsenals as being sufficient to give it security. For real or perceived reasons, both felt compelled to keep on increasing or modernising their nuclear arsenals. Nuclear strategy has never been in a steady state; it has always been changing. And one of the chief motive forces for these changes is the input from science and technology.

Ever since the nuclear arms race began in August 1949, there has been a continuous growth of nuclear arsenals, either in numbers or in quality. As far as numbers is concerned, the 1987 Stockholm International Peace Research Institute (SIPRI) *Yearbook* shows that the United States and the Soviet Union each have about 25 000 nuclear warheads.[1] More than half of them are in the category of strategic forces (See Table 21.1). Actually, the number of warheads in the United States has gone down a little since the previous year, but this has been more than compensated by the introduction of new, superior systems, like the MX, an intercontinental ballistic missile, and the B-1B heavy bomber. These are part of the large modernisation programme embarked upon by the Ronald Reagan Administration. The Soviets, too, introduced new improved weapons, the SS-25, an ICBM, and the SS-NX-23, a submarine-launched ballistic missile. The megatonnage of the warheads held by each side is at least twenty-five times greater than the amount deemed in the 1960s to be necessary to satisfy the requirements of deterrence. What is the reason for this enormous

Table 21.1 Nuclear forces in 1987 (numbers of warheads or bombers)

	USA	USSR[1]
A Strategic Forces		
Intercontinental ballistic missiles	2 290	6 900–13 000
Submarine-launched ballistic missiles	6 050	2 400– 4 100
Long-range bombers	5 343	600– 1 200
B Theatre Forces		
Missiles	6 157	4 480– 8 780
Aircraft	3 800	4 770– 4 770
Total	23 640	19 150–31 850

[1]The two sets of figures are low and high estimates of MIRV loading.

overkill capacity? The answer lies mainly in the continuous technological advances, which keep changing the quality and performance of nuclear weapons and the means of their delivery, necessitating a never-ending modernisation of the arsenals. Such technological advances will be made as long as there exist military research establishments in which thousands of scientists and engineers are employed, devoting their talents and ingenuity to invent and design ever more sophisticated devices. By its very nature the nuclear arms race feeds on the continuous input of scientific innovation and technological skill. But many analysts believe that these factors have acquired a momentum of their own, that they have become the masters instead of being the tools. It is said that nowadays technology dictates policy, that new weapon systems emerge not because of any military or security requirements, but because of the sheer impetus of the technological process.

The role of the scientists in the military research establishments has been eloquently described by Lord Zuckerman, who has been for many years the chief scientific adviser to the British Government:

when it comes to nuclear weapons, the military chiefs of both sides – who by convention are the official advisers on national security – usually serve only as a channel through which the men in the laboratories transmit their views. For it is the man in the laboratory – not the solider, or sailor, or airman – who at the start proposes that for this or that arcane reason it would be useful to improve an old or to devise a new nuclear warhead. And if a new warhead, then a new missile; and given a new missile, a new system within which it has to fit. It is he, the technician, not the commander in the field, who is at the heart of the arms race, who starts the process of formulating a so-called military nuclear need. It is he, who has succeeded over the years in equating, and so confusing, nuclear destructive power with military strength, as though the former were the single and sufficient condition of military success. The men in the nuclear weapons laboratories of both sides have succeeded in creating a world with an irrational foundation, on which a new set of political realities has in turn had to be built. They have become the alchemists of our times, working in secret ways which cannot be divulged, casting spells which embrace us all. They may never have been in battle, they may never have experienced the devastation of war; but they know how to devise means of destruction.[2]

These are strong words, but they have to be taken seriously, coming from a scientist who had been closely associated with both military and political leaders, and was directly involved in decision-making on matters of national security. Similar sentiments were expressed by other knowledgeable people such as Herbert York, the first Director of the Lawrence Livermore National Laboratory.[3]

What is the main motivation of these scientists and technicians? For this we should listen to the person who actually designed weapons at the Los Alamos National Laboratory. Theodore Taylor said recently: '. . . the most stimulating factor of all was simply the intense exhilaration that every scientist or engineer experiences when he or she has the freedom to explore completely new technical concepts and then to bring them into reality.'[4]

Of course, science and technology are not the only factors which contribute to the continuing existence of the nuclear arms race. Primarily, the arms race is the product of political forces, and there are many factors, often interacting with each other, that enter into play. There exist strong institutional pressures and the ubiquitous vested interests, commonly described as the 'military–industrial complex'. A serious driving force is the 'asymmetry of perception', each side feeling threatened by the technical progress apparently made by the opponent and believing that the other side had achieved superiority. Nevertheless the fact remains that new military gadgetry is frequently the trigger for new stategies. New weapon systems tend to act as catalysts for new political and strategic doctrines, as we saw, for example, in the change of the strategy of deterrence from retaliatory to war-fighting, after the introduction of Multiple Independently-targetable Re-entry Vehicles (MIRVs) and the achievement of great accuracy in hitting a target. Technology also provides serious obstacles to the attempts to halt the arms race by negotiations. Often new advances are made for the explicit purpose of serving as bargaining chips in these negotiations, to be given up in an exchange for a concession by the other side. But usually the bargaining chip becomes an integral element of the weaponry, thus adding another spiral to the continuing arms race. As long as scientific innovation is available there will be no end to the arms race.

Of course, any kind of research at universities and learned institutions may result in the development of means of destruction, as the discovery of fission has illustrated so dramatically. Nevertheless it is the existence of dedicated research laboratories, specifically designated to work on military problems, that is directly responsible for fuelling the arms race, with all the dangers inherent in it. The two main establishments of this type in the United States are the Los Alamos National Laboratory in New Mexico and the Lawrence Livermore National Laboratory in California. Between them they employ some 16 000 people, more than half of them on explicitly military projects. Despite their academic cloak, through the association with the University of California, they are the breeding grounds for new generations of nuclear weapons and new sophisticated methods for their deployment. Thus the Los Alamos Laboratory says in one of its annual statements: 'Our ongoing responsibility since the founding of the Laboratory in 1943 has been the design, testing, and maintenance of nuclear warheads... In 1980 the Laboratory was responsible for the design of all new strategic warheads entering the stockpile.'[5]

The Lawrence Livermore National Laboratory is vying with Los Alamos in these achievements. Its workers take great pride in having developed the MX missile, which they insist on calling the 'Peacemaker', and in designing its warhead, the W87, as well as the W84 warhead for ground-launched Cruise missiles, and the B83 warhead for the new heavy bomber, to mention just a few of the recent additions. Such laboratories also exist in other countries, for example, the Atomic Weapons Research Establishment at Aldermaston, Great Britain, and of course there are others in the Soviet Union. There is not much information about the latter, and no doubt they play as important a role as those in the West, but there are reasons to believe that the main impetus of the nuclear arms race comes from the United States. At least this has been so in the past, as can be seen in John Holdren's Table of 'Milestones in the Nuclear Arms Race' (see Table 21.2) Thirteen of the fourteen events listed in it occurred first in the United States; the Soviet Union lagged behind by about five years on the average.[6]

The overall effect of military technology, then, is to make the situation less and less stable. One of the main outcomes of technological innovation is to erode the value of the existing deterrent force, making it vulnerable. This creates the need for continuous updating of the nuclear arsenals in order to overcome a genuine or perceived fallibility of existing arsenals. The British Prime Minister, Margaret Thatcher, put it succinctly, when defending the replacement of Polaris by Trident: 'Unless you modernize the strategic deterrent, it soon ceases to be a deterrent.'

The quest for a less vulnerable deterrent carries with it a great

Table 21.2   Milestones in the nuclear arms race

Milestone	Year of Achievement by: USA	USSR
Test of atomic (fission) bomb	1945	1949
Deployment of intercontinental bomber	1948	1956
Deployment of jet bomber	1951	1954
Test of practical hydrogen bomb	1954	1955
Deployment of tactical nuclear weapons in Europe	1954	1957
Deployment of nuclear artillery	1954	1980
Strategic reconnaissance for targeting	1955	1962
Test of intercontinental ballistic missile	1958	1957
Deployment of ICBM	1959	1960
Deployment of submarine-launched ballistic missile	1960	1964
Deployment of solid-fuel ICBM	1963	1968
Deployment of swing-wing, supersonic bomber	1967	1974
Test of multiple independent re-entry vehicles	1968	1973
Deployment of MIRV	1970	1975

danger: it increases the probability of a nuclear war started accidentally or inadvertently. There are several reasons for this. One is the greatly reduced time interval – especially if Strategic Defense Initiative (SDI) were to be implemented – in which a decision has to be made about a response to a perceived attack. Another is the ever-increasing reliance on computers and other electronic gadgetry. Indeed one may have to do away altogether with human intervention in the decision-making process, and leave the decision to pre-programmed highly sophisticated sensors and their associated computer hardware and software; but, as recent disasters have shown, highly sophisticated technology can go wrong. With the ever-decreasing time available for decisions and the ever-increasing complexity of equipment, sooner or later a malfunction or misinterpretation is bound to occur, and civilisation will be wiped out. To sum up: increasing technological advances, made by scientists in the military research establishments, keep eroding the value of the deterrent, necessitate continuous modernisation of nuclear arsenals, and increase the danger of an inadvertent nuclear war. The deterrent, far from being stable, inexorably leads to our destruction.

These perils to mankind, inherent in current strategic policies, have been generally recognised. Hence the efforts to reach agreements in international negotiations on the reduction of military arsenals of all kinds, nuclear, chemical, conventional, as well as on confidence-building measures and the moderation of confrontational postures. The main difference of opinion appears to be about the ultimate objectives of these negotiations, particularly in relation to nuclear arms: is it our aim to reduce greatly the nuclear arsenals but retain permanently nuclear weapons at the level of minimum deterrence, or do we want to see the complete elimination of nuclear weapons? Curiously, both leaders of the superpowers have come out in favour of the latter; in Reykjavik they discussed the elimination of nuclear weapons within ten years. But this has not been taken seriously in the West, where many leaders, notably Thatcher, consider a nuclear-free world as a Utopian dream.

Despite the fundamental differences of approach between those who want to maintain the nuclear deterrent at the minimum level and those who want to eliminate nuclear weapons, in practice the difference may not be significant during the foreseeable future. But before going into this it is important to note the great change in the climate of public debate that has taken place recently, and which makes it possible to put on the agenda ideas which would have been considered completely unrealistic even a year ago. This is largely due to the radically changed situation inside the Soviet Union. The changes that became identified in the West by the Russian words *glastnost* and *perestroika* have already helped to reduce hostile propaganda, and have created an atmosphere conducive to the signing of international treaties. After decades of sterile arms control negotiations, which produced much rhetoric but not a single, mutually agreed measure of actual reduction of nuclear arsenals, a breakthrough has finally happened. The agreement to eliminate completely one class of weapons is justifiably hailed as an historical event. In terms of numbers, the dent put in the nuclear arsenals by the removal of intermediate-range missiles is very small, but it signifies a halt in the arms race and has revealed the existence of a reverse gear. It is also gratifying that the superpowers are discussing a test ban treaty, in addition to deep cuts in strategic nuclear forces.

We must not, however, allow euphoria about these events to blind us to realities. It would be folly to underestimate the powerful forces that act in the opposite direction: vested interests will certainly put up every possible obstacle in the way to a world with fewer arms. But we are right to feel elated that a new approach, the new way of thinking which many scientists have been exhorting for so long, has at last taken hold in the political arena.

This encourages us to look further ahead, to examine where we want to go from here. What do we have in mind when calling for radical reductions in nuclear arsenals? What should be the limit aimed at in these reductions? Should it be zero, or a finite value, referred to as the minimum deterrent? The latter is often defined as the number of weapons just needed to ensure the necessary retaliatory capacity, but there are differences of opinion about its actual numerical value; some say that 1000 warheads would be sufficient, others fear that this may be too low, while others believe that we could go down to a few hundred. But more important than numbers is the actual concept of the minimum deterrent. Its protagonists argue that nuclear weapons are here to stay: we have to live with the bomb – they say – but at a level that would reduce to a minimum the chances of its use by accident or miscalculation. There are some basic difficulties in the idea of a permanent nuclear deterrent, even at a low level. These difficulties are quite apart from the moral objections to the very concept of nuclear deterrence, cogently expressed by such diverse authorities as the Catholic Bishops in the United States and the General-Secretary of the Soviet Communist Party. For one thing, who will be entitled to possess the minimum deterrent? The two superpowers only? Would this be acceptable to Great Britain and China, let alone to France? And what about other countries, some of which may already have more than the minimum, although undeclared? If the idea is perpetuated that the security of a country demands the possession of the nuclear deterrent, how can one deny other countries the same security? According to Article VI of the Non-Proliferation Treaty, the nuclear weapon states undertook to pursue the aim of nuclear disarmament (as part of the ultimate objective of General and Complete Disarmament), not to settle for a permanent – even if very low – level of nuclear armament. Horizontal proliferation of nuclear weapons has been kept under control so far, but this cannot go on for ever.

Another troubling question is whether one can really conceive that the deterrent will be maintained at its constant minimum value. We are contemplating going back to a situation, in terms of numbers, that existed some thirty years ago, but is there reason to believe that this time the arsenals will not start growing as they did then? As long as either side has nuclear weapons, however small in number, there will be the need to maintain them. A core of scientists and engineers will be necessary for this purpose. And, as is their habit, they will think of ways of making their own weapons more effective, and those of the other side more vulnerable. A nuclear arms race will start all over again. In short, the minimum deterrent is unlikely to stay at the minimum for long.

In view of these and other objections, one should look seriously at the prospects of getting rid of nuclear weapons altogether, within the ultimate objective of General and Complete Disarmament. We should look at this problem in the light of the recent changes, the new way of thinking which encourages bolder approaches. One argument against complete nuclear disarmament, frequently put forward, is that nuclear weapons cannot be disinvented: we cannot erase from our memories the knowledge of how to make them. It is argued that, even if all existing weapons were dismantled, it would not take long to reintroduce them, should a military conflict occur. This is of course true, although it is puzzling that the same argument is not used against a ban on chemical weapons – which everybody seems to desire – even though it is much easier to reintroduce chemical than nuclear weapons. A reply to this argument is that, even if the incentive to use nuclear weapons did arise in a nuclear-free world, the situation would still be much safer than if nuclear weapons were ready in the arsenals. If the military research establishments have halted their weapons projects – and this should be an essential element of nuclear disarmament – it would take some months for either side to manufacture the weapon, particularly if no weapon-grade plutonium or uranium were kept in storage, and this would provide valuable time for negotiations to end the conflict by peaceful means.

The main difficulty about complete nuclear disarmament, however, is the worry about cheating. The destructive potential of nuclear weapons is so enormous that the concealment of even a small number of such weapons could give the side that has cheated an overwhelming advantage. Therefore we have to address the problem of verification and compliance with undertakings during the process of nuclear disarmament.

Obviously we cannot get rid of nuclear weapons overnight. It will have to be a gradual process, strictly correlated with verification effectiveness. The connection between reduction of nuclear arms and verification is depicted on the so-called Wiesner graph, first put forward at a Pugwash Conference (see figure 21.1). The vertical axis on the left is a measure of the status of armaments, for example, the

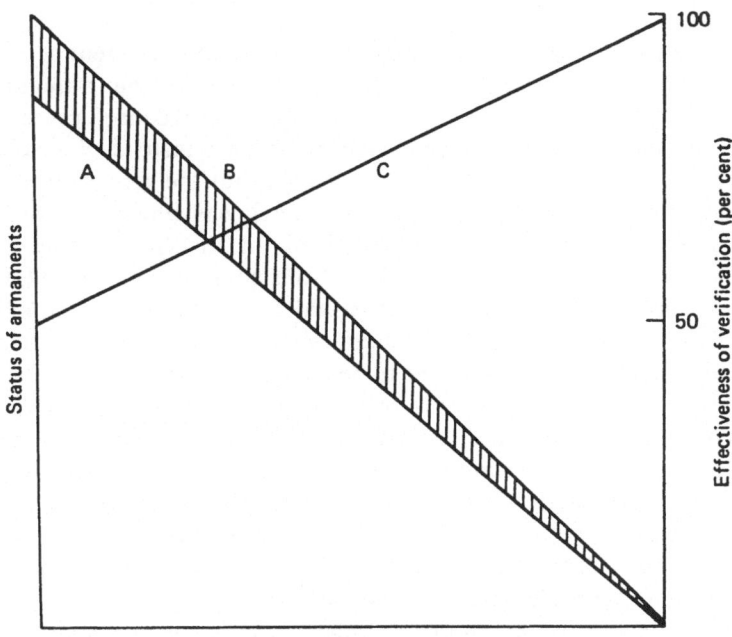

Figure 21.1

number of nuclear weapons held by one side, or some other quantity characterising the potential of these weapons. The horizontal scale gives the time from the start of the disarmament process. Both these scales are arbitrary. Let line *A* indicate the number of weapons held at a given time by our adversary, as declared by him. Natural caution will make us assume that he is cheating, that he managed to conceal some weapons. Let line *B* be the real number held by him. Note that at the beginning, when the arsenals are still large, even a considerable concealment would not matter, but the situation changes as the arsenals become smaller. Even a small amount of cheating could then be decisive. Therefore, the necessary condition to proceed to low levels of arsenals is effective verification of compliance. The state of necessary verification is indicated by line *C*, with its percentage scale on the vertical axis on the right. The lower the arsenals the greater must be the verification capability, and we can go down to complete abolition only when verification is 100 per cent effective. I propose a different definition of the minimum deterrent: it is the number of weapons needed to balance a possible deception by the adversary. The stripes between lines *A* and *B* denote the minimum deterrent according to this definition. As is clear, it is not a constant, but a quantity gradually decreasing to zero. This is of course an idealised presentation. In practice there would be no straight lines; instead of a steady decrease there would be a step-wise change. But the general trend is what matters. According to this scheme, the time in which to complete the disarmament process would depend on the time needed for verification to become fully effective. If this could be accelerated, the time to achieve full disarmament would be shortened accordingly; and this brings one back to the military research establishments. As already stated, these would have to be closed down if a new arms race is to be avoided, but they could be kept going if they were converted to work entirely for peace, specifically to improve verification techniques. Some such research is already going on, but it should be considerably expanded. The scientists and engineers should be asked to use their talents to discover new methods of surveillance, new ways to reduce the possibilities of undetected cheating, thus shortening the time scale for complete nuclear disarmament.

In the present discussion the political aspect of the problem has so far been ignored, although it is the most important. Indeed, it has been suggested that what we should really work for is a climate of trust in the world in which nuclear weapons would become an irrelevancy. As an illustration it is pointed out that the Americans are not worried about the Btitish having nuclear weapons, and vice versa. It is doubtful whether this is true. Anyway, the aim would be a world in which people would be able to say: 'Nuclear weapons, who cares?' In my opinion this expects far too much; it envisages a biblical world where 'the wolf shall dwell with the lamb'. It implies a complete change not only in political thinking but also in human nature; that nobody, not even a bunch of maniacs or terrorists, would bother to pick up the existing nuclear weapons and hold us all to ransom.

The political aspect is closely interrelated with the technogical aspect. The mistrust and hatred that dominated our lives during the past decades were to a large extent caused by the very existence of nuclear weapons which enable either side to destroy the other in an instant. We have to justify the huge expenditure on arms by creating the stereotype of an enemy bent on our destruction; we have to generate an atmosphere of fear and antagonism. But the very event of starting to reduce the arsenals will encourage the belief that a catastrophe can be avoided, and this in turn will be conducive to

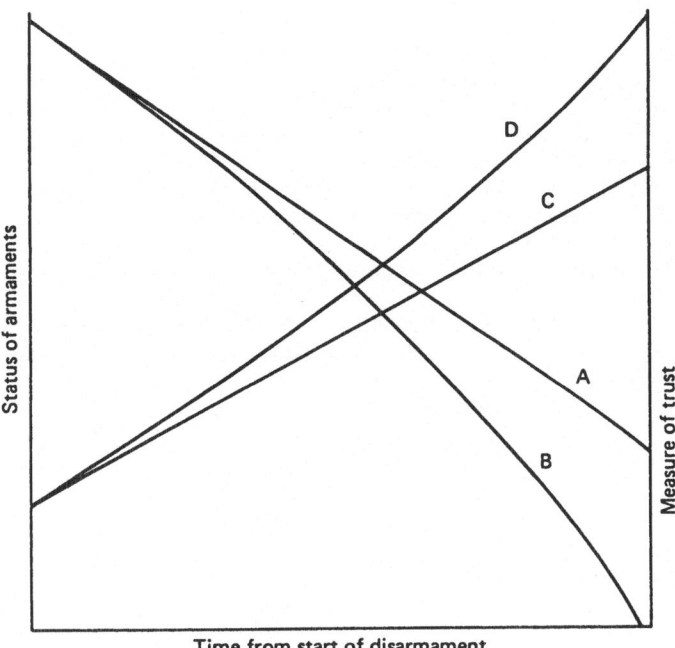

Figure 21.2

taking further steps towards disarmament. It is a process of positive feedback and its effect is depicted in Figure 21.2. Line *A* is the same as on Figure 21.1, showing the reduction in arms without taking into account the political factor, while curve B shows the effect of the improved political climate. Line *C* and curve *D* denote the increase in goodwill created and enhanced by the process of disarmament. In calculating these curves I used a very simple mathematical model, but this is probably sufficient to illustrate the trend. Reducing the arsenals reduces fear and bolsters confidence, which in turn makes it easier to implement further reductions, which create a better climate, and so it goes on, resulting in a shorter time needed to achieve complete nuclear disarmament. How many years it will take to achieve this one cannot tell, but the sooner we start the sooner we will get there. What is important is to have the ultimate objective in front of us.

Notes

1. Stockholm International Peace Research Institute, *World Armaments and Disarmament: SIPRI Yearbook 1987* (Oxford, 1987).
2. Lord Zuckerman, *Science Advisers and Nuclear Weapons* (London, 1980).
3. Herbert York, *Race to Oblivion* (New York, 1970).
4. T. B. Taylor, 'Third-Generation Nuclear Weapons', *Scientific American* (April 1987).
5. Los Alamos National Laboratory, *Annual Report 1980* (Los Alamos, 1980).
6. J. P. Holdren, 'The Dynamics of the Nuclear Arms Race: History, Status, Prospects', Energy and Resources Group Working Paper no. 83–5, University of California, 1983.

22 Experimental Nuclear Explosions and the Arms Race

Francesco Lenci

It is highly significant that, ever since the first few years of experimental explosions of thermonuclear devices, the three nuclear powers of the time (the United States, Great Britain and the Soviet Union) have grappled with the problem of suspending their nuclear tests. This awareness of the importance that the discontinuance of programmes to develop military nuclear technology could have for international security and peace led, in the summer of 1958, to the convocation of a conference of experts of the Eastern and Western countries to tackle the technical questions linked with the detection of experimental nuclear explosions. The conclusions of those discussions were that it would be possible to detect and identify nuclear explosions in the atmosphere above the power of 1 kiloton (kt; 1 kt = 1000 tons of TNT), and to detect, with a reliability of approximately 90 per cent, underground nuclear tests of more than 5 kt. The monitoring network necessary for this purpose would have required a system of around 160 to 170 control stations installed on the ground and about ten appropriately-equipped ships.

Also in 1958, the Soviet Union, the United States and Great Britain inaugurated a 'voluntary' moratorium and began negotiations for the ending of all experimental nuclear explosions. On account of the difficulties that these negotiations encountered, the multilateral voluntary moratorium came to an end, and in 1961 first the Soviet Union and then the United States resumed their tests.

On the one hand, the radioactive contamination caused by the nuclear tests in the atmosphere, and, on the other hand, the moments of great tension that characterised the international situation in those years, created a state of alarm and preoccupation in public opinion which, in turn, gave rise to political pressure on the superpowers to reach an agreement banning all nuclear tests.

On 5 August 1963, the United States, the Soviet Union and Great

Britain thus signed the Limited – or Partial – Test Ban Treaty (LTBT or PTBT), which prohibits carrying out nuclear tests in the atmosphere, in outer space and under water. The partial nature of the resulting agreement is clearly recognised in the preamble to the Treaty, in which it is declared that 'seeking to achieve the discontinuance of all test explosions of nuclear weapons for all time, [the signatories] determined to continue negotiations'. And, even if intermittently, the negotiations did in fact continue over the years, always with the ultimate goal of drawing up a Comprehensive Test Ban Treaty (CTBT). However after 1963 the Soviet Union, the United States and Great Britain pursued their test programmes by carrying out nuclear explosions underground. The LTBT, by allowing underground tests to be conducted, sets no limit to the development of nuclear technologies for the acquisition of new weapons; it merely seeks to prevent contamination from radioactive fall-out due to tests in the atmosphere. (France and the People's Republic of China, not having adhered to the LTBT, conducted numerous tests in the atmosphere up to 1974 and 1980, respectively.)

However, thanks to this commitment by the United States, the Soviet Union and Great Britain to strive for agreement on the CTBT, over the years following 1963 various other treaties were negotiated and completed, both with the aim of greatly limiting the areas in which nuclear arms could be installed or tested (the Treaty on Outer Space and that of Tlatelolco relating to Latin America in 1967, the Sea Bed Treaty in 1972 and the Treaty of Rarotonga relating to the Pacific Ocean in 1986) and in order to try to avoid the horizontal proliferation of nuclear weapons (the Non-Proliferation Treaty in 1970). Also in the Non-Proliferation Treaty, the signatory states undertake to 'pursue negotiations in good faith on effective measures relating to cessation of the nuclear arms race at an early date and to nuclear disarmament, and on a treaty on general and complete disarmament' (Article VI).

In 1974, returning to the commitments of the LTBT of 1963, the United States and the Soviet Union, in order to continue to negotiate a CTBT and with the intention of facilitating the attainment of this objective, signed a treaty prohibiting underground nuclear tests of more than 150 kt, namely the Threshold Test Ban Treaty (TTBT). To allow satisfactory reciprocal verification of respect for the Treaty by National Technical Means (NTM), the Treaty and the Protocol to the Treaty provided that the United States and the Soviet Union exchange precise information on the geological and geophysical characteristics of the ground in which the tests were carried out (such as density, seismic velocity, degree of humidity and porosity). It was also stipulated that the two parties should exchange detailed, complete data on two tests to be used for the preliminary calibrations. The TTBT was never ratified, but it has been substantially respected by both parties since 1976. The reciprocal accusations of violation of the Treaty, mainly directed against the Soviet Union by the Ronald Reagan Administration, have always been contested as unwarranted, not only by the Soviets but also by numerous American experts and by organisations such as the Swedish National Institute for Defence Research.[1]

In 1977 the trilateral negotiations for a CTBT between the United States, the Soviet Union and Great Britain seemed to be headed towards the final stage. The Soviet Union seemed willing to accept voluntary on-site inspections and the United States appeared to agree on the 'non-obligatoriness' of such inspections. An accord had also been reached on the number of national seismic stations to be installed on the territory of the other party, and it had been established that between 10 and 15 of these stations would be supplied with cryptological systems that would guarantee the data were transmitted without interruption and could not be altered or modified. It seems that the negotiations were so promising as to induce the Jimmy Carter Administration not to bother with the ratification of the TTBT, which had already been approved by the Senate Foreign Relations Committee in 1977. However various external factors slowed the course of the negotiations in a way that later proved to be disastrous: unforeseen difficulties in the second phase of the Strategic Arms Limitation Talks II (SALT II) , the kidnapping of the American hostages in Iran, and the Soviet intervention in Afghanistan.[2]

The negotiations to reach agreement on the CTBT were broken off immediately after the election of Reagan as President of the United States, and were resumed only in September 1987. Quite an important contribution to the resumption of the negotiations on the CTBT came from the unilateral initiative of the Soviets to suspend their nuclear test programme from 6 August 1985, the fortieth anniversary of the dropping of the atomic bomb on Hiroshima, to 1 January 1987, despite the United States's continuing to conduct nuclear tests. The Reagan Administration justified the non-adherence of the United States to the moratorium by claiming that the Soviets' initiative was merely propagandistic and that in reality the Soviets had brought their test programme to a conclusion with a long series of nuclear explosions in 1984 and the first half of 1985. Consequently the United States, according to the official declarations, would take the possibility of a moratorium into consideration only after having completed its own test programme.

In reality this unilateral initiative had the great value of constituting concrete proof of the Soviets' 'new way of thinking', and appreciable results were seen immediately. First, in May 1986 the Academy of Sciences of the Soviet Union agreed on collaboration with the Natural Resources Defense Council (NRDC), an American environmental protection organisation, in the installation of seismic detection stations, run jointly by Soviet and American scientists, in the area of Semipalatinsk and in the desert of Nevada, where the nuclear tests of the Soviet Union and the United States are carried out. This collaboration is unequivocal proof that it is possible to find adequate solutions to the technical problems connected with the verification of a CTBT provided there is the political determination to consider international agreements for the control, limitation and reduction of armaments to be the decisive instrument for mutual security and peace.[3] Secondly, in July 1986 an International Forum of Scientists was held in Moscow to analyse the problems linked with the verification and control of a complete ban on nuclear tests and to assess the reliability of the possible technical solutions. On that occasion the first results obtained by the US–Soviet 'Verification Team' in the detection stations installed in the Semipalatinsk area were made public. The final document of the Forum also reaffirmed the technical feasibility of adequately verifying a CTBT and clearly evidenced the contribution that a CTBT, by preventing the development of new weapons systems, could make to stopping, or at least slowing, the arms race. Finally, on 7 August 1986, the group of the six countries that constitute the 'Initiative of the Five Continents' (Argentina, Greece, India, Mexico, Sweden and Tanzania) made a public appeal for the CTBT in which, among other things, the six countries declare themselves to be 'prepared to participate in co-operative efforts together with the USA and the USSR and also to take certain steps on our own to facilitate the achievement of adequate verification arrangements'.

At the end of February 1987, however, the Soviet Union too began conducting experimental nuclear explosions again, although stating its willingness to suspend them as soon as the United States did the same. The Soviet Union justified this decision with the usual arguments regarding the necessity of safeguarding its security and not finding itself in a position of strategic and military inferiority vis-à-vis the United States. In reality this decision to resume conducting nuclear tests could be interpreted as a concession to those Soviet military and political sectors that consider the 'new way of thinking' of the Soviet leadership a continual, unjustified backing down and a declaration of weakness towards the Reagan Administration.

Substantially, underground nuclear tests must be detected by means of techniques of the seismological type, whose sensitivity is a crucial factor. A significant part of the energy released by a nuclear explosion which takes place underground is in fact transmitted to the earth (approximately: 0·01 per cent if the explosion occurs in cavity, 0·1 per cent if it takes place in dry alluvial soil and 1 per cent in the case of granitic rocks), thus generating seismic waves.

One of the most delicate problems in monitoring underground experimental nuclear explosions is discriminating them from natural seismic phenomena: repeated 'false alarms' due to earthquakes would lead to rapid deterioration of the reliability of the verification system and therefore of the credibility of the treaty itself. Today agreement among experts is nearly unanimous that the technologies now available can permit detection of weak seismic signals produced by underground nuclear tests of a power of around 1·0 kt, even in the event that the explosion is 'decoupled'.[4] Putting adequate verifiability into doubt thus appears to be a pretext so as to avoid arriving at an agreement on a total ban on nuclear tests. Finally, it should be kept in mind that the nuclear tests of less than two kilotons are a decidedly small fraction of the total number of tests, since tests of such low power have little significance or demonstrative worth if it is desired to design and develop new nuclear weapons. Indeed, tests of new arms must have explosive powers of at least half or a third that of the weapon one wants to produce.

The questionable utility of low-power tests for the development of new nuclear weapons has led some to suggest the advisability of negotiating very rapidly a ban on tests of a power above a quite low level, namely a Low Threshold Test Ban Treaty. The strongest arguments against this alternative solution to the CTBT may be, on the one hand, the possibility of perfecting new weapons systems anyway, and on the other hand, the plethora of allegations and accusations of having carried out 'above-threshold' tests that the parties might make against each other.

In the middle of the month of September 1987, the Soviet Union and the United States decided to start a negotiating process that gradually, beginning with the ratification of the TTBT of 1974 and through intermediate limitations on the number and power of the nuclear tests, could lead to the achievement of the ultimate goal of banning all nuclear tests. The verification measurements during this negotiation will probably be performed by means of on-site controls using both seismological techniques as well as those of the 'CORRTEX' type (Continuous Reflectometry for Radius vs. Time Experiment).

Experimental nuclear explosions have basically three aims: a study of the effects of nuclear weapons; the development of new nuclear weapons; and control of the efficiency and security of nuclear weapons. As far as the effects of nuclear weapons are concerned, they have been the subject of much in-depth study, and can be easily simulated.

It is a widely held opinion, however, that the greatest obstacles to agreement on a CTBT are created by the determination to pursue development programmes for new nuclear weapons. And this is precisely the reason why a total ban on nuclear tests, although certainly not in itself a panacea, would bring nuclear weapons technology at least partially under control and make a decisive contribution to avoiding the modernisation of already existing arms systems and the development of new ones, thus significantly slowing down the arms race.

New and ever more deadly nuclear weapons can certainly be designed and acquired. One such possibility is the N-bomb, foreseen as usable as an 'anti-man' bomb on the battlefield. In it the fraction of energy emitted in the form of fast neutrons is as high as possible, while the power of the atomic triggering bomb and the quantity of fusionable material are reduced to the minimum. Another is the high Electromagnetic Pulse (EMP) production bomb. The elecromagnetic pulse generated by the asymmetric distribution of positive and negative charges is formed because of air ionisation produced by gamma rays; to maximise the pulse, asymmetric shielding could be used to make the emission of gamma radiation anisotropic.

Among the weapons systems that require nuclear tests in order to be perfected, the best known is perhaps the x-ray laser (the Excalibur Program), one of the components of the arsenal planned for the US Strategic Defense Initiative (SDI). This laser would be 'pumped' by the energy emitted in the first few microseconds following a thermo-nuclear explosion, consisting, to about 70 per cent, of x-radiation. The x-rays could be collimated by thin metallic bars from which x-rays would then be emitted in phase in beams directed against the missiles to be shot down. It is estimated that another 10 to 15 nuclear tests are necessary to validate the technical feasibility of the Excalibur Program, while at least 100 to 200 additional experimental explosions would be needed to allow development of the weapon.[5]

In order to make it easier to destroy hardened objectives, such as missile silos, a warhead is being studied at the Lawrence Livermore National Laboratory which, installed on intercontinental missiles and bombers, would be able to penetrate into the earth and explode underground, thus damaging the target much more effectively. These Earth Penetrating Warheads would considerably increase the proba-bility of success of a counterforce attack and therefore would make the enemy's nuclear weapons more vulnerable, with a decidedly destabilising effect.

One of the most important goals sought in all new arms projects seems to be that of directional channelling of the energy released by a nuclear explosion in order to concentrate it on selected objectives.[6] Weapons that emit microwaves (wavelength between 3 cm and 1m) in a relatively narrow angle could much more efficiently damage electri-cal installations and electronic systems, such as communications and data transmission networks, for instance. According to an estimate reported by Theodore Taylor, if 5 per cent of the energy released by a 1-kt blast could be converted into microwaves, energy fluxes of some 800 J/m^2 could be achieved (enough to seriously damage many kinds of electronic equipment) over an area of approximately 250 km^2, assuming that the explosion took place at a distance of some 30 000 kilometres from the earth (for example, by detonating, at the desired moment, a device installed on a satellite place in a geosynchronous orbit).[7] The energy flux would rise to something like $5\,000\,000 \text{ J/m}^2$ if the explosion occurred 400 kilometres away from the earth.

For the reliability and the security of nuclear weapons, the United States Department of Defense specifies the principal military charac-teristics that must be controlled in nuclear warheads (for example, those of the MX missile): security against triggering of the nuclear explosive in case of accident; compatibility of the dimensions and weight envisaged for the warhead by the Department of Energy with the characteristics of the launching and warhead transport systems; security from dispersion of plutonium in case of accident; efficiency of the triggering systems, also with regard to the high-power conventional explosive; real power of the warhead; state of conservation of the nuclear material (plutonium and tritium, for example); and resistance of the warhead.[8]

As far as the security of the warheads is concerned, the entire world community can only hope that this is absolutely guaranteed for all operative nuclear weapons. As for their reliability, the researchers of the Lawrence Livermore National Laboratory report that over a third of the nuclear weapons introduced from 1958 to the present have displayed problems of reliability, 75 per cent of which were solved thanks to nuclear tests. The continuation of the tests would thus be necessary for reasons of national security. Of a decidedly contrary opinion are numerous scientists and nuclear arms experts, who hold that 'continued nuclear testing is not necessary in order to insure the reliability of the nuclear weapons in our stockpile'. 'The best way to confirm reliability,' they continue, 'is to disassemble sample weapons and to subject components to non-nuclear tests'.[9]

Even many of those who today assert that it is necessary to control the reliability of the nuclear arsenals by means of tests concur that a drastic reduction in the number of nuclear weapons is a condition that could facilitate reaching agreement on a CTBT. Actually, in a situation of minimum deterrence, if not subjecting nuclear arms to control tests truly lowered their reliability and created uncertainties as to their functioning, the danger that one of the two parties tried a first strike attack could also be less. Such a first strike would, in fact, be a completely irrational decision, both owing to the uncertainties regarding one's own arsenal and to the necessity of assuming the arsenal of the adversary to be perfectly efficient ('worst case hypothesis'), and hence capable of retaliating with a devastating counter-attack.

Notes

1. L. R. Sykes and D. M. Davis, 'The Yields of Soviet Strategic Weapons', *Scientific American* (April 1987) pp. 21–9; and A. Krass, 'Recent Developments in Arms Control Verification Technology', in Stockholm International Peace Research Institute, *World Armaments and Disarmament: SIPRI Yearbook 1987* (Oxford, 1987) pp. 431–46.
2. H. F. York, 'U.S.–Soviet Negotiations and the Arms Race', in W. F. Hanreider (ed.), *Technology, Strategy and Arms Control* (Boulder, Colorado, 1985) pp. 1–14.

3. See Thomas B. Cochran, 'The US National Resources Defense Council/Soviet Academy of Sciences Nuclear Test Ban Verification Project', below, pp. 354–62.
4. J. F. Evernden, C. B. Archambeau and E. Cranswick, 'An Evaluation of Seismic Decoupling and Underground Nuclear Test Monitoring Using High Frequency Seismic Data', *Review of Geophysics*, vol. xxiv (1986) pp. 143–215.
5. J. A. Stein, 'Nuclear Tests Mean New Weapons', *Bulletin of the Atomic Scientists* (November 1986) pp. 8–11.
6. T. B. Taylor, 'Third-Generation Nuclear Weapons', *Scientific American* (April 1987) pp. 30–9.
7. Ibid.
8. Lawrence Livermore National Laboratory, *Energy and Technology Review* (Washington, DC, 1987).
9. H. Bethe, N. Bradbury, R. Garwin, S. M. Keeny Jr, W. Panofsky, G. Rathjens, H. Scoville Jr and P. Warnke, in *Bulletin of the Atomic Scientists* (November 1985) p. 11.

23 The US National Resources Defense Council/Soviet Academy of Sciences Nuclear Test Ban Verification Project

Thomas B. Cochran

INTRODUCTION

The first week in September 1987 was an extraordinary one for arms control verification. As part of the co-operative Test Ban Verification Project of the Natural Resources Defense Council (NRDC) and the Soviet Academy of Sciences, fourteen American scientists from the Scripps Institution of Oceanography (at the University of California-San Diego), University of Nevada-Reno and the University of Colorado went to the region of the Soviet's principal nuclear test site near Semipalatinsk. Together with their Soviet counterparts from the Institute of Physics of the Earth (IPE) in Moscow, they fired off three large chemical explosions. The purpose of these explosions was to demonstrate the sensitivity of the three seismic stations surrounding the test site, to study the efficiency with which high-frequency seismic waves propagate in the region, and to study differences between chemical explosions, nuclear explosions and earthquakes in order more firmly to establish procedures for verification of a nuclear test ban. Before reviewing the results of these experiments, a brief update on the status of the joint project will be offered, followed by a review of the significance of high frequency seismic data to test ban verification.

PROJECT STATUS

In May 1986, NRDC and the Soviet Academy of Sciences agreed to

354

establish and jointly staff three seismic monitoring stations near each of the principal nuclear weapons test sites in the United States and Soviet Union. The objectives of this scientific exchange are to conduct research on nuclear test ban verification and to demonstrate that verification is not an obstacle to a nuclear test ban or moratorium.

By September 1986 the American field team from Scripps, together with their Soviet counterparts from IPE, had established and were jointly operating temporary surface seismometers at Karkaralinsk, Bayanaul, and Karasu, the three stations around the Soviet nuclear weapons test site in the Republic of Kazakh (see Figure 23.1). The stations are each about 200 kilometres from the centre of the test site.

Between September and November 1986 the IPE team constructed at each site permanent facilities, including instrument and housing trailers, an underground vault for three-component short- and inter-mediate-period surface seismometers, and a 100 metre borehole for housing three-component high frequency seismometers. During the same period the Scripps team began ordering some $600 000 worth of American-made seismic equipment to be installed at these new facilities.

In November 1986, IPE seismologists came to the United States to join the Scripps team in selecting locations for three seismic monitor-

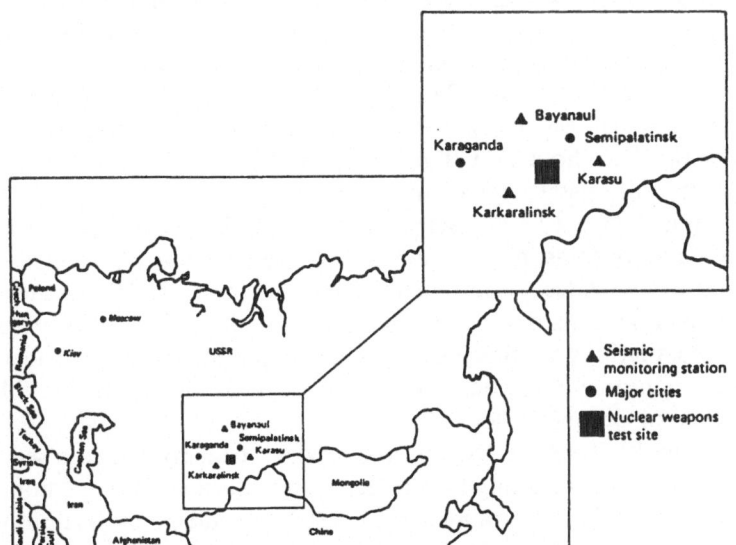

Figure 23.1 Seismic monitoring stations in the Soviet Union

ing stations around the Nevada Test Site and to examine the new equipment being manufactured. Owing to restrictions placed on their visas by the Reagan Administration, the visit by the Soviets was restricted to seven days and they were not able to visit the proposed station locations. The Scripps and IPE teams were permitted to meet briefly at Scripps in La Jolla, California. There, using maps, slides and geological samples, the locations of the US stations were selected. These stations are at Troy Canyon and Nelson, Nevada, and Deep Springs, California (see Figure 23.2).

In February 1987, the Scripps team began installing at the three Kazakh stations the more sophisticated American-made seismic equipment, including the high-frequency borehole instruments. These state-of-the-art seismometers are capable of detecting explosions as small as a few tons anywhere on the test site.

February 1987 also marked the end of the Soviet's nineteen-month unilateral testing moratorium, at which time the Soviet Government asked NRDC and the Soviet Academy to turn off the monitoring stations during each subsequent nuclear test. Owing to the difficulty of getting the stations back on line this resulted in the equipment being operated for only about 50 per cent of the time, and it delayed completion of the installation of the new equipment until May 1987. Nevertheless the stations have continued to record seismic data from earthquakes, industrial explosions in the region and nuclear tests outside the Soviet Union. These data, previously unavailable in the United States, are being made available to both governments and to the scientific community. The data are being used to improve estimates of Soviet test yields, useful in verification of the 150-kilton limit under the Threshold Test Ban Treaty, and to study the transmission properties of seismic signals in the region.

Also in February 1987, scientists from the University of Nevada–Reno and Scripps began monitoring the Nevada Test Site with temporary surface seismometers at the three NRDC/Academy stations in Nevada and California. Construction of the three permanent installations has now been completed and the more sophisticated seismic equipment identical to that placed in the Soviet union will be installed in October 1987.

In June 1987 the NRDC and the Soviet Academy signed a second year agreement extending the project for an additional fifteen months. It was agreed that beginning in early 1988 two additional stations will be added in the Soviet Union increasing the size of the network to five. The three seismic monitoring stations already in Kazakh will be

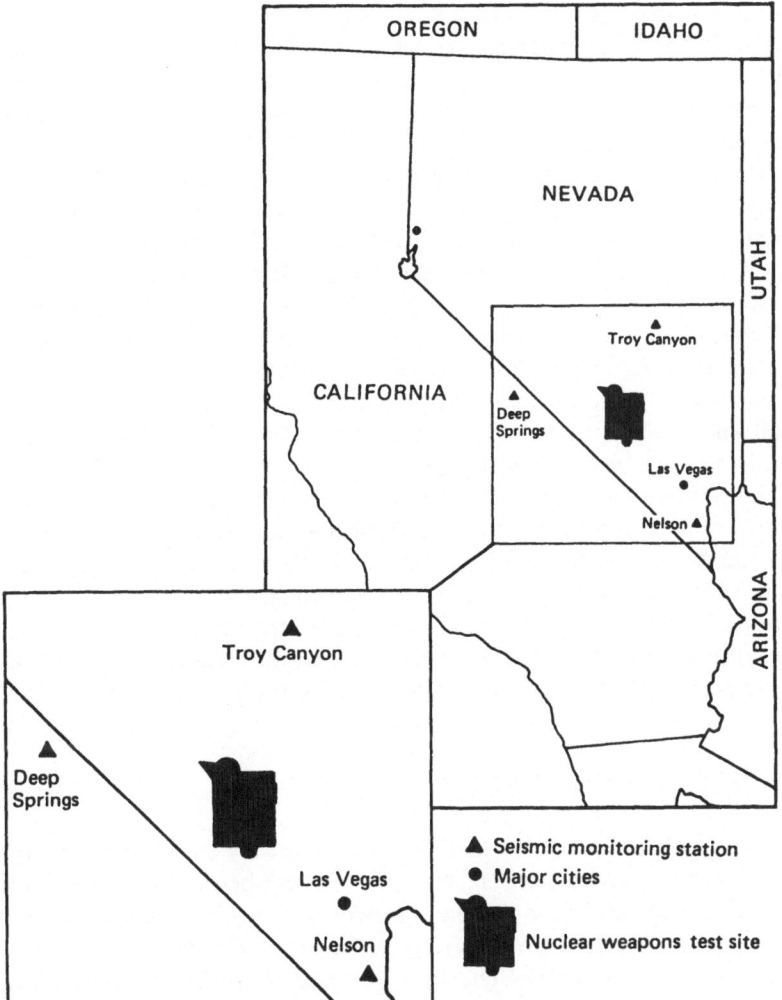

Figure 23.2   Seismic monitoring stations in the United States

moved out to a distance of about 1000 kilometres from the Kazakh
Test Site in order to focus the research agenda on a one-kiloton
threshold test ban. At this time, according to the agreement, the
NRDC/Soviet Academy stations will begin recording Soviet nuclear
tests. This will be the first time Americans have ever recorded Soviet
tests at stations in the Soviet Union.

It was also agreed that NRDC and the Soviet Academy would conduct the above-mentioned chemical explosion experiments near the Kazakh Site and a reciprocal set of chemical explosion experiments near the Nevada Test Site. The latter are now tentatively scheduled for early 1988.

Plans for a Soviet Academy field team to return to the United States to participate in the monitoring of the Nevada Test Site have been postponed, owing to visa problems. The US Government will permit the scientists to work on the project in the United States only if they first agree to witness a nuclear explosion and demonstration of CORRTEX, another method for estimating test yields. The recently announced agreement between the US and Soviet Governments to begin formal negotiations on testing may resolve the visa problem; we will know when the Soviet seismologists from IPE apply for visas to go to Scripps to assist in the analysis of recently collected data.

In part to overcome the barrier to Soviet participation in the monitoring programme in the United States, NRDC and the Soviet Academy agreed in June 1987 to establish a satellite 'hot line' for rapid transmission of seismic data between the Soviet Union and the United States. Seismic stations and data centres in both countries will be connected by dedicated telecommunications links for close-to-real-time exchange of seismic recordings, including nuclear tests.

HIGH FREQUENCY SEISMIC DATA

To appreciate the significance of high-frequency signal propagation to test ban verification, it may be useful, before describing the experimental results, to review the two most probable methods that have been suggested for cheating under a low threshold or comprehensive test ban treaty. The first of these is to set off the nuclear explosion in a large underground cavity in order to muffle the sound of the blast. The second is to hide the sound of the explosion in the coda, or background signal of a large earthquake.

The first scenario is referred to as the 'decoupling' scenario because the shock wave energy from the blast is not efficiently coupled into the surrounding medium. If the cavity is of sufficient size the low-frequency seismic signals – those of the order of one hertz – are reduced by a factor of about one hundred over what they would be if the explosion was well tamped, that is, if no cavity was used (Figure 23.3). With seismometers operating in this low-frequency range,

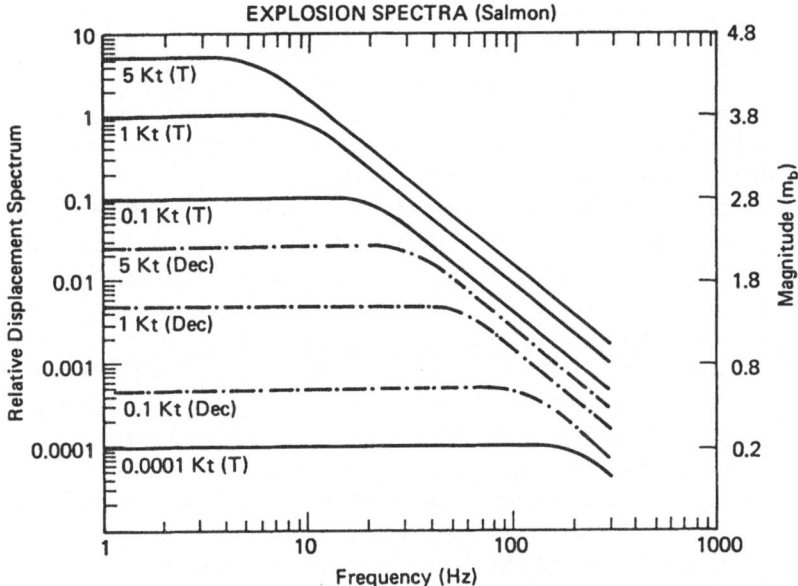

EXPLOSION SPECTRA (Salmon)

Notes: Predictions based upon the Sharpe (1942) model. Left-hand scale: normalised displacement. Right-hand scale: m_b magnitude scale (calibrated for explosions in shield areas of the United States and the Soviet Union). T: tamped; Dec: fully decoupled.

Source: J. F. Evernden, C. B. Archambeau, and E. Cranswick, 'An Evaluation of Seismic Decoupling and Underground Nuclear Test Monitoring Using High-Frequency Seismic Data', *Reviews of Geophysics*, vol. xxiv, (1986) p. 146.

Figure 23.3   Predicted displacement source spectra for tamped and decoupled explosions in strong salt

typically monitored in earthquake studies, a small nuclear explosion could be lost in the background noise, and if detected it would be indistinguishable from a large chemical blast.

In recent years several seismologists studying the test ban verification have noted that the decoupling efficiency is greatly reduced at higher frequencies. For example, it can be as low as a factor of ten in the 30–50 hertz region (Figure 23.3). Thus, if as they predict, high-frequency seismic signals propagate efficienctly in most of the Soviet Union, a few in-country high-frequency seismic stations would be adequate to monitor a low-threshold test ban. In fact, Evernden,

Archambeau and Cranswick have predicted that 25 in-country stations would suffice to monitor a one-kiloton threshold in the Soviet Union.[1] Two uncertainties in their model are the efficiency of high-frequency signal propagation and the high-frequency background noise levels in the regions of interest. We have set out to measure both of these parameters in the Soviet Union.

The second scenario is of somewhat less concern than the first; but here again the solution lies in the high-frequency regime. Explosions are much richer in high frequencies relative to low frequencies as compared to earthquakes. Modern digital recording techniques permit us to analyse the frequency spectra of seismic events. Thus again, if the high-frequency signals propagate far enough to be detected above the background noise, spectral analysis or filters can be used to detect an explosion even against the background of an earthquake.

With regard to both of these scenarios, it can be argued that one can always ensure detection of the high-frequency signals by increasing the density of seismic stations. There is, however, a practical limit. To monitor a low threshold test ban in the Soviet Union with twenty-

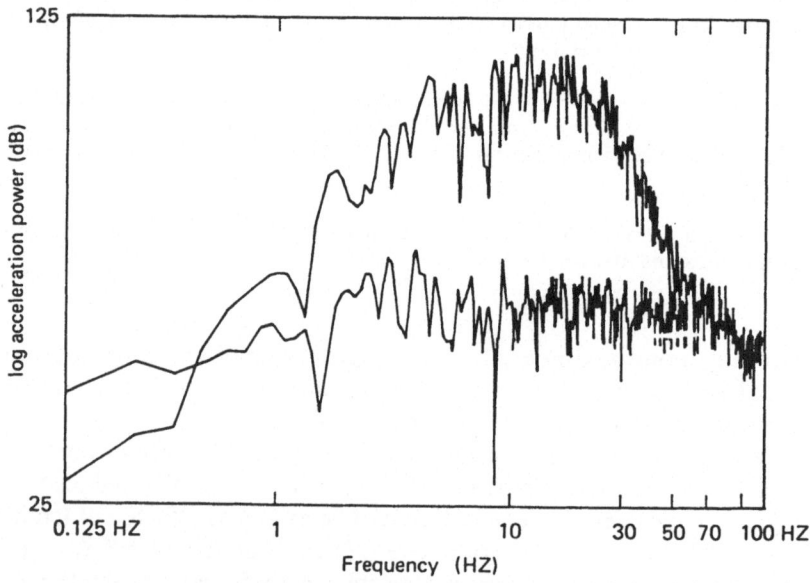

Figure 23.4 Acceleration power spectra of vertical component of P wave signal from Kazakh explosion no. 2 at Karkarlinsk Station (146 km), with noise acceleration power-surface seismometer (GS–13)

five stations, the mean distance between stations would be about 1000 kilometres.

In order to measure the efficiency of high-frequency signal propagation near the Kazakh Test Site, two ten-ton TNT explosions were detonated by IPE at Osakarovka (50°16'50"N, 72°10'20"E) near the city of Karaganda in central Asia, and one 20-ton explosion on the western edge of the test site (50°00'05"N, 77°20'10"E). The seismic signals from all three explosions were recorded by each of the three permanent seismic stations jointly established by NRDC and the Academy of Sciences at Karkaralinsk (49°20'N, 75°23'E), Bayanaul '50°49'N, 75°33'E) and Karasu (49°57'N, 81°05'E), and by four additional temporary stations set out specifically for this experiment at 30-kilometre intervals along the road between Karaganda and Karkaralinsk. The most distant station, Karasu, was 400 miles (630 km) from the explosion site at Osakarovka. While the data have not yet been quantitatively analysed, qualitatively they appear to be consistent with the assumptions on the model of Evernden *et al.*[2] Strong high-frequency signals (up to 50 hertz) were recorded. (See, for example, Figure 23.4). Thus we are confident a low threshold test ban

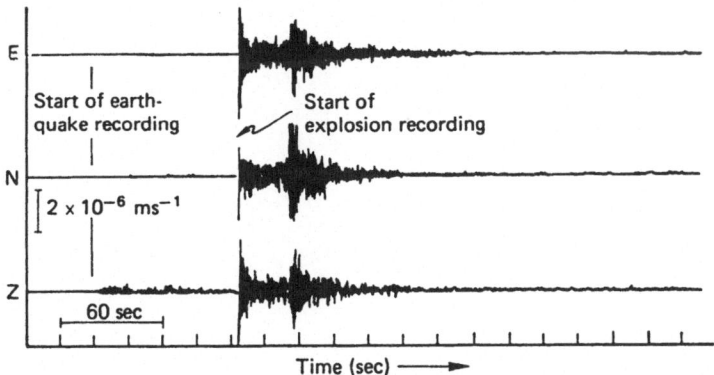

Note: Seismic signal amplitudes shown are proportional to ground velocity at the receiver. Three components of ground motion are shown, vertical (Z), East–West (W) and North–South (N). Earthquake was at a distance of 127° in McQuarrie Islands, south of New Zealand, surface wave magnitude 7·2.

Figure 23.5 Simultaneous recording of a teleseismic earthquake and a 10-ton chemical explosion (no. 3) near Karaganda, USSR, recorded at Karkarlinsk, USSR (explosion distance 255 km); surface seismometer (GS–13) – $T_0 = 1$ sec; explosion origin time: 3 September 1987, 7:00:00 GMT

can be adequately verified with a reasonable number of in-country seismic stations.

In a remarkable coincidence, a large earthquake occurred just before the last scheduled explosion, the second ten-ton explosion at Osakarovka. The earthquake was subsequently located in the Macquarrie Islands between New Zealand and Antarctica and had a surface wave magnitude of 7·2. All the seismic stations recorded both events. In spite of the presence of the large earthquake signal, the explosion signals were clear and unmistakable, with their high-frequency content much greater than that of the earthquake (see Figure 23.5). Thus with high-frequency stations there appears to be little chance that a country could successfully hide nuclear shots in the background of distant earthquakes.

Notes

1.   J. F. Evernden, C. B. Archambeau, and E. Cranswick, 'An Evaluation of Seismic Decoupling and Underground Nuclear Test Monitoring Using High-Frequency Seismic Data', *Reviews of Geophysics*, vol. xxiv, (1986) 143–215.
2.   Ibid.

Index

ABM defence systems 34, 67, 94, 288
ABM Treaty 60, 62–3, 68, 76, 77, 80–2, 219, 234, 289–91, 301, 312, 319, 325
  clarification 90–3
  constraints 93–5
  goals 82–90
  threshold limits 81, 85, 87–8, 90–3
Abrahamson, General James A. 67, 70, 78
ACCHAN 242
Accident Measures Agreement 298
ACE 242, 246
ACE HIGH communications system 251, 252
ACLANT 242
active countermeasures SDI 70
ADA 102
Aeritalia 46
aeronautical industry 101–2
Afghanistan 211, 213, 214, 292, 347
Africa 239
AGIs 254
air defence, NATO 108
air defence radar 109–10, 129–30
airbases, NATO 110
airborne alert, US 220
Airborne Command Post, USINCEUR 255
Airborne Laser Experiment 89
Airborne Optical System 88–90
Airbus 11, 43
aircraft 33, 43
  SIGINT 254
  technological development 21
aircraft runways 153–4
  TBM targets 137–47
aircraft systems, integrated 231–2
AirLand Battle 232, 233
ALCMs 83, 114
alert levels, NATO 256–8

American Physical Society 59, 68, 171
Antarctic Treaty 296
anti-missile systems
  alternative regimes 85–90
  deployment 82–3
anti-satellite weapons 302, 322
anti-simulation 69
Anti-submarine Warfare 246, 262
AOC Strike Command 246
apertures, sensor systems, limits 94
Apollo project 6, 7
Archambeau, C.B. 359–60
Argentina 180, 181, 302, 318, 348
Ariane 11
Ariane 5 100
Arkin, William 171, 274
arms control 161, 326
  achievements 286–99
  interaction with nuclear technology options 191–4
  international negotiations 319–21, 337–8
  modelling 234–6
  priorities 299–307
  purposes 285–6
Arms Control and Disarmament Agency 321
arms industry, Italy 46
arms race 34, 48, 162–6, 219, 310–11, 316–18, 329, 333
  nuclear 336–7
  space 324–5
  steps to contain 319–23
  technological 226–7
Armstrong 33
artificial intelligence 228–32
ASAT weapons 69, 70, 72, 77
astronauts 96
ATBMs 108, 111
atomic bomb 31
Atomic Weapons Research Establishment (UK) 336
Attached Pressurized Module 100

AUTODIN 252
Autokonetz 247
AUTOVON 252
*Aviation Week and Space
Technology* 274
AWACS 89, 279

B-1 bomber 227
B-1B bomber 333
Bacon, Francis 97
balance of trade, US 26
Ball, Desmond 274, 275, 278, 279
ballistic missile defence
technology 80, 82, 84
ballistic missiles
elimination 301–2
launchers 290
Baneberry event 288
Bangladesh 318
BAOR 247
Bardeen, John 64
bargaining power, nuclear
weapons 162
battle management 90, 228, 232
battlefield, automated 232–3
battlefield nuclear weapons 327
Belgium 241
Bell 36
Bertram, Christoph 301
beryllium 212
Bhutto, Zulfikar Ali 210
biological agents, TBM delivery
121–2
Biological Weapons Convention
78, 319
biomass 14–15
Blair, Bruce 237
blast, munitions 116–18
BMEWS station 262
Boeing 707 aircraft 43
Boeing 767 aircraft 88–9
boiling water reactors 172
bombers
intermediate-range 123
US 83
boost-phase defence 61
Bracken, Paul 237

Brazil 5, 165, 318
nuclear technology 180, 181
Brezhnev, Leonid 294
British Space Centre 99
broadcasting, satellites 75
Brown, Paul S. 285
Buccaneer aircraft 246
Bulgaria 254
bunkers
above-ground, TBM
targets 134–7
hardened 117
semi-hard 127–8
TBM targets 148–50
Business Week 24–5, 44

C³ I 39, 228, 232–4, 237–40, 256,
257, 260
Europe: complexity 240–8;
vulnerability 248–56
NATO 263, 275–82; critical
points 268–74
strategic 261–2, 264
Campbell, Duncan 274
Canada 98, 179, 210
CANDU 173, 183
Carnegie 33
Carter, President Jimmy 210–11,
213, 292, 322
CCIS 247
Cemetery Net HF System 251
Center for International Security
and Arms Control, Stanford
University 65
Challenger 8
Charles, Daniel 241
chemical explosions, seismic
monitoring 358
chemical weapons 61, 121–2, 302,
325, 328
China 165
nuclear power 164, 179, 181,
200, 201, 202, 287, 289, 318,
346
choice, of technology 21
Chopra, V.D. 212
Churchill, Winston 29, 31, 286
CIA 262

CINCAFCE 246
CINCEASTLANT 242, 246, 247
CINCEUR 259, 260
CINCHAN 242, 246, 247, 249
CINCLANT 242
CINCUKAIR 246, 249
CINSAC 237
Circle of Equal Probability 113,
 128, 131, 134, 135, 136, 138,
 142, 143, 144, 149, 152
CISC 249
civil activities, space 75–6
civil aircraft 43
civilian spin-offs 33
 developing countries 5
 emerging countries 5–6
 microelectronics 7
 military R&D 18, 24–7, 47, 51–2
 military technology 3–4
 negative 42–3, 48
 new materials 8
 Strategic Defense Initiative 9–10
Classic Wizard 261
cobalt 212
Cochran, Thomas B. 171
Cold War 332
Columbus project 98, 99, 100
COMCENTLANT 246
COMMAIRCHAN 246
COMMAIREASTLANT 246
command centres, NATO 110,
 249–51
command headquarters, NATO
 268–71
Commander Allied Forces South
 Norway 249
Commander Iberian Atlantic
 Area 249
commercialisation, technological
 advance 22
Commission on Industrial
 Competitiveness (US) 25
common security 311–12, 314
communication facilities, TBM
 targets 148–50
communications, satellites 75
communications centres, NATO
 272–4

communications networks 251–3
COMNORCHAN 246
COMNORLANT 246
competition, military technology
 4–5
competitiveness, space technology
 and 103
COMPLYMCHAN 246
Comprehensive Test Ban 163, 164,
 166, 287, 302, 306, 327
Comprehensive Test Ban Treaty
 194, 346, 347, 348, 350, 352
computer chips 227
computers 25, 33, 227–8
 fifth generation 11–12
 optical 102
COMSEC 255–6, 264
Concorde 37
Conference of the Committee on
 Disarmament 319
Conference on Confidence and
 Security-building Measures in
 Europe 298–9
Conference on Disarmament 328
Conference on Security and
 Cooperation in Europe 329
confidence-building measures 321,
 329
Congressional Research Service
 115, 123
controlled escalation 238
CONUS 240
conventional forces, Europe 277
conventional weapons 4
 disarmament 328
Coot A ELINT aircraft 254
CORRTEX 350, 358
cost–benefit effectiveness, military
 technology 6
costs
 military R&D 52
 nuclear bomb materials 174
 space programmes 99
Cousins, N. 45
Cranswick, E. 359–60
craters, runways 138, 140
crisis management 238, 298
Cross Fox HF system 251–2

Cruise missiles   110–11, 146–7, 153,
    326, 336
  characteristics   112–16
  delivery capabilities   122–6
Cuban Missile Crisis   288
Cunningham, Chris   71, 73
Czechoslovakia   254, 258

DANASAT   70
DARPA   228, 230
De Grasse, R.   38
decision-making artificial
    intelligence and   228–30
  NATO   258
  nuclear war   337
decoupling, seismic   358–60
decoys   69, 110, 111, 129
DEFCON THREE   256
defence market, nature of   39, 40–1
defence needs   52–3
Defence Planning Committee   241,
    242
defences, TBMs   108
Defense Advanced Reseach Project
    Agency   100
defensive weapons, space   66–73
Delhi-6   302, 348
denaturant, Pu-239   186
deterrence   60, 162, 276, 285, 332–4
Deutch, John   301
deuterium   187–8, 203
developing countries   4
  civilian spin-offs   5
  food production   13–14
Dickens, Charles   324, 325
directed-energy weapons   59, 93–4
disarmament   286, 324–31
divisional command posts, TBM
    targets   150–3
DMAIN   150–1
Dornier   102
drones   130
DSCS   251, 252, 261
DTOC   151
Duchin, F.   32
Dumas, Lloyd J.   17

E-2C Hawkeye aircraft   213

E-3 AWACS aircraft   89
E-3A AWACS aircraft   213
early deployment   62
early warning radar   248
early warning sites, NATO   271–2
early warning systems   262
Earth Penetrating Warheads   351
earthquakes   360, 362
economic role, technology   19–20
Edison, Thomas   22
Eighteen Nations Disarmament
    Committee   288, 297, 319
Eisenhower, President Dwight D.
    60, 64–5, 321
electricity generation, nuclear   171–
    8, 206
Electromagnetic Pulse   37, 227, 252,
    350
electromagnetic rail guns   37
electronic industry, Great
    Britain   39, 42
Elettronica   46
ELINT aircraft   254
ELINT ocean surveillance satellite
    station   261
ELINT satellites   262
Ellis, General Richard C.   238, 253
Emergency Action Messages   251
emerging countries, civilian spin-
    offs   5–6
EMSI   100
energy biomass   14–15
enrichment, isotopes   181, 182
Enzing, Christien   10
escalation   276, 278
Esprit   11
Eureka   11, 102–3
Eureka, B   100
Europe   11
  C³I: complexity   240–8;
      vulnerability   248–56
  security   314
  space technology   98–101
European Military Telephone
    Network   252
European Space Agency   34–6, 98–
    100
Evernden, J.F.   359–60
Excalibur Program   350–1

exo-atmospheric intercept 67
expenditure, military R&D,
    Italy 45–6
expert systems 229–30, 235
explosions, high frequencies 360
extended deterrence 162
externalities, military R&D 17–19

F–15 interceptors 137, 140, 144
F-16 fighters 140, 144
F-16 launchers 125
F-16C fighters 213
Fairchild 38
FAS Public Interest Report 31
Federation of American
    Scientists 81
Feiveson, Harold 200
Fieldhouse, Richard W. 274
fifth generation computers 11–12
fighters 137, 138, 140
    US, for Pakistan 213–14
firms, large, defence industry 44
first use, nuclear weapons 163–4
First World War 33
fissile materials, production 191,
    192
fission energy systems, vulnerability
    to proliferation 181–7
fission weapons 202, 203
Fleet Ballistic Missile submarines
    256
flexible response 238, 256, 277
FOFA 233
food production, developing
    countries 13–14
Ford, Daniel 238
Forum for a Nuclear-free World
    305
Fourth Allied Tactical Air Force
    249
Foxbat D ELINT aircraft 254
fragmentation weapons 118-19
France 9, 165, 179–80, 210, 235,
    241, 297
    nuclear power 164, 179, 181,
        200, 202, 287, 288, 346
    plutonium recycling 207
    SDI participation 102
    space technology 98

Frontal Aviation 107, 108, 109,
    112, 122–6, 137, 140
Frosi, M 46
fuel cycle, sensitive parts 186
fuel reprocessing 172
Fuel–Air Explosives 116, 121
fusion reactors, vulnerability to
    proliferation 188–91
fusion-fission hybrid reactors
    187–8

gallium arsenide 227
GALOSH interceptors 70
gamma rays 186, 350
garages, space 72
gas centrifuge plants 210
General and Complete
    Disarmament 339, 340
General Dynamics F-16
    fighter 140, 144
General Staff, Soviet 222
Geneva Conference on
    Disarmament 325
Geneva negotiations 81, 82, 300,
    301, 319, 332
German Democratic Republic 254
German Federal Republic *see* West
    Germany
Gibson, Roy 99
glasnost 321, 338
GLCM missiles 114, 257, 259, 336
Glenn, John 214
Global Positioning System 75
Gorbachev, Mikhail 81, 86, 302,
    303–4, 321, 322, 326
graphite-moderated reactors 170
gravity bombs 83
Great Britain 9, 24, 241, 297
    electronic industry 39, 42
    military expenditure 44
    nuclear power 164, 179, 181,
        200, 288
    nuclear tests 345–6, 347
    plutonium recycling 207
    SDI participation 102
    space technology 98, 99, 100–1
Greece 241, 302, 348
Gregory, Shaun 274
Grenada, invasion 239–40

Gromyko, Andrei 85
ground-based interceptors 61
ground-based lasers 75–6
GRU 255

Haig, Alexander 256
Halley, Edmund 97
hardware, computers 227
Harpoon missiles 214
Harrison, John 97
Hart, Gary 85
Harvard Avoiding Nuclear War
    Project 298
Harvard Negotiating Project 298
heat-seeking sensors 61
heavy water 179
heavy-water moderated
    reactors 170, 173
Helsinki process 313, 315
Heriot-Watt University 102
Hermes shuttle 100
HEROS 247
High Frequency communications
    systems 248
high-frequency seismic data 358–62
high-temperature gas-cooled
    reactors 183
Hiroshima 57–8
Hoenig, M.M. 171
Holdren, John P. 336
Homer 104
homing sensors 91
horizontal proliferation 200–2, 204,
    289, 318
hostage crisis, Iran 347
Hot Line Agreement 297
HOTOL 100–1
House of Representatives Armed
    Services Committee 44
Human Frontier programme 15
Hungary 254
hybrid reactors 187–8

IBM 36
ICBMs 67, 69, 70, 73, 74, 83, 111,
    286, 290, 292, 333
Iceland summit 86
IDEA 14–15
IFF systems 239, 247

Il-18 ELINT aircraft 254
image processing 234
imaging, satellites 75
implosion fission weapons 203
India 5, 165, 302, 316, 348
    confrontation with Pakistan 318
    nuclear power 179, 181, 201,
        212, 213, 214, 289, 318
    war with Pakistan 210, 212
INE 300
inertial guidance, missiles 113
INF agreement 300, 312, 314
INF negotiations 219
INF weapons 280
informatics 7
information technology 226–36
infra-red telescopes 88, 89, 92
Instrument Pointing System 98
integrated circuits 38
Integrated Communications
    System 247
Intel 39
intelligence agencies 221
intelligence satellites 84
intelligence stations, NATO 272
interceptors 91, 94, 138, 140, 289
intermediate-range missiles 338
International Atomic Energy
    Agency 180, 186–7, 192, 195,
        210, 211
international negotiations, arms
    control 319–21, 337–8
international reactions,
    proliferation 164
International Satellite Monitoring
    Agency 235
international trade, USA 11
inventions 22
investment
    military R&D 29–34, 51
    military technology 3–4
ion sources, star weapons 102
Iran 347
isotopes, enrichment 181, 182
Israel 44, 180, 181, 201, 318
Italy 180
    military research and
        development 45–7
    NATO forces 241

SDI participation   29, 46
  space technology   98
Italy. Defence Ministry   45, 46
Italy. Ministry of Industry   46

Japan   11, 24, 98, 99, 224, 323
  industrial policy   44
  plutonium recycling   207
  research and development   54
Japanese Agency of Industrial
    Science and Technology   15
jet aircraft   33
Joint Chiefs of Staff
    Organization   221–2
Joint Consultative Commission,
    PNET   294
Joint Council of Engineers   43
JIF–120   239
JUMP   151

KANNUP   210, 211
Karpov, Victor   85
Kashmir   212
Kelleher, Catherine   258
Kennedy, President John F.   18, 322
Keyworth, Jay   64
KGB   225
Khan, Dr A.Q.   210, 212–13
Khrushchev, Nikita   321–2
Kieler Institut für
    Weltwirtschaft   103
Killian, Jim   65
kinetic energy   9
Kissinger, Henry   276
KKVs   291
Kohl, Helmut   102
Krupp   33
Kvitsinsky, Yuli   85–6

laboratory, space   100
labour costs   20
Lance missiles   113
large-scale integration   6
lasers   23–4, 36, 37, 70, 102, 227
  brightness limits   92, 93–4
  ground-based   75–6
  orbiting   74
  X-ray   37, 70, 350-1

Last Talk HF System   251
latent proliferation   200, 204
Latham, Donald C.   238–9
Latin America   346
Latin American Nuclear-Free Zone
    Treaty   296
Lawrence Livermore National
    Laboratory   335, 336, 351, 352
LEO   78
Leontiev, W.   32
LERTCOM   251, 255, 257
light bulb, invention   22
light water reactors   206
Limited Test Ban Treaty   234, 287–8, 302, 328, 346
Liquid Metal Fast Breeder
    Reactor   188
Livermore project   37
living standards, USA   26
Long, F.A.   32
long-range interceptors   84
long-rod penetrators   74–5
longitude, determination   97
Los Alamos National Laboratory
    335, 336
Lovins, L. Hunter   191
Low Earth Orbit   72
Low Threshold Test Ban Treaty   349
lunar tables   97

M-1 tanks   214
M-60 tanks   214
McDonald–Douglas F-15 fighter
    140, 144
Macmillan, Harold   59
Maddox report   39
magnetic-confinement, approaches
    to fusion energy   191
Man-Tended Free Flyer   100
maraging steel   211–12
markets, leading production   12–13
Massachusetts Institute of
    Technology   43
massive retaliation   238
materials, new, civilian spin-offs   8
Measure of Economic Welfare   322
Melman, Seymour   43
Mexico   302, 348

Meyer, Tobias 97
microelectronics 6, 7
microprocessors 39
microwave links 251
microwave weapons 351
Middle East 297
Midgetman missiles 227
MIG-23/27 aircraft 123
MIG-25R ELINT aircraft 254
Miggiano, Paolo 45
military advantages, nuclear
 weapons 162
military aircraft 43
military applications, space
 technology 101–2
Military Committee, NATO 242
military contracts 38–42
The Military Engineer 139
military expenditure 44, 310, 317,
 322–3
military forces, levels 316–17
military operations, routine 219–25
military personnel, relations with
 defence industry 45
military reactors 42
military research and development
 313, 316, 323
 attributes 52–4
 civilian spin-offs 51–2
 diversion of resources 24–7
 externalities 17–19
 investment 29–34
 Italy 45–7
 limitations 36–7
 scientists in 334–6
military technology
 cost–benefit effectiveness 6
 investment 3–4
 non-military justifications 50–4
 space 97
military uses, space technology 98,
 99
military-industrial complex 335
minimum deterrent 339–40, 341–2
Ministry of Defence (UK) 39, 42
Minuteman II missiles 38
MIRVs 34, 67, 83, 290, 335
missile launchers 117
Mitterrand, François 102

mobile command posts 263
Mobile Ground Terminals 263
Mobile Subscriber Equipment 247
molybdenum 212
monitoring systems 92
Moon, landings 96, 105
MOS 38–9
Mostek 39
Mowery, D.C. 39
MRVs 83
Mujaheddin 213
munitions, conventional, effects
 against targets 116–22
mutual assured security 312–14
Mutual Interim Restraint 292, 293
MX missiles 227, 333, 336, 351

Nagasaki 57–8
Namboodiri, P.K.S. 212
NARS 252
National Aeronautics and Space
 Administration 8, 34
National Science Foundation 25
National Technical Means 287, 346
National Test Bed 232, 234
NATO 50, 102, 122, 239, 313, 314,
 316, 317
 airbases 137–47, 153–4
 alert levels 256–8
 C³I 263, 275–82; complexity,
  240–8; critical points, 268–74;
  vulnerability, 248–56
 extended deterrence 162
first use policy 163–4
non-nuclear threats to 106–12
North Atlantic Treaty Organisation:
  Facts and Figures 243, 244,
  245
 nuclear release 258–60
 war-fighting doctrines 233
 Warsaw Pact attack 127
Natural Resources Defense
 Council 302, 348, 354–62
naval forces, nuclear weapons
 326–7
navigation, satellites 75
Navstar–GPS 75
NCA 263
'need to know' principle 221

Netherlands 210, 241, 297
neutron bombs 203, 350
neutron reflectors 212
Nevada test site 288
new technologies
  strategic defence 62
  testing 63
Newton, Sir Isaac 97
nickel 212
NICS 247
Nike–Zeus system 89
Nimrod aircraft 246
Nitze, Paul H. 68, 76, 81
Nixon, President Richard 294
no-first use 313
non-national proliferation 200
non-nuclear strategic weapons
  74–5
Non-Proliferation Treaty 165, 186,
  187, 194–5, 201–2, 288–9, 302,
  319, 339, 346
NORAD RAPIER system 263
North Korea 239
Northrup F-5E/F fighter 140
NSA 262
NSDD 172 68, 73
NTMs 294
nuclear bombers, US 220
nuclear bombs, Pakistan 211–12
nuclear disarmament 119–200, 340
nuclear energy 33, 51, 161
nuclear energy facilities 171–8
Nuclear Engineering International
  177
nuclear explosions 37
  experimental 345–52
  underground 294–6, 346
nuclear forces, US 224
nuclear fuels, reprocessing 172
nuclear materials dismantled
  warheads 205–8
  stockpiles 204–5
Nuclear News 176
nuclear operations, routine 219–25
nuclear power plants, dismantled
  warheads as fuel for 205–8
nuclear reactors 169–70
  commercial 193–4
nuclear release, NATO 258–60

nuclear submarines 42, 220
nuclear targets, countries as 164
nuclear technology, arms control
  and 191–4
nuclear tests 203–4, 324
  atmospheric, banning 287–8
  banning 327–8
  limiting 302–6
  moratorium 321–2
  suspension 345–52
Nuclear Tests Experts' Meetings
  96, 303, 304
nuclear weapons 4, 58, 66, 317
  acquisition 162–5, 167–9, 194;
    nuclear energy
    facilities 171–8
  arms race 162–6
  clandestine production 191–2
  dedicated facilities 169–71
  deterrence 332–4
  dispersal, Europe 257
  first use 163–4
  naval forces 326–7
  negotiations 324
  orbiting 74
  security 351–2
  selective use 275–6
  stockpiles 199
  technology 202–4, 334–5
nuclear-free zones 296–7, 315, 346

offensive weapons, space 74–5
Office of Naval Research (US) 43
Office of Technology Assessment
  (US) 170, 171, 174
Okean exercise 255
on-site inspections 299
Oppenheimer, Robert 58
optical computers 102
Outer Space Treaty 296, 325

Pacific Ocean 346
Pakistan 165, 316
  confrontation with India 318
  nuclear technology 180, 181,
    210, 210–14, 318
Palme Commission 311, 327
PALs 257, 279–80
PARR 211

particle beam weapons   227
passive countermeasures, SDI   69
passive sensors   91
Patriot air defence radar   109, 129
Peaceful Nuclear Explosions
    Treaty   294–6, 302, 303, 304–5
perestroika   338
Perimeter Aquisition Radar   89
Pershing I missiles   113
Pershing II missiles   110, 113, 114,
    128, 143, 144, 257, 259
phased-array radar   76–7, 94–5
Pindar bunker   249
Pioneer   10, 96
planar technology   38
plutonium   172–3, 173, 174, 179,
    192–3, 202, 203
  nuclear warheads   206
  production   211
  recycling   206–7
  reprocessing plant   175
  stockpiles   204
Poland   254
Polaris   336
Politburo   222
Political Action Committee, defence
    industry   44
Poseidon submarine   247
pre-emptive strikes   107, 238
President's Science Advisory
    Council   64–5
pressurised water reactors   42, 172,
    183
Prevention of Nuclear War
    Agreement   298
production, efficiency   20
proliferation   161–2, 200–2, 204,
    289, 318, 213–4
  nuclear arms race and   162–6
  vulnerability to: fission energy
    systems   181–7; fusion and
    hybrid energy systems
    187–91
Ptarmigan   247
PTT networks, Europe   252–3
Pu-236   188
Pu-239   94, 173, 187, 188
  denaturant   186
Pu-240   172, 174, 202

Pu-242   172
pumped energy lasers   350–1

radar   25, 51, 67, 84
  air defence   109–10
  large   90–1
  phased-array   76–7
  targets   117, 129–30
  types of   89
radio-electronic combat   252
radioactivity   287–8, 345, 346
RAF Germany   246
re-entry vehicles, tracking and
    identification   88
reactor-grade plutonium   172–3,
    174, 175, 202
reactors, dual-purpose   200
Reagan Doctrine   213
Reagan, President Ronald   25, 59,
    64, 66, 67, 68, 106, 211, 292,
    302, 303, 304, 326, 347
rearmament, Second World
    War   31–2
Recherche, La   35
recycling, plutonium   204
Reforger exercise   256
remote sensing   9
Reppy, J.   32
reprocessing plant, plutonium   175
research and development
  goal-setting   23
  see also military research and
    development
Reviews of Modern Physics   68, 70
Reykjavik summit   301, 302, 303,
    304, 314
Richelson, Jeffrey, T.   274
risk money   53–4
  military R&D   52
Risk Reduction Centres   298
RITA   247
robotics, battlefield   232
rockets
  ground-based energy sources
    75–6
  mass equations   71–3
Rogers, General Bernard   259–60
Royal Air Force   246, 249

Royal Navy 246, 249
Rusk, Dean 321–2

SA-12 system 129
SACEUR 242, 246, 247, 259–60
SACHERTS system 263
SACLANT 242, 246, 247
Sakharov, Andrei 58, 64, 305–6, 321
SALT Agreements 234
SALT I 60, 289–91, 297, 298, 325
SALT II 291–4, 298, 302, 312, 325
Sanger project 101
SATCOM network 251, 252
satellite ground stations 248
satellites 61–2, 96
  arms control 234–5
  civil uses 75
  early-warning 67
  intelligence 84
  monitoring systems 92
SCF 261–2
Schelling, Thomas 300, 301
Schott 102
science, breakthroughs 21
scientists
  military R&D 334–6
  SDI and 57–65
Scowcroft, Brent 301
Seabed Arms Control Treaty 296
Second Allied Tactical Air Force 246, 247
Second World War 25, 33
  rearmament, 31–2
secrecy
  defence industry 44
  military R&D 25–6, 45
  SDI projects 102, 103
security 311–15, 329
security guarantees 162–3, 164
seismic data, high-frequency 358–62
seismic monitoring 295, 302, 354–8
seismological techniques, underground test detection 349
seismometers 355, 356
Selenia 46
Sematech 12

Semiconductor Industry Association 12
semiconductors 38, 39
Senate Armed Services Committee 44
Senate Committee on Foreign Relations 77
Senate Defense Appropriations Committee 44–5
sensor systems, apertures, limits 94
sensors 9, 67, 88, 89–90, 233–4
  verification 90–1
Sentinel/Safeguard system 89
SHAPE 242, 249
shaped-charge warheads 116, 119–20
Shevardnadze, Eduard 298, 300, 301, 304, 305
Shultz, George 304
SIGINT 248, 251, 262, 264
  Soviet Union 253–6
Sims, Jennifer 10
Site Defence system 89
SKYNET 251, 252
SLBMs 67, 70, 83, 247, 333
SLCMs 247
SNDVs 292
Sochaczewski, Major General Joachim 253
social process, technology as 20–1
software, military 227–8
SOSUS 262
South Africa 179–80, 201
South America 239
South Korean airliner incident 222
South-west Asia 239
Soviet Academy of Sciences 45, 302, 348, 354–62
Soviet Navy 221
Soviet Union 24, 31, 50, 163, 316
  ABM Treaty and 290
  changes within 338
  early-warning radar 76–7
  invasion of Afghanistan 211, 213, 292, 347
  military expenditure 44
  military R&D 29
  military technology 4, 5, 42
  NPT violation 202

Soviet Union—*cont.*
  nuclear balance 318
  nuclear power 164, 179, 181,
    200, 288
  nuclear tests 287, 345–50, 356
  nuclear weapons 66, 333
  routine nuclear operations 219,
    220, 223, 224, 225
  SIGINT activities 253–6
  space technology 96
  strategic relationship with
    USA 59
  strategic weapons, cuts 63
  suspension of nuclear tests 347–8
  TBMs 106–12
  Treaty compliance 292–3, 295-6
Spa 46
space
  arms race 324–5
  civil activities 75–6
  defensive weapons 66–73
  industry 9
  manufacturing in 75
  non-militarisation 313
  offensive weapons 74–5
  testing in 86
space mines 69, 70
space programmes, spin-off 7–8,
  34–6
space research, European 11
Space Shuttle, accident 229
space shuttles 78
space station, European 99
space technology 96–8
  Europe 98–101
  military applications 101–2
  motivation behind 103–5
space weapons 85
Space-based Interceptors 71–3
space-based testing 86–7
space-track radar 77
Spacelab 98
Spector, Leonard S. 177
spent fuel, reactors 204, 205, 207
Spetsnaz 238, 251
SPOT satellite 235
Sputnik, I 96
SRAMs 83
SS-1 launchers 123

SS-1c SCUD B missiles 116
SS-12 launchers 123
SS-12 mod missiles 106, 116, 123,
  138
SS-13 missiles 292
SS-18 missiles 70, 73, 300
SS-21 missiles 106, 113, 114
SS-22 missiles 106, 113, 123, 138
SS-23 missiles 106, 116, 123, 138
SS-25 missiles 73, 286, 292, 333
SS-NX-23 missiles 333
SSBNs 262
Stalin, J.V. 31
Standing Consultative
  Commission 82, 292–3
Stanford University 65, 298
START 219, 292, 300
Stealth bomber 227
steel industry 33
Steinbruner, John 237
Stern, Der 211
Stockholm International Peace
  Reseach Institute 30, 45–6,
  333
STOMA/SIGMA 247
strategic ballistic missiles 68
Strategic Bombing Survey, US
  57–8
strategic C^3I 261–2, 264
Strategic Computing Initiative 228,
  230–2
strategic defence, United States
  60–5
strategic defence systems, command
  and control 233–4
Strategic Defense Initiative 23–4,
  34, 50–1, 67, 68, 106, 302, 330,
  337
  ABM Treaty and 80–1, 290–1
  battle management 232
  civilian spin-offs 9–10
  costs 29
  criteria 68
  effects on mainstream
    research 36
  European participation 101–3
  halting 313
  Italian participation 46
  lasers 37

scientists and 57–65
Soviet concern 300, 301
testing under 88
weapons for 226–7
X-ray lasers 350–1
Strategic Defense Initiative Office
(US) 67, 97–8
strategic forces, provocative
moves 223–4
Strategic Rocket Forces 221
strategic weapons 66–7, 83
cuts 63
Strike Command 246, 249
SU-17 aircraft 123
SU-24 aircraft 123
sub-munitions 112, 116, 120–1,
130–7, 144–7, 152
submarines
FBM 256
Soviet 224
US 247
superconductivity 9–10
superpowers 4, 224, 299–300, 326,
333
supersonic aircraft 37, 100
Sweden 9, 224, 302, 329, 348
Swedish National Institute for
Defence Research 347
Symington Amendment 211

Tactical Ballistic Missiles 106–12
army divisional command
posts 150–3
characteristics 112–16
delivery capabilities 122–6
effectiveness against aircraft
runways 137–47
effectiveness against
communications and
command facilities 148–50
munitions 116–22
single-warhead, effectiveness
against targets 127–30
submunitions, effectiveness against
targets 130–7
tactical nuclear weapons 328–9
tanks, US, for Pakistan 214
Tanzania 302, 348
Taylor, Theodore B. 335, 351

technological progress
military needs and 33
nuclear weapons 334–5
technology
economic role 19–20
misconceptions 20–4
technology transfer
military R&D 47
military–civilian 8–9
telescopes 84, 85
infra-red 88, 89, 92
Tercom guidance systems 126
terminal sensing systems 126
terrain sensing ballistic missiles
113, 114
Test Ban Verification Project
354–62
testing, new technologies 63
Texas Instruments 38
Th-232 187, 188
Thatcher, Margaret 336
theatre nuclear war 237–40, 261–2
Third World *see* developing
countries
thorium 202
threshold limits 81
ABM Treaty 85, 87–8
new 90–3
Threshold Test Ban Treaty 294–6,
302, 303, 304–5, 346–7, 350, 356
Time 211, 305
Tirman, J. 32
Tomahawk SLCMs 247
Tornado aircraft 279
Tow-2 missiles 214
TRANSIT satellites 75
Treaty on Outer Space 346
Treaty for the Prohibition of
Nuclear Weapons in Latin
America 296–7
Treaty of Rarotonga 346
Treaty of Tlatelolco 346
Tri-tak 247
Trident 336
tritium 189–90, 203
stockpiles 205
tropospheric scatter links 251, 252
trucks, targets 117
TU-16 bombers 123

TU-22 bombers 123
TU-22M bombers 123
Turkey 201, 241

U-232 187
U-233 182, 186, 188, 202
U-235 94, 172, 175, 182, 186, 205, 206
U-238 172, 186, 187, 188, 193, 194
UHF communications systems 239
underground command centres 249–50
underground explosions 294–6
underground structures, hardened 110;
unilateral arms control initiatives 320, 321–3
Union of Concerned Scientists 77
United Kingdom *see* Great Britain
United Nations 43, 314, 319
  General Assembly 77
United States 163, 180, 201, 241, 297, 316
  ABM Treaty and 290–1
  aid to Pakistan 210–11, 213–14
  Air Force 238
  Army 239
  Army, Field Manual FM 100–5 258, 259
  arms control legislation 302–3
  Army. Training and Doctrine Command 259
  budget deficit 317
  C³I systems 237–40
  Congress 322
  Department of Defense 12, 38–9, 239, 250, 351
  Department of Energy 176
  General Accounting Office 9
  House of Representatives 302–3
  military expenditure 44
  military R&D 29, 47
  military technology 4, 5
  Navy 238, 239
  NPT violation 202
  nuclear balance 318
  nuclear policy 285
  nuclear power 164, 179, 181, 200, 288

nuclear testing 287–8, 345–50
nuclear weapons 66, 333, 336–7
routine nuclear operations 219, 220, 223, 224, 225
Senate 303
SALT II 292–4
space technology 96, 98, 99, 100
State Department 68
Strategic Bombing Survey 57–8
strategic defence 60–5
strategic offensive weapons 83
strategic relationship with Soviet Union 59
technological progress 24–5, 26
technology sector 44
USAEUR 242, 246
USAFE 242, 251
USCINCEUR 242
USEUCOM 242, 246, 249–50, 255
USINCEUR 255
USNAVEUR 242, 249
university research, government orientation 43
uranium 175, 192
  enrichment 42
  highly-enriched 202, 203;
    dilution 205–6;
    stockpiles 204–5
uranium enrichment 169, 170, 172, 182, 201, 210
uranium-plutonium oxide fuel 206
USSR *see* Soviet Union

V2s 96
Vanunu, Mordechai 201
verification 340–1
  AMB testing 84–5
  criteria 92–3
  CTBT 348
  dismantling nuclear warheads 207–8
  joint experiments 304–5
  sensors 90–1
  technology 234–6
  TTBT 295–6, 346–7
vertical proliferation 200–2, 289
Very High Speed Integrated Circuits 227

Very Large Scale Integration 227
very low frequency communications
    systems 248
Vickers 33
Vietnam War 232
violations, TTBT 347 SALT II 347
von Hippel, Frank 192
VSTOL aircraft 111

W84 warhead 336
W87 warhead 336
wages, rising 19–20
war, agreements to reduce the risk
    of 297–8
warheads 116
    dismantled 205–8
    high-fragmentation 118–19
    superpowers 333
Warsaw Pact 4, 50, 108, 122, 123,
    127, 249, 256, 276, 281, 314,
    316, 317
Washington Post 211
WAVELL 247
weapon launchers 90
weapon systems, intelligent 230–2
weapons 90, 91
    information technology and
        226–8
    technological, understanding of
        59
weapons package, US, for
    Pakistan 213–14
weapons stockpiles 192
weapons-grade plutonium 173

weather observation, satellites 75
Weinberger, Caspar W. 73, 250
Weizsäcker, Carl-Friedrich von 315
West Germany 24, 180, 323
    NATO forces 241
    plutonium recycling 207
    PTT system 253
    research and development 54
    SDI participation 102
    space programme 98
    space technology 98, 101
Westmoreland, General William
    232
White Book 1985 45, 46
White Cloud 261
White House Science Council 64
Wickham, General John A. 240
Wiesner graph 340–1
WINTEX Command Post Exercises
    257, 258
Woolsey, R. James 301

X-30 supersonic aircraft 100
X-ray lasers 37, 70, 350–1

Yom Kippur War 225
York, Herbert 335

Zegveld, Walter 10
Zeiss 102
Zia ul-Haq, President 210, 211, 214
Zodiac 247
Zuckerman, Lord 334